TECHNICAL MATHEMATICS

Robert Donovan

Prentice Hall
Englewood Cliffs, New Jersey   Columbus, Ohio

Library of Congress Cataloging-in-Publication Data

Donovan, Robert.
    Technical mathematics/Robert Donovan.
        p.   cm.
    Includes index.
    ISBN 0-13-440694-X
    1. Mathematics.   I. Title.
QA39.2.D644   1996   95-9505
510′.246—dc20

Cover Photo: Michel Tcherevkoff/The Image Bank
Cover Designer: Brian Deep
Editor: Stephen Helba
Developmental Editor: Carol Hinklin Robison
Production Editor: Christine M. Harrington
Production Manager: Deidra M. Schwartz
Project Management and Text Design: Elm Street Publishing Services, Inc.

This book was set in 10/12 Times Roman by Carlisle Communications, Inc. and was printed and bound by R. R. Donnelley & Sons Co. The cover was printed by Phoenix Color Corp.

©1996 by Prentice-Hall, Inc.
A Simon & Schuster Company
Englewood Cliffs, New Jersey 07632

All rights reserved. No part of this book may be reproduced, in any form or by any means, without permission in writing from the publisher.

Photos in text courtesy of the following: BK Precision, MAXTEC International Corp.; Ohmite Manufacturing Company; James Millen, Andover, MA; Vitramon, Inc.; Clarostat Sensors and Controls; Leader Instruments Corporation; Illinois Capacitor, Inc.; Johnson Manufacturing; Bell Industries; Microtan; Eveready; Catalyst Research; Dale Electronics; Philips.

Printed in the United States of America

10 9 8 7 6 5 4 3 2 1

ISBN: 0-13-440694-X

Prentice-Hall International (UK) Limited, *London*
Prentice-Hall of Australia Pty. Limited, *Sydney*
Prentice-Hall of Canada, Inc., *Toronto*
Prentice-Hall Hispanoamericana, S. A., *Mexico*
Prentice-Hall of India Private Limited, *New Delhi*
Prentice-Hall of Japan, Inc., *Tokyo*
Simon & Schuster Asia Pte. Ltd., *Singapore*
Editora Prentice-Hall do Brasil, Ltda., *Rio de Janeiro*

To my family
Sue, Kyle, and Kimme

Preface

Students are often anxious to begin their studies in technology but are not so enthused about mathematics. This book is written with those students in mind to mainstream them into their chosen fields quickly. This book is designed as a one-semester, stand-alone mathematics text for technology students at the community college, proprietary school, or Vo-Tech center. It assumes no prerequisite courses in math at the college level. A high school algebra background is desirable and will suffice.

The primary purpose of this book is to teach mathematics by providing a solid foundation in algebra and trigonometry. Examples and exercises are designed to allow the student to apply the basics to practical problems. It is in the process of working these exercises that the student prepares for the technical courses for which this course is a prerequisite. It is a goal of this book to develop in the student the diligent work habits and thought processes essential for success in the workplace.

Many incoming students have some notions about algebra, but most are inexperienced in trigonometry. This book first prepares the student with a solid algebra foundation and then presents trigonometry. Every opportunity is taken throughout the remainder of the text to enhance this knowledge and to build a working knowledge of trig principles. It is important that the student feels comfortable with trigonometry. In several places in the text, material is presented in a sequence that builds toward a culminating application complex enough to cause the student to "break a sweat." It is this growth that causes excitement in students and develops a sense of accomplishment.

■ INSTRUCTIONAL FEATURES

As shown in the accompanying figure, each chapter opener consists of three parts:

- **Chapter Outline**
- **Objectives**
- **New Terms to Watch For**

The **Outline** lists the sections within the chaper, and the **Objectives** provide a numbered list of the performance-based objectives to be accomplished during the course of the chapter. **New Terms to Watch For** lists each key term that will be encountered within the chapter. When each term is first presented, it is highlighted in **boldface** color print.

Each section has been developed to include numerous instructional pedagogical elements. The intent of each of these elements is to strengthen the instructional value of the section by clarifying content and providing opportunities for interaction for the student. Section-level features include:

vi ■ PREFACE

CHAPTER OPENER

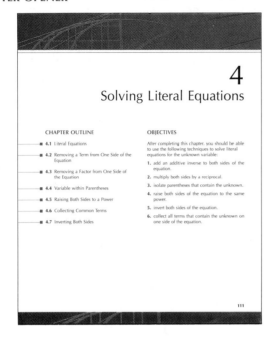

EXAMPLE AND EXERCISE

- ☐ A clear, concise **introductory discussion of the theory or concept** being developed helps students bridge content throughout the chapter.
- ☐ **Numerous examples with solutions** displayed in detail and **answers** provided allows students to practice skills and reaffirms comprehension of content.
- ☐ A set of **exercise problems** concludes each section. These exercises are similar to the examples and provide immediate feedback to the student on his or her progress. The answers to all exercises within the chapter are listed at the end of that chapter, while the complete solution to each of the exercises is included in the Instructor's Resource Manual.

A typical example and exercise is shown in the figures on p. vi.

Each chapter ends with a **Summary,** the **Answers to the Exercises,** and a wide assortment of **Problems** from all sections of the chapter. The **Solutions to Odd-Numbered Problems** are listed at the end of the text. The complete solutions to all of the chapter-end problems are given in the Instructor's Resource Manual.

Instruction in the use of calculators is integrated throughout the text. Due to the wide variety of calculators used by students, no particular model is presented. Calculator sequences are given to introduce principles and generate ideas. These sequences, which can be modified by the student to fit the calculator being used, are not intended to replace the user's manual that comes with the calculator.

■ SUPPLEMENTS TO THE TEXT

The *Instructor's Resource Manual* contains valuable information for the instructor, information that can make the class flexible enough to meet a variety of situations. It includes:

1. Complete solutions to all the exercises.
2. Complete solutions to all chapter-end problems.
3. Additional problems and solutions for some of the topics.
4. Additional applications in some areas. For example, geometry is applied to simple and compound gear trains, and trig is applied to the optics of light traveling down a fiber optic cable.
5. Thirty transparency masters in the form of black line-art, which illustrate difficult-to-duplicate figures.
6. A sample test for each chapter.

■ CHAPTER DESCRIPTIONS

Chapters 1 through 4 provide a foundation in basic algebra to prepare the student for the work with literal equations that follows.

Chapter 1 is a review of arithmetic. It concentrates on operations containing fractions, the order of operations, and exponents. No literal expressions are included in this chapter.

Chapter 2 builds skills in handling large and small numbers with scientific notation and engineering notation. It also introduces concepts in significant digits. These concepts are applied throughout the remainder of the text.

Chapter 3 parallels Chapter 1. Many of the same topics are covered with the inclusion of signed numbers and variables.

Chapter 4 puts the basic algebra skills to work at solving literal equations for a specified unknown. It is essential that students master the material in this chapter.

Chapter 5 introduces U.S. and SI units of measure for a wide variety of quantities. It also presents the dimensional analysis method of covering or reducing units. This chapter contains material that serves as a reference for the rest of the book.

Chapter 6 presents an eight-step systematic approach to solving technical problems. This approach is used in most of the examples in the remainder of the text. A variety of topics are presented and used as models for problem solving.

Chapter 7 presents general background information in geometry. It covers the Pythagorean theorem, area, radian, degrees, minutes, seconds, angular velocity, and volume.

Chapter 8 discusses linear equations and graphing. The trig functions are introduced as an exercise in graphing data obtained from a calculator.

Chapter 9 presents systems of equations and their solutions. Cramer's rule is studied and applied to loop equations in DC circuits.

Chapter 10 presents trigonometry. It begins a sequence of material based on trigonometry that culminates in Chapter 13 with the application of vectors. Many of the following chapters revisit trig in some form to entrench those concepts into the students' minds and to build on the basics.

Chapter 11 presents addition and subtraction of vectors, graphically and algebraically. Phasors are used to represent sine waves with various amplitudes and phase shifts.

Chapter 12 presents complex numbers. Here the student learns to manipulate complex numbers in both rectangular and polar form and to convert from rectangular coordinates to polar and vice versa using a calculator.

Chapter 13 applies much of the preceding material to applications in surveying, statics, and airplane navigation. Solutions to the problems involve trig, vectors, complex numbers, and simultaneous equations.

Chapter 14 introduces common and natural logarithms. The fundamentals are applied to sound level, power gain, voltage gain, and RC time constants.

Chapter 15 presents four methods for solving quadratic equations. Quadratic equations are applied to dish antennas.

Chapter 16 presents the binary and hexadecimal number systems, including 1's and 2's complement subtraction, binary coded decimal, and signed 2's complement numbers.

■ SYLLABUS OPTIONS

An ideal program allots enough time to work through this text from cover to cover. If that is not the case, here are some suggestions for streamlining this text.

Chapter 1, Arithmetic: A pretest can determine whether your students can skip the bulk of the material. If so, cover only the resistor color code material at the end of the chapter.

Chapter 3, Algebra: If your students have a background in algebra, a pretest can determine whether this chapter can be skipped.

Chapter 4, Solving Literal Equations: It is imperative that students master this chapter. Be cautious about skipping any of this material.

Chapter 5, Units of Measure: Students should master the dimensional analysis approach to unit conversion. The information about units of each of the quantities can be treated as reference material to be used as needed.

Chapter 6, Solving Problems: This chapter presents an eight-step approach to solving problems that should not be skipped. If necessary, omit a few of the topics presented.

Chapter 7, Geometry: The Pythagorean theorem is used extensively in the remaining chapters, as are radians. Assign these sections:

7.1 Angles

7.2 Triangles

7.3 Pythagorean Theorem

7.7 Circles

7.9 Radian Measure

7.12 Angular Velocity

Chapter 8, Linear Equations and Graphing: The fundamentals at the beginning of the chapter should be covered, as should the graphing techniques at the end of the chapter. These sections can be omitted:

8.6 Slope or Gradient

8.7 Midpoint of a Line

Chapter 9, Systems of Linear Equations: The applications in Chapter 13 require the use of simultaneous equations, so the fundamentals need to be covered. If necessary, omit this section:

9.7 Loop Equations

Chapter 10, Trigonometry: The fundamentals of trig functions and inverse trig functions are essential. If necessary, omit one or both of these applications:

10.3 Snell's Law

10.4 Angles of Elevation and Depression

Section 10.5, Surveying, is the foundation for one of the applications in Chapter 13 and should not be omitted.

Chapter 12, Complex Numbers: At the end of the chapter, in section 12.11, complex numbers are applied to series AC circuit analysis. This application can be omitted if necessary.

Chapter 14, Logarithms: If necessary, omit some of these applications:

14.4 Sound Level

14.5 Power Gain (G_p)

14.7 Voltage Gain (G_v)

14.8 Log Graph Paper

14.11 Universal Time Constant Chart

Chapter 15, Quadratic Equations: If necessary, omit this section:

15.6 Applications of Quadratic Equations

Chapter 16, Number Systems: If necessary, omit these sections:

16.15 Signed 2's Complement Numbers

16.16 Binary Coded Decimal (BCD)

To challenge students who move quickly through the material in this text, take advantage of the additional applications and problems in the *Instructor's Resource Manual*. Assign these extra applications to the class or to individual students.

ACKNOWLEDGMENTS

While much of the labor in writing a book is of a solitary nature, it takes the efforts of numerous individuals who provide guidance, support, and attention to a variety of tasks and details to assure the highest quality in the final product.

I wish to thank the following reviewers: Miriam Alexander, College of Albermarle; Robert Ho, Mt. San Antonio College; Donald King, ITT Technical Institute; O. Ray Miller, Oklahoma State University Technical Branch; Charles Oster, Sauk Valley Community College; Jill Rugare, DeVry Institute of Technology, Columbus; and Molly Summer, Pikes Peak Community College.

In addition, I would like to thank Carol Robison and Steve Helba for their continuing editorial support and guidance; Barb Lange and Nancy Shanahan for managing the production details; and Nancy Moudry for the fine copyediting of the manuscript.

Contents

Chapter 1   **ARITHMETIC**   1
     1.1   ARITHMETIC OPERATIONS   2
         1.1.1   *Addition*   2
         1.1.2   *Subtraction*   3
         1.1.3   *Multiplication*   3
         1.1.4   *Division*   4
     1.2   PRIME FACTORS   4
     1.3   REDUCING FRACTIONS   5
     1.4   RAISING FRACTIONS TO HIGHER TERMS   6
     1.5   ADDING AND SUBTRACTING LIKE FRACTIONS   7
     1.6   LEAST COMMON MULTIPLE   7
     1.7   LEAST COMMON DENOMINATOR AND ADDING AND SUBTRACTING UNLIKE FRACTIONS   8
     1.8   MULTIPLYING FRACTIONS   10
     1.9   DIVIDING FRACTIONS   11
     1.10   COMPLEX FRACTIONS   12
     1.11   POSITIVE INTEGER EXPONENTS   13
         1.11.1   *Calculator y^x Key, or x^y Key, or $\wedge$ Key*   14
         1.11.2   *Calculator x^2 Key*   14
         1.11.3   *Calculator +/− Key*   15
         1.11.4   *Calculator x^3 Key*   15
     1.12   NEGATIVE INTEGER EXPONENTS   15
     1.13   RATIONAL EXPONENTS   16
     1.14   RULES FOR OPERATIONS WITH EXPONENTS   18
         1.14.1   *Rule 1—Multiplying Exponential Expressions*   18
         1.14.2   *Rule 2—Dividing Exponential Expressions*   18
         1.14.3   *Rule 3—Raising a Power to a Power*   19
         1.14.4   *Rule 4—Raising a Product to a Power*   20
         1.14.5   *Rule 5—Raising a Quotient to a Power*   21
     1.15   ORDER OF OPERATIONS   23
     1.16   PERCENT   24
         1.16.1   *Converting a Decimal to a Percent*   25
         1.16.2   *Converting a Fraction to a Percent*   25

		1.16.3 Converting a Percent to a Decimal 25

            1.16.3   Converting a Percent to a Decimal   25
            1.16.4   Converting a Percent to a Fraction   25
            1.16.5   Applying Percent to Resistor Values   27

Chapter 2   SCIENTIFIC NOTATION, ENGINEERING NOTATION, AND SIGNIFICANT DIGITS   42
    2.1   DECIMAL NOTATION   43
    2.2   SCIENTIFIC NOTATION   44
        2.2.1   *Converting From Scientific Notation to Decimal Notation*   44
        2.2.2   *Converting From Decimal Notation to Scientific Notation*   45
        2.2.3   *Using a Calculator*   47
                  +/− Key   47
                  EE, EEX, or EXP Key   47
        2.2.4   *Using Scientific Notation in Arithmetic Operations*   48
    2.3   ENGINEERING NOTATION   50
        2.3.1   *Converting From Decimal Notation to Engineering Notation*   50
        2.3.2   *Converting From Engineering Notation to Decimal Notation*   51
    2.4   SIGNIFICANT DIGITS   52
        2.4.1   *Accuracy*   59
        2.4.2   *Precision*   61
    2.5   ORDER OF OPERATIONS   63

Chapter 3   ALGEBRA   72
    3.1   SIGNED NUMBERS   74
    3.2   ADDITION OF SIGNED NUMBERS   74
    3.3   SUBTRACTION OF SIGNED NUMBERS   75
    3.4   MULTIPLICATION OF SIGNED NUMBERS   77
    3.5   DIVISION OF SIGNED NUMBERS   78
    3.6   ADDITION AND SUBTRACTION OF MONOMIALS   79
    3.7   MULTIPLICATION OF MONOMIALS   79
    3.8   DIVISION OF MONOMIALS   81
    3.9   POWERS AND ROOTS OF MONOMIALS   83
    3.10   MULTIPLICATION OF BINOMIALS   84
    3.11   MULTIPLYING A POLYNOMIAL TIMES A POLYNOMIAL   86
    3.12   FACTORING   87
    3.13   FACTORING TRINOMIALS   89

- 3.14 FACTORING OTHER TRINOMIALS 91
- 3.15 REDUCTION OF FRACTIONS 91
- 3.16 RAISING FRACTIONS TO HIGHER TERMS 93
- 3.17 LEAST COMMON MULTIPLE, LCM 93
- 3.18 ADDING AND SUBTRACTING FRACTIONS 94
- 3.19 MULTIPLICATION OF FRACTIONS 96
- 3.20 DIVISION OF FRACTIONS 97
- 3.21 COMPLEX FRACTIONS 98

Chapter 4 SOLVING LITERAL EQUATIONS   111
- 4.1 LITERAL EQUATIONS 112
- 4.2 REMOVING A TERM FROM ONE SIDE OF THE EQUATION: REVIEW OF TERMINOLOGY 113
- 4.3 REMOVING A FACTOR FROM ONE SIDE OF THE EQUATION: REVIEW OF TERMINOLOGY 116
- 4.4 VARIABLE WITHIN PARENTHESES 120
- 4.5 RAISING BOTH SIDES TO A POWER: REVIEW OF TERMINOLOGY 123
- 4.6 COLLECTING COMMON TERMS: REVIEW OF TERMINOLOGY 128
- 4.7 INVERTING BOTH SIDES 132

Chapter 5 UNITS OF MEASURE   142
- 5.1 SYSTEMS OF UNITS OF MEASURE 144
- 5.2 QUANTITY—LENGTH, SYMBOL—s 144
- 5.3 QUANTITY—AREA, SYMBOL—A 146
- 5.4 QUANTITY—VOLUME, SYMBOL—V 147
- 5.5 QUANTITY—TIME, SYMBOL—t 149
- 5.6 QUANTITY—VELOCITY, SYMBOL—v 149
- 5.7 QUANTITY—ACCELERATION, SYMBOL—a 150
- 5.8 QUANTITY—MASS, SYMBOL—m 151
- 5.9 QUANTITY—FORCE, SYMBOL—F 151
- 5.10 QUANTITY—DENSITY, SYMBOL—p 152
- 5.11 QUANTITY—PRESSURE, SYMBOL—P 152
- 5.12 QUANTITY—TEMPERATURE, SYMBOL—T 154
- 5.13 QUANTITY—CHARGE, SYMBOL—Q 154
- 5.14 QUANTITY—CURRENT, SYMBOL—I OR i 154
- 5.15 QUANTITY—VOLTAGE OR ELECTROMOTIVE FORCE (EMF), SYMBOL—E OR V_s 154
- 5.16 QUANTITY—RESISTANCE, SYMBOL—R 156
- 5.17 QUANTITY—CONDUCTANCE, SYMBOL—G 157
- 5.18 QUANTITY—CAPACITANCE, SYMBOL—C 157
- 5.19 QUANTITY—INDUCTANCE, SYMBOL—L 159

5.20 QUANTITY—FREQUENCY, SYMBOL—f   159
5.21 QUANTITY—ENERGY, SYMBOL—W   159
5.22 QUANTITY—POWER, SYMBOL—P   161

Chapter 6   **SOLVING PROBLEMS**   169
6.1 THE EIGHT-STEP APPROACH TO PROBLEM SOLVING   170
6.2 SERIES CIRCUITS   171
6.3 GAGE AND ABSOLUTE PRESSURE   172
6.4 OHM'S LAW   173
6.5 TORRICELLI'S THEOREM   174
6.6 TANK CIRCUIT   175
6.7 COEFFICIENT OF FRICTION   177
6.8 TWO RESISTORS IN PARALLEL   178
6.9 TWO OR MORE RESISTORS IN PARALLEL   180
6.10 SUPERELEVATION   181
6.11 MASS AND WEIGHT   182
6.12 °FAHRENHEIT,° CELSIUS   183
6.13 AIRPLANE STALL SPEED   184
6.14 PRESSURE EXERTED BY A COLUMN OF LIQUID   185
6.15 WAVE-LENGTH, PERIOD, FREQUENCY, AND VELOCITY OF ELECTROMAGNETIC WAVES   186

Chapter 7   **GEOMETRY**
7.1 ANGLES   194
7.2 TRIANGLES   196
7.3 PYTHAGOREAN THEOREM   198
7.4 AREA OF TRIANGLES   200
7.5 QUADRILATERALS   203
7.6 AREA OF QUADRILATERALS   204
7.7 CIRCLES   207
7.8 DEGREES, MINUTES, AND SECONDS   211
7.9 RADIAN MEASURE   212
7.10 AREA OF A SECTOR   213
7.11 ARC LENGTH   214
7.12 ANGULAR VELOCITY   215
7.13 ANGULAR DISPLACEMENT   216
7.14 LINEAR VELOCITY   216
7.15 VOLUME   217

Chapter 8   **LINEAR EQUATIONS AND GRAPHING**   233
8.1 CARTESIAN COORDINATE SYSTEM   234
8.2 DISTANCE BETWEEN TWO POINTS   236

	8.3	SLOPE OF A LINE 238
	8.4	PARALLEL LINES 241
	8.5	PERPENDICULAR LINES 242
	8.6	GRADE OR GRADIENT 243
	8.7	MIDPOINT OF A LINE SEGMENT 244
	8.8	LINEAR EQUATIONS 245
	8.9	POINT-SLOPE FORM OF EQUATION 249
	8.10	SLOPE-INTERCEPT FORM OF EQUATION 250
	8.11	GENERAL FORM OF EQUATION 252
	8.12	GRAPHING EXPERIMENTAL DATA 254
	8.13	GRAPHING TRIG FUNCTIONS 260
	8.14	TRIG IDENTITIES 263

Chapter 9 SYSTEMS OF LINEAR EQUATIONS 281

- 9.1 GRAPHICAL SOLUTION 282
- 9.2 SOLUTION BY SUBSTITUTION METHOD 284
- 9.3 SOLUTION BY ADDITION METHOD 287
- 9.4 GENERAL SOLUTION OF TWO EQUATIONS-TWO UNKNOWNS 289
- 9.5 MATRICES AND DETERMINANTS 291
- 9.6 CRAMER'S RULE 292
- 9.7 LOOP EQUATIONS 295

Chapter 10 TRIGONOMETRY 307

- 10.1 TRIG FUNCTIONS 308
- 10.2 INVERSE TRIG FUNCTIONS 317
- 10.3 SNELL'S LAW 323
- 10.4 ANGLES OF ELEVATION AND DEPRESSION 329
- 10.5 SURVEYING 332

Chapter 11 VECTORS AND PHASORS 344

- 11.1 ADDING VECTORS—GRAPHICAL SOLUTION-DAISY CHAIN METHOD 345
- 11.2 ADDING VECTORS—GRAPHICAL SOLUTION-PARALLELOGRAM METHOD 347
- 11.3 SUBTRACTING VECTORS—GRAPHICAL SOLUTION 350
- 11.4 RESOLVING A VECTOR INTO HORIZONTAL AND VERTICAL COMPONENTS 352
- 11.5 COMBINING HORIZONTAL AND VERTICAL COMPONENTS INTO A RESULTANT VECTOR 355
- 11.6 ADDING VECTORS ALGEBRAICALLY 359

	11.7	SUBTRACTING VECTORS ALGEBRAICALLY 363
	11.8	PHASORS 368

Chapter 12 COMPLEX NUMBERS 392

- 12.1 COMPLEX NUMBERS 393
- 12.2 GRAPHING COMPLEX NUMBERS 394
- 12.3 RECTANGULAR NOTATION 395
- 12.4 POLAR NOTATION 396
- 12.5 CONVERTING FROM POLAR COORDINATES TO RECTANGULAR COORDINATES 396
  - 12.5.1 *Calculator P>R Key* 397
- 12.6 CONVERTING FROM RECTANGULAR TO POLAR COORDINATES 399
  - 12.6.1 *Calculator R>P Key* 400
- 12.7 ADDING COMPLEX NUMBERS 402
  - 12.7.1 *Graphical Interpretation* 402
- 12.8 SUBTRACTING COMPLEX NUMBERS 403
  - 12.8.1 *Graphical Interpretation* 404
- 12.9 MULTIPLYING COMPLEX NUMBERS 405
  - 12.9.1 *Polar Form* 405
  - 12.9.2 *Rectangular Form* 405
- 12.10 DIVIDING COMPLEX NUMBERS 406
  - 12.10.1 *Polar Form* 406
  - 12.10.2 *Rectangular Form* 407
- 12.11 SERIES AC CIRCUIT ANALYSIS 407

Chapter 13 VECTOR APPLICATION 422

- 13.1 SURVEYING 423
  - 13.1.1 *Bearing of Perpendicular Vectors* 423
  - 13.1.2 *Bearing of a Vector in the Opposite Direction* 424
  - 13.1.3 *Traverse* 425
  - 13.1.4 *Closure (Closed Traverse)* 425
- 13.2 MECHANICS 429
  - 13.2.1 *Free Body Diagram* 430
  - 13.2.2 *Moment* 432
- 13.3 NAVIGATION 437

Chapter 14 LOGARITHMS 450

- 14.1 LOGARITHMIC NOTATION 451
- 14.2 COMMON LOGS 452
- 14.3 SIGNIFICANT DIGITS 453
- 14.4 SOUND LEVEL 455

	14.5	POWER GAIN (G_P) 457
	14.6	PROPERTIES OF LOGARITHMS 460
	14.7	VOLTAGE GAIN (G_V) 461
	14.8	LOG GRAPH PAPER 463
	14.9	NATURAL LOGARITHMS 467
	14.10	UNIVERSAL TIME CONSTANT CHART 468
Chapter 15	**QUADRATIC EQUATIONS**	**485**
	15.1	QUADRATIC EQUATIONS 486
	15.2	FINDING THE ROOTS BY FACTORING 487
	15.3	FINDING THE ROOTS BY COMPLETING THE SQUARE 488
	15.4	FINDING THE ROOTS BY QUADRATIC FORMULA 491
	15.5	GRAPHING PARABOLAS 494
	15.6	APPLICATIONS OF QUADRATIC EQUATIONS 501
Chapter 16	**NUMBER SYSTEMS**	**513**
	16.1	COUNTING IN BINARY 514
	16.2	CONVERTING FROM BINARY TO DECIMAL 516
	16.3	CONVERTING FROM DECIMAL TO BINARY 518
	16.4	COUNTING IN HEXADECIMAL 520
	16.5	CONVERTING BINARY TO HEXADECIMAL 522
	16.6	CONVERTING HEXADECIMAL TO BINARY 523
	16.7	BINARY ADDITION 525
	16.8	BINARY 1'S COMPLEMENT SUBTRACTION 526
	16.9	BINARY 2'S COMPLEMENT SUBTRACTION 528
	16.10	SIGNED 2'S COMPLEMENT NUMBERS 530
	16.11	BINARY CODED DECIMAL (BCD) 535
Appendix	**SOLUTIONS TO END OF CHAPTER PROBLEMS**	**548**
Index 579		

1
Arithmetic

CHAPTER OUTLINE

- **1.1** Arithmetic Operations
- **1.2** Prime Factors
- **1.3** Reducing Fractions
- **1.4** Raising Fractions to Higher Terms
- **1.5** Adding and Subtracting Like Fractions
- **1.6** Least Common Multiple
- **1.7** Least Common Denominator and Adding and Subtracting Unlike Fractions
- **1.8** Multiplying Fractions
- **1.9** Dividing Fractions
- **1.10** Complex Fractions
- **1.11** Positive Integer Exponents
- **1.12** Negative Integer Exponents
- **1.13** Rational Exponents
- **1.14** Rules for Operations with Exponents
- **1.15** Order of Operations
- **1.16** Percent

OBJECTIVES

After completing this chapter, you should be able to

1. factor a composite number into its prime factors.
2. reduce fractions to lowest terms.
3. raise fractions to higher terms.
4. add and subtract like fractions.
5. calculate the least common multiple.
6. add and subtract unlike fractions.
7. multiply fractions.
8. divide fractions.
9. evaluate complex fractions.
10. evaluate exponential quantities.
11. apply the rules for operations with exponents.
12. calculate percent.
13. use percent to calculate upper and lower bounds on resistors.

NEW TERMS TO WATCH FOR

addend	common multiple
sum	least common multiple (LCM)
commutative	unlike fractions
associative	least common denominator (LCD)
subtrahend	reciprocal
minuend	positive integer
difference	base
multiplicand	exponent
multiplier	cubed
product	squared
factor	exponential quantity
divisor	negative integer
dividend	index
quotient	radicand
numerator	radical
denominator	square root
prime number	cube root
composite number	rationalizing the denominator
prime factors	percent
like fractions	radix

1.1 ■ ARITHMETIC OPERATIONS

Even though you already know how to add, subtract, multiply, and divide, you might not remember the terminology involved in these operations. These terms are important because they are used in this chapter and the following chapters to introduce new concepts. Review the definitions here and refer back to this section as needed.

1.1.1 Addition

In addition, each number being added is called an **addend**. The result is called the **sum**. The first or top number is also called the *augend*.

$$\text{addend} + \text{addend} = \text{sum}$$

$$\text{augend} + \text{addend} = \text{sum}$$

It is not necessary to give different names to each addend because the order in which they are written does not matter; the sum will be the same. In other words, addition is **commutative**.

$$a + b = b + a$$

Addition is also **associative**. If three addends are being added, it does not matter which two are added first. The process can begin with the last two numbers or with the first two numbers.

$$(a + b) + c = a + (b + c)$$

For example,
$$2 + (5 + 7) = 2 + 12 = 14$$
$$(2 + 5) + 7 = 7 + 7 = 14$$

1.1.2 Subtraction

In subtraction, the number being subtracted is called the **subtrahend.** The number being subtracted from is called the **minuend.** The result is called the **difference.**

$$\text{minuend} - \text{subtrahend} = \text{difference}$$

Here, order is important. Subtraction is not commutative.

$$a - b \neq b - a \,(\neq \text{ means ``is not equal to''})$$

For example, $9 - 7 = 2$ but $7 - 9 = -2$.

1.1.3 Multiplication

In multiplication, the numbers being multiplied are called the **multiplicand** and the **multiplier.** The result is called the **product.**

$$(\text{multiplicand})(\text{multiplier}) = \text{product}$$

Multiplication can be expressed in a variety of ways. The multiplicand and multiplier can be enclosed in parentheses (), brackets [], or braces {}. They can be separated by an ×, an asterisk, or a dot, or the factors can be written with nothing separating each of them.

$$(a)(b) = [a][b] = \{a\}\{b\} = a \times b = a * b = a \cdot b = ab$$

Multiplication is commutative; the order in which the numbers are written is not important.

$$ab = ba$$

Since order is not important, it does not matter which is the multiplier and which is the multiplicand. Each of the numbers being multiplied is called a **factor** of the product.

$$(\text{factor})(\text{factor})(\text{factor}) = \text{product}$$

Multiplication is also associative. If three factors are being multiplied, it does not matter which two are multiplied first. The process can begin with the last two numbers or with the first two numbers.

$$(a \cdot b) \cdot c = a \cdot (b \cdot c)$$

For example,
$$2 \cdot (5 \cdot 7) = 2 \cdot 35 = 70$$
$$(2 \cdot 5) \cdot 7 = 10 \cdot 7 = 70$$

1.1.4 Division

In division, the number being divided by is called the **divisor** and the number being divided into is called the **dividend**. The result is called the **quotient**. The number being divided by can also be called the **denominator**, and the number being divided into can be called the **numerator**.

$$\frac{\text{dividend}}{\text{divisor}} = \text{quotient}$$

$$\frac{\text{numerator}}{\text{denominator}} = \text{quotient}$$

Most technical expressions and equations are written in single-line format.

$$\text{dividend/divisor} = \text{quotient}$$
$$\text{numerator/denominator} = \text{quotient}$$

If you have a problem visualizing or understanding equations or expressions in single-line format, rewrite them in fraction form.

Division is not commutative.

$$a/b \neq b/a$$

For example, $4/2 = 2$ but $2/4 = 1/2$.

SECTION 1.1 EXERCISES

Answer True or False.

1. $2 + 12 + 9 = 9 + 12 + 2$
2. $732 - 981 = 981 - 732$
3. $98 \cdot 27 = 27 \cdot 98$
4. $101/39 = 39/101$
5. The answer to a division problem is called the difference.
6. If $a - b = c$, b is called the subtrahend.
7. The answer to a multiplication problem is called the multiplicand.
8. Subtraction and division are not commutative, but addition and multiplication are.

1.2 ■ PRIME FACTORS

A number that has only two factors, itself and 1, is called a **prime number**. 11 and 17 are each prime numbers. A number that has more than two factors is called a **composite number**. A composite number can be written as a product of **prime factors**.

EXAMPLE 1.1 Write 168 as a product of prime factors.

Solution $168 = 2 \cdot 84 = 2 \cdot 2 \cdot 42 = 2 \cdot 2 \cdot 2 \cdot 21 = 2 \cdot 2 \cdot 2 \cdot 7 \cdot 3$

When factoring numbers, watch for these situations to occur.

1. If a number ends in 0, it is divisible by 10.
2. If a number ends in 5, it is divisible by 5.
3. If a number is even, it is divisible by 2.
4. If the sum of the digits of a number is divisible by 3, the number is divisible by 3.

EXAMPLE 1.2   Factor 513.

Solution   Check the sum of the digits. $5 + 1 + 3 = 9$. 9 is divisible by 3; so is 513.

$$513 = 3 \cdot 171$$

$1 + 7 + 1 = 9$. 9 is divisible by 3; so is 171.

$$513 = 3 \cdot 171 = 3 \cdot 3 \cdot 57$$

$5 + 7 = 12$. 12 is divisible by 3; so is 57.

$$513 = 3 \cdot 3 \cdot 3 \cdot 19$$

19 is prime.
The prime factors of 513 are 3, 3, 3, and 19.

SECTION 1.2 EXERCISES   For problems 1 through 5, write each number as a product of its prime factors.

1. 18
2. 24
3. 125
4. 30
5. 42
6. 330
7. 256
8. 204
9. 1020
10. 7700
11. 361
12. List all the prime numbers below 100.

1.3 ■ REDUCING FRACTIONS

Reducing a fraction means finding a fraction with smaller numbers that is equivalent to (has the same value as) the original fraction. Prime factors are used to reduce fractions to the lowest possible terms. To reduce a fraction to lowest terms, factor the numerator into prime factors, factor the denominator into prime factors, and cancel factors that are common to both numerator and denominator. Multiply the remaining factors in the numerator to form the numerator of the result, and multiply the remaining factors in the denominator to form the denominator of the result.

EXAMPLE 1.3   Reduce 66/105.

Solution

$$\frac{66}{105} = \frac{3 \times 22}{5 \times 21} = \frac{\cancel{3} \times 2 \times 11}{5 \times \cancel{3} \times 7} = \frac{22}{35}$$

EXAMPLE 1.4 Reduce 70/154.

Solution
$$\frac{70}{154} = \frac{2 \times 35}{2 \times 77} = \frac{2 \times 7 \times 5}{2 \times 7 \times 11} = \frac{5}{11}$$

SECTION 1.3 EXERCISES Reduce each fraction to lowest terms.

1. 15/20
2. 49/63
3. 24/30
4. 54/66
5. 65/85
6. 112/126

1.4 ■ RAISING FRACTIONS TO HIGHER TERMS

Raising fractions to higher terms is the opposite process from reducing a fraction to lowest terms. The denominator of the resulting fraction is a multiple of the original denominator. To raise a fraction to higher terms, determine what number the original denominator must be multiplied by to become the new denominator. Multiply both the numerator and the denominator by this same number so that the value of the original fraction will not be altered.

EXAMPLE 1.5 Express 2/7 as a fraction with a denominator of 63.

Solution 7 must be multiplied by 9 to get 63. Multiply the numerator and the denominator by 9.

$$\frac{2 \times 9}{7 \times 9} = \frac{18}{63}$$

EXAMPLE 1.6 Express 25/17 as a fraction with a denominator of 136.

Solution 17 divides into 136 eight times. Multiply the numerator and the denominator by 8.

$$\frac{25}{17} = \frac{25 \times 8}{17 \times 8} = \frac{200}{136}$$

This process will be used for combining fractions with unlike denominators.

SECTION 1.4 EXERCISES
1. Express 5/7 as a fraction with a denominator of 105.
2. Express 2/3 as a fraction with a denominator of 39.
3. Express 6/11 as a fraction with a denominator of 132.
4. Express 4/5 as a fraction with a denominator of 105.

5. Express 9/10 as a fraction with a denominator of 710.
6. Express 8/13 as a fraction with a denominator of 390.

1.5 ■ ADDING AND SUBTRACTING LIKE FRACTIONS

Two or more fractions whose denominators are identical are called **like fractions**. To add or subtract like fractions, simply add or subtract the numerators and use the like denominator as the denominator of the result.

$$\frac{a}{d} + \frac{b}{d} - \frac{c}{d} = \frac{a + b - c}{d}$$

EXAMPLE 1.7 Add these fractions: 12/5 + 3/5 + 6/5 = .

Solution The denominators are like. Add the numerators.

$$\frac{12}{5} + \frac{3}{5} + \frac{6}{5} = \frac{21}{5}$$

EXAMPLE 1.8 Subtract these fractions: 12/17 − 9/17 = .

Solution The denominators are like. Subtract the numerators.

$$12/17 - 9/17 = 3/17$$

SECTION 1.5 EXERCISES Combine these fractions. Reduce the result to lowest terms.

1. 15/4 + 6/4 + 9/4 =
2. 5/7 + 12/7 + 8/7 =
3. 34/12 − 8/12 =
4. 82/3 − 32/3 + 14/3 =

1.6 ■ LEAST COMMON MULTIPLE

If a quotient is a whole number (no remainder), then the dividend is a *multiple* of the divisor. For example, 7 divides evenly into 14, 21, 28, and 35; so 14, 21, 28, and 35 are multiples of 7.

The product of two or more factors is a **common multiple** of each of the factors. For example, 54 is a common multiple of 6 and 9. However, the product of two or more factors may not yield the **least common multiple** (**LCM**) of the numbers. 54 is not the LCM of 6 and 9 since 6 and 9 will each divide into a smaller number, 18. To find an LCM of two or more numbers, break each number into its prime factors. Each prime factor must be a factor in the LCM the maximum number of times that it appears in any of the numbers.

EXAMPLE 1.9 Find the least common multiple of 108, 42, and 98.

Solution Break 108, 42, and 98 into their prime factors.

$$108 = 2 \cdot 2 \cdot 3 \cdot 3 \cdot 3$$
$$42 = 2 \cdot 3 \cdot 7$$
$$98 = 2 \cdot 7 \cdot 7$$

2 appears twice as a factor of 108, 3 appears three times as a factor of 108, and 7 appears twice as a factor of 98. The factors of the LCM are two 2s, three 3s, and two 7s.

$$\text{LCM} = 2 \cdot 2 \cdot 3 \cdot 3 \cdot 3 \cdot 7 \cdot 7 = 5292$$

Note that the prime factors of each number appear in the LCM. 5292 is the smallest number that 108, 42, and 98 will divide into.

EXAMPLE 1.10 Find the LCM of 36, 108, 40, and 200.

Solution:

$$36 = 2 \cdot 2 \cdot 3 \cdot 3$$
$$108 = 2 \cdot 2 \cdot 3 \cdot 3 \cdot 3$$
$$40 = 2 \cdot 4 \cdot 5$$
$$200 = 2 \cdot 4 \cdot 5 \cdot 5$$

The LCM contains two 2s, three 3s, one 4, and two 5s.

$$\text{LCM} = 2 \cdot 2 \cdot 3 \cdot 3 \cdot 3 \cdot 4 \cdot 5 \cdot 5 = 10\ 800$$

SECTION 1.6 EXERCISES

Find the LCM for each of the following sets of numbers.

1. 2, 8, 12
2. 5, 10, 15
3. 49, 14, 21, 98
4. 44, 140, 385
5. 30, 42, 54
6. 770, 132, 60, 210

1.7 ■ LEAST COMMON DENOMINATOR AND ADDING AND SUBTRACTING UNLIKE FRACTIONS

Two or more fractions whose denominators are not identical are called **unlike fractions**. Before unlike fractions can be added or subtracted, the least common multiple of the denominators must be found. The least common multiple of the denominators is called the **least common denominator (LCD)**. Once the LCD is determined, each fraction is raised to higher terms with the LCD as the denominator of each.

1.7 LEAST COMMON DENOMINATOR AND ADDING AND SUBTRACTING UNLIKE FRACTIONS

EXAMPLE 1.11 Add 2/5 + 9/14 + 5/4 = .

Solution 2/5, 9/14, and 5/4 are unlike fractions. Find their LCD.

$$5 \text{ is prime.}$$
$$14 = 2 \cdot 7$$
$$4 = 2 \cdot 2$$

The LCD contains one 5, one 7, and two 2s.

$$\text{LCD} = 5 \cdot 7 \cdot 2 \cdot 2 = 140$$

Raise each fraction to higher terms with 140 as a denominator. To raise 2/5 to higher terms, 5 must be multiplied by the remaining factors of the LCD. Multiply the numerator and the denominator by 7·2·2 or 28.

$$2/5 = (2 \cdot 28)/(5 \cdot 28) = 56/140$$

To raise 9/14 to higher terms with 140 as a denominator, multiply the numerator and the denominator by 10.

$$9/14 = (9 \cdot 10)/(14 \cdot 10) = 90/140$$

To raise 5/4 to higher terms with 140 as a denominator, 4 must be multiplied by the remaining factors of the LCD. Multiply the numerator and the denominator by 5·7 or 35.

$$5/4 = (5 \cdot 35)/(4 \cdot 35) = 175/140$$
$$2/5 + 9/14 + 5/4 = 56/140 + 90/140 + 175/140 = 321/140$$

EXAMPLE 1.12 Subtract 31/150 − 31/210 = .

Solution The fractions are unlike. Find the LCD.

$$150 = 2 \cdot 3 \cdot 5 \cdot 5$$
$$210 = 2 \cdot 3 \cdot 5 \cdot 7$$

The LCD contains one 2, one 3, two 5s, and one 7.

$$\text{LCD} = 2 \cdot 3 \cdot 5 \cdot 5 \cdot 7 = 1050$$

To raise 31/150 to higher terms with 1050 as the new denominator, 150 must be multiplied by the factors of the LCD not included in 150. Multiply the numerator and the denominator by 7.

$$(31 \cdot 7)/(150 \cdot 7) = 217/1050$$

Likewise, to raise 31/210 to higher terms, multiply the numerator and the denominator by 5.

$$(31 \cdot 5)/(210 \cdot 5) = 155/1050$$
$$31/150 - 31/210 = 217/1050 - 155/1050 = 62/1050$$

Both the numerator and the denominator are even, so divide them by 2.

$$31/150 - 31/210 = 31/525$$

1.7.1 Calculator $a^{b/c}$ Key

Some calculators have an $a^{b/c}$ key that is used for operations involving fractions. It allows the operator to enter data in fractions and yields an answer in fraction form. The $a^{b/c}$ key is used to inform the calculator that a numerator has been entered and that a denominator will follow. For example, to enter the last example, press 31 $a^{b/c}$ 150 − 31 $a^{b/c}$ 210 = . 31/525 should appear in the display.

SECTION 1.7 EXERCISES

1. 2/3 + 3/4 =
2. 5/6 + 3/8 + 5/12 =
3. 13/33 − 1/3 =
4. 91/110 + 61/66 − 19/30 =
5. 23/1125 − 17/1000 =
6. 16/245 − 13/343 =

1.8 ■ MULTIPLYING FRACTIONS

Compared to addition and subtraction of unlike fractions, multiplication of fractions is easy. First reduce the fractions to lowest terms. This makes numbers smaller and easier to work with. The reduction can take place between the numerator of one fraction and the denominator of another. That is, a factor in the numerator of one fraction can cancel the same factor in the denominator of another fraction. When the fractions have been reduced to lowest terms, multiply numerator times numerator to form the numerator of the result, and multiply denominator times denominator to form the denominator of the result.

$$\frac{a}{b} \times \frac{c}{d} = \frac{a \times c}{b \times d}$$

EXAMPLE 1.13 Multiply these fractions: 12/21 · 14/9

Solution First, reduce to lowest terms.

$$\frac{4 \times \cancel{3}}{7 \times \cancel{3}} \times \frac{2 \times \cancel{7}}{3 \times 3}$$

Now multiply numerator times numerator to form the numerator of the result, and multiply denominator times denominator to form the denominator of the result.

$$\frac{4}{1} \times \frac{2}{3 \times 3} = \frac{8}{9}$$

Resist the temptation to grab the calculator to convert to decimal fractions. Often, fractions will cancel and leave an exact result, whereas converting to decimal form leaves an approximation. The result of the last example as worked on a calculator is 12/21 · 14/9 = 0.8889. The same process applies to the multiplication of more than two fractions.

EXAMPLE 1.14 Multiply 60 · 2/5.

Solution Think of 60 as 60/1.

$$12\,\frac{\cancel{60}}{1} \times \frac{2}{\cancel{5}} = 12 \times 2 = 24$$

EXAMPLE 1.15 Multiply these fractions: 15/11 · 35/33 · 66/45

Solution

$$\frac{\cancel{3}\times\cancel{5}}{11} \times \frac{7\times 5}{\cancel{3}\times\cancel{11}} \times \frac{2\times\cancel{3}\times\cancel{11}}{3\times\cancel{3}\times\cancel{5}} = \frac{5\times 7\times 2}{11\times 3} = \frac{70}{33}$$

SECTION 1.8 EXERCISES

1. 15/4 · 4/55 =
2. 66/75 · 25/42 =
3. 5/6 · 3/50 · 21/2 =
4. 8/15 · 80/3 · 14/3 · 5/672 =
5. 2/3 · 7/9 · 3 · 81/84 =

1.9 ■ DIVIDING FRACTIONS

To divide by a fraction, multiply by its **reciprocal**. A reciprocal is formed by inverting the fraction; the numerator becomes the denominator and vice versa.

$$\frac{\frac{a}{b}}{\frac{c}{d}} = \frac{a}{b} \times \frac{d}{c} = \frac{a\times d}{b\times c}$$

EXAMPLE 1.16 Divide 13/42 by 81/28.

Solution Multiply 13/42 by the reciprocal of 81/28.

$$\frac{\frac{13}{42}}{\frac{81}{28}} = \frac{13}{42} \times \frac{28}{81}$$

Proceed as in the multiplication of fractions.

$$\frac{13}{\cancel{2}\times\cancel{7}\times 3} \times \frac{2\times\cancel{2}\times\cancel{7}}{3\times 3\times 3\times 3} = \frac{13\times 2}{3\times 3\times 3\times 3\times 3} = \frac{26}{243}$$

EXAMPLE 1.17 Divide 3/8 by 5.

Solution Think of 5 as 5/1.

$$\frac{3}{8} / \frac{5}{1} = \frac{3}{8} \times \frac{1}{5} = \frac{3}{40}$$

SECTION 1.9 EXERCISES

1. $20/21 \div 5/14 =$
2. $3/26 \div 22/13 =$
3. $22/15 \div 19/10 =$
4. $231/20 \div 154/95 =$
5. $32/35 \div 14/15 =$
6. $343/3 \div 7/12 =$

1.10 ■ COMPLEX FRACTIONS

Complex fractions have more than one division line, for example $(4/5)/7$ or $(1 - 2/3)/(1 + 5/4)$. Simplifying complex fractions involves applying the principles of fractions studied so far.

EXAMPLE 1.18 Evaluate $4/5 \,/\, 7$.

Solution 7 is the same as 7/1.

$$4/5 \div 7/1 = 4/5 \cdot 1/7 = 4/35$$

EXAMPLE 1.19 Simplify $(3/4 - 2/3)/(1 + 5/4) =$.

Solution The numerator consists of two unlike fractions. Find their LCD and raise each fraction to higher terms using the LCD as the new denominator. Write the denominator as the sum of two like fractions by writing 1 as 4/4.

$$\frac{\frac{3}{4} - \frac{2}{3}}{1 + \frac{5}{4}} = \frac{\frac{9}{12} - \frac{8}{12}}{\frac{4}{4} + \frac{5}{4}} = \frac{\frac{1}{12}}{\frac{9}{4}}$$

Now invert 9/4 and multiply.

$$\frac{\frac{1}{12}}{\frac{9}{4}} = \frac{1}{12} \times \frac{4}{9} = \frac{1}{\cancel{2} \times \cancel{2} \times 3} \times \frac{\cancel{2} \times \cancel{2}}{9} = \frac{1}{27}$$

The next example will be solved two ways. In the first approach the numerator is combined into one fraction, the denominator is combined into one fraction, and then the numerator is divided by the denominator. In the second approach all fractions in the numerator and the denominator are eliminated by multiplying both numerator and denominator by the LCD of all the fractions in the numerator and denominator.

EXAMPLE 1.20 Simplify.

$$\frac{\frac{2}{3} - \frac{1}{5}}{\frac{3}{4} + \frac{1}{2}}$$

Solution 1 Combine the numerator into one fraction and the denominator into one fraction. Divide the numerator by the denominator.

$$\frac{\frac{2}{3}-\frac{1}{5}}{\frac{3}{4}+\frac{1}{2}} = \frac{\frac{10}{15}-\frac{3}{15}}{\frac{3}{4}+\frac{2}{4}} = \frac{\frac{7}{15}}{\frac{5}{4}} = \frac{7}{15} \times \frac{4}{5} = \frac{28}{75}$$

Solution 2 Find the LCD of all four fractions in the numerator and the denominator.

$$4 = 2 \cdot 2$$

2, 3, and 5 are prime.

$$\text{LCD} = 2 \cdot 2 \cdot 3 \cdot 5 = 60$$

Multiply both numerator and denominator by 60 to eliminate the four fractions.

$$\frac{60\left[\frac{2}{3}-\frac{1}{5}\right]}{60\left[\frac{3}{4}+\frac{1}{2}\right]} = \frac{40-12}{45+30} = \frac{28}{75}$$

SECTION 1.10 EXERCISES

Simplify.

1. $$\frac{1-\frac{7}{9}}{\frac{17}{3}-3} =$$

2. $$\frac{3-\frac{3}{7}}{\frac{2}{5}+\frac{2}{7}} =$$

3. $$\frac{\frac{2}{15}-\frac{3}{35}}{\frac{1}{60}+\frac{2}{105}} =$$

4. $$1 + \frac{1}{1 + \frac{1}{1 + \frac{1}{6}}}$$

1.11 ■ POSITIVE INTEGER EXPONENTS

Positive whole numbers (1, 2, 3, . . .) are called **positive integers**. This section investigates the result of using a positive integer as an exponent. Exponents are used to represent repeated multiplication. 3^4 represents the product $3 \cdot 3 \cdot 3 \cdot 3$. 3 is called the base and 4 is called the **exponent**. The exponent represents the number of times the base is to be used as a factor. 3^4 is read "three raised to the fourth power" or simply "three to the fourth." A quantity raised to the third power is said to be **cubed**. 4^3 is read "four cubed" and equals 4·4·4 or 64. A quantity raised to the second power is said to be **squared**. 12^2 is read "twelve squared" and equals 12·12 or 144. Expressions written with exponents are called **exponential quantities**.

EXAMPLE 1.21   Evaluate 2^{10}.

Solution $2^{10} = 2 \cdot 2 \cdot 2 \cdot 2 \cdot 2 \cdot 2 \cdot 2 \cdot 2 \cdot 2 \cdot 2 = 1024$

EXAMPLE 1.22   Evaluate 10^4.

Solution $10^4 = 10 \cdot 10 \cdot 10 \cdot 10 = 10{,}000$

1.11.1 Calculator y^x Key, or x^y Key, or ^ Key

Scientific calculators have a y^x key that is used to raise a quantity to a power. To raise a number to a power, enter the base, press the y^x key (or x^y), enter the exponent, and press =. The correct result should appear in the display. The ^ key, pronounced "caret key," is used in the same way. Enter the base number, press ^, enter the exponent, and press =.

If you are uneasy about a particular key on your calculator, try a simple example that you know the answer to and then proceed to a more difficult problem.

EXAMPLE 1.23   Use your calculator to evaluate 2^4.

Solution   Enter this sequence: **2** x^y **4 =** or **2 ^ 4 =**
16 should appear in the display. If your calculator appears to use a different method of handling exponents, consult your operator's guide.

EXAMPLE 1.24   Evaluate the expression 1.4^6.

Solution   Enter this sequence: **1.4** x^y **6 =** or **1.4 ^ 6 =**

The display should show 7.529536.

If the result of this operation yields a large enough number, the calculator will display the result in scientific notation. This format will be covered in a later chapter.

1.11.2 Calculator x^2 Key

The x^2 key is used to square whatever number appears in the display. To square a number, simply enter the number and then press the x^2 key.

EXAMPLE 1.25   Use your calculator to evaluate π^2.

Solution   Find the π key. If π is written above one of the keys, it is called a **second function,** and the 2nd key must be pressed first, followed by π. Then press the x^2 key. The result appears without hitting the = key.

1.11.3 Calculator +/− Key

Many calculators have a +/− key used to toggle (change) the sign of the number in the display. To enter −6, press 6 and then the +/− key. Other calculators have a − key that is separate and different from the subtract key. To enter a negative number, press − and then the number.

EXAMPLE 1.26   Use your calculator to evaluate $(-4.11)^2$.

Solution   Since the − is inside the parentheses, it should appear in the display of your calculator before you press the x^2 key. Enter **4.11 +/− x^2 or −4.11 x^2**. The result should be 16.8921.

Note that when a negative number is raised to an even power, the result is positive; and when raised to an odd power, the result is negative.

1.11.4 Calculator x^3 Key

Some calculators also have an x^3 key that is used to cube the number in the display. To cube a number, enter the number and press the x^3 key. The result will appear without pressing the = key.

EXAMPLE 1.27   Use your calculator to evaluate $(-2.615)^3$.

Solution   Since the − sign is within the parentheses, enter **2.615 +/− x^3. −17.881958** should appear in the display.

SECTION 1.11 EXERCISES

Use your calculator to evaluate these expressions.

1. 1.86^2
2. -13.6^3
3. $\pi^{3.5}$
4. $(-3.07)^3$

1.12 ■ NEGATIVE INTEGER EXPONENTS

The whole numbers less than zero (−1, −2, −3, . . .) are called **negative integers.** This section investigates the use of negative integers as exponents. Negative exponents are used to represent reciprocal quantities. An expression with a negative exponent can be rewritten as the reciprocal of the same expression with a positive exponent.

$$a^{-n} = 1/a^n \text{ or } 1/a^{-n} = a^n$$

For example, $5^{-3} = 1/5^3 = 1/(5 \times 5 \times 5) = 1/125$.

EXAMPLE 1.28 Evaluate 10^{-3}.

Solution $10^{-3} = 1/10^3 = 1/1000$ or 0.001

EXAMPLE 1.29 Evaluate $1/10^{-4}$.

Solution $1/10^{-4} = 10^4 = 10{,}000$

The y^x or x^y key on your calculator will handle negative exponents.

EXAMPLE 1.30 Evaluate 5.6^{-3}.

Solution Enter this sequence on your calculator: **5.6 y^x 3 +/− =**. 0.00569424 should appear.

SECTION 1.12 EXERCISES

Express as decimal numbers.

1. 2^{-5}
2. 3^{-4}
3. $1/5^{-2}$
4. $1/2^{-8}$
5. 4.2^{-3}
6. 8.2^{-2}
7. $1/10^{-6}$
8. 10^{-5}

1.13 ■ RATIONAL EXPONENTS

In the expression $\sqrt[3]{8}$, 3 is called the **index**, 8 is called the **radicand**, and the symbol $\sqrt{}$ is called the **radical**. This notation is used to mean "what number times itself three times equals 8?" $2 \cdot 2 \cdot 2 = 8$, so $\sqrt[3]{8} = 2$. 2 is the third root of 8 or cube root of 8. In addition to the radical form of notation, this same information can be written in exponential form: $8^{1/3} = 2$. The index becomes the denominator of the exponent. In general, $\sqrt[b]{a} = a^{1/b}$. The second root and third root are given special names. If no index is written above the index, the index is understood to be 2. A radical with no index or an exponent of 1/2 indicates second root or **square root**. $\sqrt{a} = a^{1/2}$. An index of three or an exponent of 1/3 indicates third root or **cube root**. $\sqrt[3]{a} = a^{1/3}$. Indices above 3 are not given names. They are referred to as fourth root, fifth root, and so on.

1.13 RATIONAL EXPONENTS

EXAMPLE 1.31 Evaluate $100^{1/2}$.

Solution $100^{1/2}$ means $\sqrt{100}$ or what number times itself equals 100? $100^{1/2} = 10$.

EXAMPLE 1.32 Evaluate $27^{1/3}$.

Solution $27^{1/3}$ means $\sqrt[3]{27}$ or what number times itself three times equals 27? $27^{1/3} = 3$.

The numerator of an exponent of a number is the power that the number is raised to. $a^{b/c}$ means raise a to the b power and take the c root of the result, or take the c root of a first and then raise the result to the power b. The order does not matter.

EXAMPLE 1.33 Evaluate $27^{2/3}$.

Solution 1 $27^{2/3} = (27^{1/3})^2 = 3^2 = 9$

Solution 2 $27^{2/3} = (27^2)^{1/3} = (729)^{1/3} = 9$

The answer is the same either way, but the numbers in Solution 1 are easier to handle.

EXAMPLE 1.34 Evaluate $25^{-5/2}$.

Solution "Use up" the negative sign by moving 25 into the denominator.

$$25^{-5/2} = 1/25^{5/2} = 1/(25^{1/2})^5 = 1/5^5 = 1/3125$$

Some calculators have an $x^{a/b}$ key that allows a fractional exponent to be entered directly. If your calculator does not have such a key, use the x^y key with parentheses enclosing the exponent.

EXAMPLE 1.35 Evaluate $32^{4/5}$.

Solution The calculator sequence is **32 x^y (4 ÷ 5) =** . The result is 16.

EXAMPLE 1.36 Evaluate $1.95^{5/7}$.

Solution The calculator sequence is **1.95 x^y (5 ÷ 7) =** . The result is 1.61127.

18 ■ CHAPTER 1 ARITHMETIC

SECTION 1.13 EXERCISES

Express in radical form.

1. $12^{1/2}$
2. $345^{1/3}$
3. $5^{2/5}$
4. $82^{5/7}$

Express in exponential form.

5. $\sqrt[3]{75^5}$
6. $\sqrt{100^3}$
7. $(\sqrt{100})^3$

Evaluate.

8. $144^{1/2}$
9. $125^{1/3}$
10. $27^{4/3}$
11. $\sqrt[3]{125^2}$

1.14 ■ RULES FOR OPERATIONS WITH EXPONENTS

1.14.1 Rule 1—Multiplying Exponential Expressions

The next example applies rules already studied. It will guide us to a general principle that applies to the multiplication of exponential quantities.

EXAMPLE 1.37 Evaluate $2^4 \cdot 2^3$ and evaluate 2^7.

Solution
$$2^4 \cdot 2^3 = 16 \cdot 8 = 128$$
$$2^7 = 128$$

The exponential expressions being multiplied have like bases. The product is the same base number with an exponent that is the sum of the original exponents. In other words,

$$2^4 \cdot 2^3 = 2^7$$

This relationship is always true. So, to multiply two or more exponential expressions that have like bases, use the common base and add the exponents.

$$a^b \cdot a^c = a^{b+c}$$

EXAMPLE 1.38 Evaluate $10^7 \cdot 10^4 \cdot 10^5$.

Solution The bases are like, so our rule applies. The product will have the same base and its exponent will be the sum of the original exponents.

$$10^7 \cdot 10^4 \cdot 10^5 = 10^{(7+4+5)} = 10^{16}$$

1.14.2 Rule 2—Dividing Exponential Expressions

The next example will lead us to a rule for evaluating quotients of exponential quantities.

1.14 RULES FOR OPERATIONS WITH EXPONENTS ■ 19

EXAMPLE 1.39 Evaluate $4^5/4^2$ and evaluate 4^3.

Solution
$$\frac{4^5}{4^2} = \frac{4 \times 4 \times 4 \times 4 \times 4}{4 \times 4} = 4 \times 4 \times 4 = 64$$
$$4^3 = 4 \times 4 \times 4 = 64$$

The exponential expressions being divided have like bases. The quotient is the same base number with an exponent that is the difference of the exponent in the numerator minus the exponent in the denominator. In other words,

$$4^5 / 4^2 = 4^{5-2} = 4^3 = 64$$

This relationship is always true. So, to divide two exponential expressions that have like bases, use the common base and subtract the exponents (exponent of the numerator minus the exponent of the denominator).

$$a^b / a^c = a^{b-c}$$

EXAMPLE 1.40 Evaluate $10^4/10^8$.

Solution $10^8/10^4 = 10^{8-4} = 10^4 = 10^4 = 10000$

EXAMPLE 1.41 Evaluate $2^9/2^5$.

Solution $2^9/2^5 = 2^{9-5} = 2^4 = 16$

EXAMPLE 1.42 Evaluate $10^6 \cdot 10^3/(10^2 \cdot 10^4) =$.

Solution $10^6 \cdot 10^3/(10^2 \cdot 10^4) = 10^{6+3-2-4} = 10^3 = 1000$

1.14.3 Rule 3—Raising a Power to a Power
Once again, an example based on previous knowledge will lead us to a rule for raising a power to a power.

EXAMPLE 1.43 Evaluate $(10^2)^4$.

Solution
$$(10^2)^4 = 10^2 \cdot 10^2 \cdot 10^2 \cdot 10^2$$
$$= 100 \cdot 100 \cdot 100 \cdot 100$$
$$= 100\,000\,000$$
$$= 10^8$$

As the example suggests, to raise a power to a power, multiply the exponents.

$$(a^b)^c = a^{bc}$$

EXAMPLE 1.44   Evaluate $(10^4)^3$.

Solution $(10^4)^3 = 10^{4 \cdot 3} = 10^{12}$

EXAMPLE 1.45   Evaluate $(10^3)^2$.

Solution $(10^3)^2 = 10^6$

1.14.4 Rule 4—Raising a Product to a Power
To raise a product to a power, raise each of the factors to the power.

$$(ab)^c = a^c b^c$$

EXAMPLE 1.46   Evaluate $(3 \times 10^2)^3$.

Solution $(3 \times 10^2)^3 = 3^3 \times (10^2)^3 = 27 \times 10^6$

EXAMPLE 1.47   Simplify $50^{1/2}$.

Solution $50^{1/2} = (5 \cdot 5 \cdot 2)^{1/2} = (5 \cdot 5)^{1/2} 2^{1/2} = 5 \cdot 2^{1/2}$ or $5(2)^{1/2}$

EXAMPLE 1.48   Simplify $48^{1/3}$.

Solution $48^{1/3} = (2 \cdot 2 \cdot 2 \cdot 2 \cdot 3)^{1/3} = (2 \cdot 2 \cdot 2)^{1/3}(2 \cdot 3)^{1/3} = 2(2 \cdot 3)^{1/3} = 2(6)^{1/3}$

The last two examples have shown that the root of a composite number can be simplified by factoring the number into its prime factors and then applying this rule. Search for factors that are duplicated the same number of times as the index of the root. That factor can be removed from the radical. For example, when taking the cube root, look for any factor that is repeated three times. Remove that factor from the radical: $(a \cdot a \cdot a)^{1/3} = a$. Factors that stand alone or that are not repeated enough times remain under the radical.

EXAMPLE 1.49　Simplify $248832^{1/4}$.

Solution

$$(2 \cdot 2 \cdot 2 \cdot 2 \cdot 2 \cdot 2 \cdot 2 \cdot 2 \cdot 2 \cdot 3 \cdot 3 \cdot 3 \cdot 3 \cdot 3)^{1/4}$$
$$= (2 \cdot 2 \cdot 2 \cdot 2)^{1/4} (2 \cdot 2 \cdot 2 \cdot 2)^{1/4} (3 \cdot 3 \cdot 3 \cdot 3)^{1/4} (2 \cdot 2 \cdot 3)^{1/4}$$
$$= 2 \cdot 2 \cdot 3 \cdot (2 \cdot 2 \cdot 3)^{1/4} = 12(12)^{1/4}$$

EXAMPLE 1.50　Simplify $1/3^{1/2}$.

Solution　It is not considered "good form" to leave a root in the denominator. Here, "simplify" means write an equivalent fraction that has no radical or root in the denominator. This rule shows us how.

$$\frac{1}{\sqrt{3}} = \frac{1}{\sqrt{3}} \times \frac{\sqrt{3}}{\sqrt{3}} = \frac{\sqrt{3}}{\sqrt{3} \times \sqrt{3}} = \frac{\sqrt{3}}{3}$$

This process is called **rationalizing the denominator**.

EXAMPLE 1.51　Rationalize $5/30^{1/2}$.

Solution

$$\frac{5}{\sqrt{30}} = \frac{5}{\sqrt{30}} \times \frac{\sqrt{30}}{\sqrt{30}} = \frac{5\sqrt{30}}{\sqrt{30}\sqrt{30}} = \frac{5\sqrt{30}}{30} = \frac{\sqrt{30}}{6}$$

EXAMPLE 1.52　Rationalize $5/35^{1/3}$.

Solution　$35^{1/3} \cdot 35^{2/3} = 35$. Multiply the numerator and the denominator by $35^{2/3}$.

$$\frac{5}{35^{1/3}} \times \frac{35^{2/3}}{35^{2/3}} = \frac{5(35)^{2/3}}{35^{\,}7} = \frac{35^{2/3}}{7}$$

1.14.5　Rule 5—Raising a Quotient to a Power
To raise a quotient to a power, raise the numerator to the power and raise the denominator to the power.

$$(a/b)^c = a^c/b^c$$

Test the rule on a problem for which you already know the answer.

EXAMPLE 1.53	Evaluate $(4/2)^3$.
Solution	$(4/2)^3 = 2^3 = 8$ Following the rule:

$$(4/2)^3 = 4^3 / 2^3 = 64/8 = 8 \quad \text{It checks!}$$

This rule also applies to negative exponents.

EXAMPLE 1.54	Evaluate $(3/2)^{-4}$.
Solution 1	In this solution, the negative exponent is applied to the numerator and the denominator, and then the negative powers are dealt with.

$$(3/2)^{-4} = 3^{-4}/2^{-4} = 1/3^{+4} \cdot 2^4 = 16/81$$

Solution 2 In this solution, the negative exponent is converted into a positive exponent by taking the reciprocal of the base. Then the positive exponent is applied to the numerator and denominator.

$$(3/2)^{-4} = 1/(3/2)^4 = 1/(3^4/2^4) = 2^4/3^4 = 16/81$$

Here is a summary of the rules of operation of exponents.

1. $a^b \cdot a^c = a^{b+c}$
2. $a^b/a^c = a^{b-c}$
3. $(a^b)^c = a^{bc}$
4. $(ab)^c = a^c b^c$
5. $(a/b)^c = a^c/b^c$
6. $a^{-n} = 1/a^n$ also $1/a^{-n} = a^n$

These rules apply to rational exponents as well as to positive and negative integer exponents.

SECTION 1.14 EXERCISES

Evaluate.

1. $10^4 \cdot 10^5 =$
2. $2^7 \div 2^3 =$
3. $(5^4)^2 =$
4. $(6 \times 10^3)^2 =$
5. $(2/7)^3 =$
6. $(3/5)^{-3} =$
7. $10^5 \cdot 10^{-3} =$
8. $10^6/10^4 =$
9. $(10^3)^3 =$
10. $(12 \times 10^2)^2 =$
11. $(3/2)^2 =$
12. $3^{-3} =$

Simplify.

13. $250^{1/2} =$
14. $2401^{1/3} =$
15. $1/42^{1/2} =$
16. $10/2500^{1/3} =$

1.15 ■ ORDER OF OPERATIONS

The order in which the mathematical operations (addition, subtraction, multiplication, division, and raising expressions to an exponent) are performed alters the result of the problem. To achieve the correct result, operations must be performed in the following order.

1. Evaluate expressions inside parentheses. You might encounter *nested* parentheses, parentheses within parentheses, or parentheses within brackets within braces: { [( )] }. Evaluate the innermost set first and work outward.
2. Raise the quantities to their powers.
3. Perform multiplications and divisions left to right.
4. Perform additions and subtractions left to right.

EXAMPLE 1.55 $3 + 4 \cdot 6 =$.

Solution Multiply $4 \cdot 6$ before adding 3.

$$3 + 4 \cdot 6 = 3 + 24 = 27$$

An algebraic calculator will give this same result when $3 + 4 \cdot 6 =$ is entered. Try it.

EXAMPLE 1.56 $2 + 4/3 =$.

Solution Divide first, then add.

$$4/3 = 1\ 1/3 \quad 2 + 1\ 1/3 = 3\ 1/3$$

Enter $2 + 4/3 =$ into your calculator and you should get the equivalent answer in decimal form, 3.3333333.

EXAMPLE 1.57 $8 + 2(25/5 + 2) =$.

Solution Within the parentheses, divide and then add.

$$(25/5 + 2) = (5 + 2) = 7$$
$$8 + 2(7) =$$

Multiply and then add.

$$8 + 14 = 22$$

This can be entered as is if your calculator has parentheses. Note that the calculator has to be told that the 2 is being multiplied by the contents of the parentheses. Here is the sequence.

$$8 + 2 \times (\,25\,/\,5 + 2\,) =$$

To enter it without using the parentheses keys, start with the contents of the parentheses and work backwards.

$$25 / 5 + 2 = \times 2 + 8 =$$

EXAMPLE 1.58 $13 + 4^2$.

Solution First raise 4 to the second power and then add 13.

$$13 + 4^2 = 13 + 16 = 29$$

EXAMPLE 1.59

$$\left[\frac{-3^2 - 28/4 + 6^2}{2 + \frac{9}{3}} - 5/2 \right]^2 + 6$$

Solution Evaluate the quantity inside the brackets, square that number, and then add 6. Inside the brackets are two fractions; simplify the numerator of the first fraction.

$$-3^2 - 28/4 + 6^2 = -9 - 7 + 36 = -16 + 36 = +20$$

Simplify the denominator of the first fraction.

$$2 + 9/3 = 2 + 3 = 5$$

Evaluate the quantity inside the brackets.

$$20/5 - 5/2 = 8/2 - 5/2 = 3/2$$

Square 3/2 and add 6.

$$(3/2)^2 + 6 = 9/4 + 24/4 = 33/4$$

SECTION 1.15 EXERCISES

Work in longhand and then confirm on your calculator.

1. $4 \cdot 2 + 3 =$
2. $3 \cdot 2 + 4/2 - 1 =$
3. $5 + 4(2 + 3/5) =$
4. $8 + 2^2 =$
5. $(8 + 2)^2 =$
6. $-2^2 =$
7. $-2^{-2} =$
8. $(-2)^2 + 6 - (12 - 4 - 6)/(3-8) + 6 =$

1.16 ■ PERCENT

Percent means per hundred. Percent (%) is a measure of quantity per hundred. 13% means 13 out of every 100.

1.16.1 Converting a Decimal to a Percent

To express a decimal fraction as a percent, move the decimal point two places to the right and add a percent sign (%). 0.32 becomes 32%. 0.00049 becomes 0.049%.

1.16.2 Converting a Fraction to a Percent

To express a fraction as a percent, convert the fraction to a decimal fraction, move the decimal point two places to the right, and add the percent sign.

EXAMPLE 1.60 What percent of 47 is 36?

Solution
$$36/47 = 0.77$$
$$0.77 = 77\%$$
$$36 \text{ is } 77\% \text{ of } 47$$

EXAMPLE 1.61 Which is a better yield—36 out of 47 or 71 out of 103?

Solution From the last example we know that 36 is 77% of 47. What percent of 103 is 71?
$$71/103 = 0.69$$
$$0.69 = 69\%$$
$$36 \text{ out of } 47 \text{ is a better yield at } 77\%$$

1.16.3 Converting a Percent to a Decimal

To convert a percent to a decimal, drop the percent sign and move the decimal point two places to the left.

$$87\% = 0.87$$
$$4.62\% = 0.0462$$

1.16.4 Converting a Percent to a Fraction

To convert a percent to a fraction, drop the percent sign and create a fraction with 100 as the denominator. Reduce the fraction to lowest terms.

EXAMPLE 1.62 Convert 34% to a fraction.

Solution Drop the percent sign. Write a fraction with 34 as the numerator and 100 as the denominator: 34/100. Reduce to lowest terms: $34/100 = 2 \cdot 17/(2 \cdot 50) = 17/50$

Try these problems.

EXAMPLE 1.63 What is 18% of 945?

Solution "Of" means multiply. $18\% \cdot 945 = 0.18 \cdot 945 = 170$

EXAMPLE 1.64 If a technician earning an annual salary of $32 450.50 receives a 4.2% raise, what is her new salary?

Solution 1 $32 450.50 is 100% of the present salary. A 4.2% increase pushes the new salary up to 104.2%.

$$104.2\% \text{ of } \$32\,450.50 = 1.042 \times \$32\,450.50 = \$33\,813.42$$

Solution 2 Find 4.2% of the present salary to find the amount of the raise. Add the amount of the raise to the present salary.

$$4.2\% \text{ of } \$32\,450.50 = 0.042 \times \$32\,450.50 = \$1\,362.92$$
$$\$32\,450.50 + \$1\,362.92 = \$33\,813.42$$

EXAMPLE 1.65 A dual trace, 50 MHz, digital storage oscilloscope has a list price of $2535. If an educational discount of 15% is given, calculate the selling price.

Solution 1 100% of the list price is $2535. The oscilloscope is being sold for 100% − 15% or 85%. Calculate 85% of the list price.

$$100\% - 15\% = 85\%$$
$$85\% \text{ of } \$2535 = 0.85 \times \$2535 = \$2154.75$$

Solution 2 Find 15% of the list price to determine the discount. Subtract the discount from the list price.

$$15\% \text{ of } \$2535 = \$380.25$$
$$\$2535 - \$380.25 = \$2154.75$$

Which procedure do you prefer?

EXAMPLE 1.66 From a 1000-foot spool of 22-gauge wire the following lengths have been cut: 33 1/3', 102 1/2', 88 1/4', and 255 3/4'. What percent of the wire remains on the spool?

Solution Calculate the amount of wire that remains on the spool by subtracting the sum of the lengths cut from 1000 feet.

$$1000' - (33\ 1/3' + 102\ 1/2' + 88\ 1/4' + 255\ 3/4') =$$
$$1000' - (33\ 4/12' + 102\ 6/12' + 88\ 3/12' + 255\ 9/12') =$$

$$1000' - 478\ 22/12'$$
$$478\ 22/12' = 478 + 1\ 10/12' = 479\ 10/12'$$
$$1000' - 479\ 10/12' = 999\ 12/12' - 479\ 10/12' = 520\ 2/12'\text{ or}$$
520.167' remain on the spool. Express 520.167' as a percent of the total.
$$520.167/1000 = 0.520167\text{ or }52.0167\%$$

Approximately 52% of the wire remains on the spool.

Are you ready for an application of arithmetic to electronics?

1.16.5 Applying Percent to Resistor Values

One method manufacturers use to denote the resistance of a resistor is a standardized color code. Four colored bands around the body of the resistor identify the nominal or face value of the resistor in ohms (Ω) and the tolerance of that value. A color-coded resistor is shown in Figure 1-1.

FIGURE 1-1
Resistor Color Code

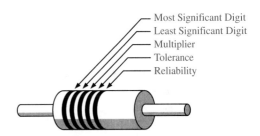

The colored stripes begin closer to one end of the resistor than the other. The stripe closest to one end identifies the first digit of the value according to a standardized color chart. Likewise, the second stripe identifies the second digit. The third stripe is the multiplier. It tells the number of zeros that follow the first two digits. Here is the color code chart for the first three bands.

Color	Value
Black	0
Brown	1
Red	2
Orange	3
Yellow	4
Green	5
Blue	6
Violet	7
Grey	8
White	9

EXAMPLE 1.67

What is the nominal value of a resistor whose first three bands are yellow, violet, and orange?

Solution Yellow is 4, violet is 7, and orange is 3. The nominal value is 47 with three zeros following. 47 000 Ω.

With the fourth band the manufacturer tells us the tolerance of that value according to this chart.

Color	Tolerance
Brown	1%
Red	2%
Gold	5%
Silver	10%
No band	20%

EXAMPLE 1.68

Suppose the resistor in the last example has a gold fourth band. Calculate its resistance.

Solution Its resistance is 47 000 + or −5% Ω. Think of 47 000 Ω as 100%. We need to calculate 95% and 105%.

$$95\% \text{ of } 47\,000 \text{ Ω} = 0.95 \cdot 47\,000 \text{ Ω} = 44\,650 \text{ Ω}$$
$$105\% \text{ of } 47\,000 \text{ Ω} = 1.05 \cdot 47\,000 \text{ Ω} = 49\,350 \text{ Ω}$$

This resistor can range from 44 650 to 49 350 Ω and be considered within tolerance (good).

EXAMPLE 1.69

Calculate the resistance of a resistor whose color bands are brown, black, black, and gold.

Solution Brown, black, and black means 10 with no additional zeros. Gold means 5%. Find 95% and 105% of 10 Ω.

$$95\% \text{ of } 10 \text{ Ω} = 0.95 \cdot 10 \text{ Ω} = 9.5 \text{ Ω}$$
$$105\% \text{ of } 10 \text{ Ω} = 1.05 \cdot 10 \text{ Ω} = 10.5 \text{ Ω}$$

EXAMPLE 1.70

A resistor measures 9297 Ω on an ohmmeter. Its color code is grey, red, red, and silver. Is the resistor good?

Solution Calculate the extremes or bounds and see whether the resistor falls within these bounds.
Grey is 8, red is 2, red is 2, and silver is 10%. The nominal value is 8200 Ω and the tolerance is + or − 10%. 100% is 8200 Ω. Find 90% and 110%.

$$90\% \text{ of } 8200 \, \Omega = 0.90 \cdot 8200 \, \Omega = 7380 \, \Omega$$
$$110\% \text{ of } 8200\% = 1.1 \cdot 8200 \, \Omega = 9020 \, \Omega$$

The resistor is out of tolerance on the high side. It is not good.

Higher precision resistors are identified by numbers stamped on the body of the resistor. See Figure 1-2. Three significant digits are given, followed by a multiplier.

FIGURE 1-2
Precision Resistors

EXAMPLE 1.71 A precision resistor is marked 3462. What is its nominal resistance?

Solution The first three digits are 346, followed by two zeros, 34 600 Ω.

On these precision resistors the tolerance is indicated by a letter according to this code.

Letter	Tolerance
F	1%
G	2%
J	5%
K	10%

EXAMPLE 1.72 Suppose the resistor in the last example has a tolerance code F. Calculate the bounds of its resistance.

Solution The resistance of the resistor is 34 600 + or − 1%. 100% is 34 600 Ω. Calculate 99% and 101%.

$$99\% \text{ of } 34\,600 \, \Omega = 0.99 \cdot 34\,600 \, \Omega = 34\,254 \, \Omega$$
$$101\% \text{ of } 34\,600 \, \Omega = 1.01 \cdot 34\,600 \, \Omega = 34\,946 \, \Omega$$

The resistance of a resistor marked 3462F can range from 34 254 to 34 946 Ω and still be considered good.

EXAMPLE 1.73 A resistor measures 345 Ω on an ohmmeter. It is stamped 3420G. Is the resistor good?

Solution 3420 means 342 with no zeros following, just 342 Ω. G means + or − 2%. Calculate 102% and 98% of 345 Ω.

$$102\% \text{ of } 345\ \Omega = 1.02 \cdot 345\ \Omega = 351.9\ \Omega$$
$$98\% \text{ of } 345\ \Omega = 0.98 \cdot 345\ \Omega = 338.1\ \Omega$$

The resistor falls within these bounds and is considered good.

An R in the code signifies **radix** or decimal point.

EXAMPLE 1.74 Calculate the resistance of a precision resistor that is labeled 7R4F.

Solution 7R4 means 7.4 Ω. F means + or − 1%. Calculate 101% and 99% of 7.4 Ω.

$$99\% \text{ of } 7.4\ \Omega = 0.99 \cdot 7.4\ \Omega = 7.33\ \Omega$$
$$101\% \text{ of } 7.4\ \Omega = 1.01 \cdot 7.4\ \Omega = 7.47\ \Omega$$

A resistor marked 7R4F can range from 7.33 to 7.47 Ω.

Some higher-quality color-coded resistors have a fifth band that signifies the reliability of the resistor. Reliability is the percent of the resistors that will fail during the first 1000 hours of operation. Here is the reliability color-code chart.

Color	Reliability (%)
Brown	1.0%
Red	0.1%
Orange	0.01%
Yellow	0.001%

EXAMPLE 1.75 In a batch of 150 000 resistors, how many can be expected to fail during the first 1000 hours of operation if their color code is brown, red, orange, gold, and yellow?

Solution The fifth band is yellow, which signifies a reliability of 0.001%. Calculate 0.001% of 150 000.

$$0.001\% \text{ of } 150\ 000\ \Omega = 0.00001 \cdot 150\ 000 = 1.5 \text{ resistors}$$

One or two resistors can be expected to fail per 1000 hours of operation.

SECTION 1.16 EXERCISES

Convert to percent.

1. 0.34
2. 1.02
3. 0.954
4. 7/10
5. 56/125
6. 250/1000
7. 34 is what % of 45?
8. 99 is what % of 207?

Convert to decimal.

9. 12.9%
10. 0.22%

Convert to fraction.

11. 45%
12. 20%
13. What is 39% of 176?
14. What is 67% of 2570?
15. If a gallon of gasoline is $1.099 per gallon and it increases to $1.159 per gallon, what is the % increase in cost?
16. If a technician earning $12.56 per hour gets a promotion and a 6.8% raise, what is the new hourly wage?
17. If a digital multimeter normally sells for $187, what will it cost during an 18% sale?

Calculate the nominal resistance.

18. Green, blue, and yellow
19. Brown, red, and green
20. 5622
21. 15R8

Calculate the upper and lower bounds of these resistors.

22. Yellow, violet, red, and gold
23. 3524G
24. 55R5F
25. Calculate the expected number of failures in a batch of 20,000 resistors that have a color code of grey, red, brown, silver, and red.

■ SUMMARY

1. In addition: addend+addend=sum.
2. In subtraction: minuend−subtrahend=difference.
3. In multiplication: factor · factor=product and multiplicand · multiplier=product.
4. In division: dividend/divisor=quotient and numerator/denominator=quotient.
5. Addition and multiplication are associative and commutative; subtraction and division are not.
6. A number that has only two factors, itself and 1, is called a prime number.
7. When factoring a composite number into prime factors, these facts are helpful:

If a number ends in 0, it is divisible by 10.

If a number ends in 5, it is divisible by 5.

If a number is even, it is divisible by 2.

If the sum of the digits of a number is divisible by 3, the number is divisible by 3.

8. To reduce a fraction to lowest terms, factor the numerator into prime factors, factor the denominator into prime factors, and cancel factors that are common to both numerator and denominator.
9. To raise a fraction to higher terms, determine what number the original denominator must be multiplied by to become the new denominator. Multiply both the numerator and the denominator by this same number so that the value of the original fraction will not be altered.
10. Fractions whose denominators are identical are called like fractions.
11. To add or subtract like fractions, simply add or subtract the numerators and use the like denominator as the denominator of the result.
12. If a quotient is a whole number (no remainder), then the dividend is a multiple of the divisor.
13. The smallest number that is the multiple of a set of numbers is called the least common multiple (LCM) of the numbers.
14. Two or more fractions whose denominators are not identical are called unlike fractions.
15. The least common multiple of the denominators of a set of fractions is called the least common denominator (LCD) of the fractions.
16. To add or subtract unlike fractions, find the least common denominator of the fractions. Raise each fraction to higher terms with the LCD as the denominator. The fractions are now like fractions and can be combined.
17. To multiply fractions, first reduce the fractions to lowest terms. The reduction can take place between the numerator of one fraction and the denominator of another. Then multiply numerator times numerator to form the numerator of the result, and multiply denominator times denominator to form the denominator of the result.
18. To divide by a fraction, multiply by its reciprocal.
19. Complex fractions have more than one division line.
20. Positive whole numbers (1, 2, 3, ...) are called positive integers.
21. Positive integer exponents are used to represent repeated multiplication.
22. Expressions written with exponents are called exponential quantities.
23. The whole numbers less than zero ($-1, -2, -3, \ldots$) are called negative integers.
24. An expression with a negative exponent can be rewritten as the reciprocal of the same expression with a positive exponent.

$$a^{-n} = 1/a^n \text{ also } 1/a^{-n} = a^n$$

25. In the expression $\sqrt[3]{8}$, 3 is called the index, 8 is called the radicand, and the symbol $\sqrt{}$ is called the radical.

26. The index of a radical can also be written as the denominator of an exponent. In general, $\sqrt[b]{a} = a^{1/b}$.
27. To multiply two or more exponential expressions that have like bases, use the common base and add the exponents.

$$a^b \cdot a^c = a^{b+c}$$

28. To divide two exponential expressions that have like bases, use the common base and subtract the exponents (exponent of the numerator minus the exponent of the denominator).

$$a^b / a^c = a^{b-c}$$

29. To raise a power to a power, multiply the exponents.

$$(a^b)^c = a^{bc}$$

30. To raise a product to a power, raise each of the factors to the power.

$$(ab)^c = a^c b^c$$

31. Writing an equivalent fraction that has no radical or root in the denominator is called rationalizing the denominator.
32. To raise a quotient to a power, raise the numerator to the power and raise the denominator to the power.

$$(a/b)^c = a^c/b^c$$

33. Perform operations in this order:
    a. Evaluate expressions inside parentheses. Evaluate the innermost set first and work outward.
    b. Raise the quantities to their powers.
    c. Perform multiplications and divisions left to right.
    d. Perform additions and subtractions left to right.
34. Percent means per hundred.
35. To express a decimal fraction as a percent, move the decimal point two places to the right and add a percent sign (%).
36. To express a fraction as a percent, convert the fraction to a decimal fraction, move the decimal point two places to the right, and add the percent sign.
37. To convert a percent to a decimal, drop the percent sign and move the decimal point two places to the left.
38. To convert a percent to a fraction, drop the percent sign and create a fraction with 100 as the denominator. Reduce the fraction to lowest terms.
39. The colored stripes on a resistor are codes that identify its resistance in ohms. The stripe closest to one end identifies the first digit of the value according to a standardized color chart. Likewise, the second stripe identifies the second digit. The third stripe is the multiplier. It tells the number of zeros that follow the first two

digits. The fourth stripe signifies the tolerance, and the fifth band signifies the reliability of the resistor.

SOLUTIONS TO CHAPTER 1 SECTION EXERCISES

Solutions to Section 1.1 Exercises

1. True
2. False Subtraction is not commutative.
3. True
4. False Division is not commutative.
5. False The answer to a division problem is called a quotient.
6. True
7. False The answer to a multiplication problem is called the product.
8. True

Solutions to Section 1.2 Exercises

1. $2 \cdot 3 \cdot 3$
2. $2 \cdot 2 \cdot 2 \cdot 3$
3. $5 \cdot 5 \cdot 5$
4. $2 \cdot 5 \cdot 3$
5. $2 \cdot 7 \cdot 3$
6. $5 \cdot 2 \cdot 11 \cdot 3$
7. $2 \cdot 2 \cdot 2 \cdot 2 \cdot 2 \cdot 2 \cdot 2 \cdot 2$
8. $3 \cdot 17 \cdot 2 \cdot 2$
9. $2 \cdot 2 \cdot 3 \cdot 5 \cdot 17$
10. $2 \cdot 2 \cdot 5 \cdot 5 \cdot 7 \cdot 11$
11. $19 \cdot 19$
12. 1, 2, 3, 5, 7, 11, 13, 17, 19, 23, 29, 31, 37, 41, 43, 47, 49, 53, 59, 61, 67, 71, 73, 79, 83, 89, 97

Solutions to Section 1.3 Exercises

1. 3/4
2. 7/9
3. 4/5
4. 9/11
5. 13/17
6. 8/9

Solutions to Section 1.4 Exercises

1. 75/105
2. 26/39
3. 72/132
4. 84/105
5. 639/710
6. 240/390

Solutions to Section 1.5 Exercises

1. 15/2
2. 25/7
3. 13/6
4. 64/3

Solutions to Section 1.6 Exercises

1. 24
2. 30
3. 294
4. 1540
5. 1890
6. 4620

Solutions to Section 1.7 Exercises

1. 17/12
2. 13/8
3. 2/33
4. 123/110
5. 31/9000
6. 47/1715

Solutions to Section 1.8 Exercises

1. 3/11
2. 11/21
3. 21/40
4. 40/81
5. 3/2

Solutions to Section 1.9 Exercises

1. 8/3
2. 3/44
3. 44/57
4. 57/8
5. 48/49
6. 196

Solutions to Section 1.10 Exercises

1. 1/12
2. 15/4
3. 4/3
4. 20/13

Solutions to Section 1.11 Exercises

1. 3.4596
2. −2515.456
3. 54.9572
4. −28.93444

Solutions to Section 1.12 Exercises

1. 1/32 = 0.03125
2. 1/81 = .012346
3. 25
4. 256
5. 0.013497
6. 0.014872
7. 1 000 000
8. 1/100 000 = 0.000 01

Solutions to Section 1.13 Exercises

1. $\sqrt{12}$
2. $\sqrt[3]{345}$
3. $\sqrt[5]{5^2}$
4. $\sqrt[7]{82^5}$
5. $75^{3/5}$
6. $100^{3/2}$
7. $100^{3/2}$
8. 12
9. 5
10. 81
11. 25

Solutions to Section 1.14 Exercises

1. 10^9
2. 2^4
3. 5^8
4. 36×10^6
5. 8/343
6. 125/27
7. $10^2 = 100$

8. $10^2 = 100$
9. 10^9
10. 144×10^4
11. 9/4
12. 1/27
13. $5(10)^{1/2}$
14. $7(7)^{1/3}$
15. $42^{1/2}/42$
16. $(20)^{2/3}/10$

Solutions to Section 1.15 Exercises

1. 11
2. 7
3. 77/5
4. 12
5. 100
6. -4
7. $-1/4$
8. 82/5

Solutions to Section 1.16 Exercises

1. 34%
2. 102%
3. 95.4%
4. $0.7 = 70\%$
5. 44.8%
6. 25%
7. 75.5%
8. 47.8%
9. 0.129
10. 0.0022
11. $45/100 = 9/20$
12. $20/100 = 1/5$
13. $0.39 \cdot 176 = 69$
14. $0.67 \cdot 2570 = 1722$
15. 5.5% increase
16. $13.41
17. $153.34
18. 560 000 Ω
19. 1 200 000 Ω
20. 56 200 Ω

21. 15.8 Ω
22. 4465 Ω to 4935 Ω
23. 3 449 600 Ω to 3 590 400 Ω
24. 54.9 Ω to 56.1 Ω
25. 20 failures

CHAPTER 1 PROBLEMS

Answer True or False.

1. $13 + 47 + 16 = 16 + 49 + 13$
2. $3 \cdot 12 \cdot 5 = 12 \cdot 3 \cdot 5$
3. $37 - 16 = 16 - 37$
4. $17 / 5 = 5 / 17$
5. The answer to a subtraction is called a quotient.
6. Numbers being added are called addends.
7. Numbers being subtracted are called subtractors.
8. The answer to a multiplication problem is called the multiplier.
9. Addition and multiplication are both commutative and associative.
10. If the sum of the digits of a number is divisible by 3, the number is divisible by 3.
11. If the sum of the digits of a number is divisible by 5, the number is divisible by 5.
12. 4 and 5 are the prime factors of 20.
13. Including 1, there are 13 prime numbers less than 30.
14. 53/15 can be reduced.
15. $5/7 = 65/91$
16. Two or more fractions whose numerators are identical are called like fractions.
17. The least common multiple for 54, 21, and 14 is 108.
18. $2/7 + 3/7 = 5/7$
19. $8/5 - 3/2 = 5/3$
20. To multiply fractions, first find the LCD.
21. One-third of one-fourth is one-twelfth.
22. Two-thirds divided by one-half is four-thirds.
23. 3/5 divided by 1/3 is 9/5.
24. A complex fraction is a fraction that is difficult to work with.
25. To subtract unlike fractions, subtract the numerators and subtract the denominators.
26. $(2/5 - 1/4) / (1/2 - 1/5) = 1/2$
27. $2^8 = 256$
28. $(10^2)(10^3) = 10^6$

29. $(2^3)^{-2} = 1/64$
30. $(-3)^{-3} = 1/27$
31. $8^{2/3} = 1/8^{3/2}$
32. $8^{-2/3} = 1/4$
33. $\sqrt[3]{27^2} = 3^2$
34. $\sqrt[a]{b^c} = b^{c/a}$

Factor into prime factors.

35. 165
36. 180
37. 81 000
38. 350
39. 715
40. 507
41. 4199

Reduce these fractions to lowest terms.

42. 175/150
43. 72/108
44. 266/154
45. 330/390
46. 23 500/24 500
47. 207/261
48. 363/1023

Raise these fractions to higher terms with the given number as the new denominator.

49. 13/11 = /143
50. 43/19 = /209
51. 5/17 = /2040
52. 7/5 = /655
53. 3/31 = /2325
54. 13/41 = /4305
55. 49/37 = /38 850

Find the least common multiple for each set of numbers.

56. 14, 21, 15
57. 12, 33, 36
58. 32, 36, 8, 4, 2
59. 63, 75, 375
60. 221, 187, 143
61. 30, 70, 315
62. 1000, 275, 88

Add these fractions.

63. 2/3 + 7/3 =
64. 17/21 + 25/21 =
65. 2/5 + 1/5 + 4/5 =
66. 21/107 + 36/107 + 91/107 =
67. 2/5 + 7/10 =
68. 22/3 + 3 =
69. 3 1/3 + 7 1/5 =
70. 1/21 + 1/28 + 1/35 =
71. 1 5/6 + 2/15 + 7/24 =
72. 5/6 + 7/90 + 7/60 =
73. 1/2 + 7/8 + 5/12 + 5/108 =
74. 1/55 + 1/33 + 1/35 + 1/21 + 1/15 + 1/77 =

Subtract these fractions.

75. 4/5 − 1/5 =
76. 3 − 3/7 =
77. 3 1/5 − 2 7/8 =
78. 7/11 − 2/13 =
79. 7/8 − 1/12 =
80. 20/21 − 7/55 =

Multiply.

81. $1/2 \cdot 4/7 \cdot 3/2 =$ **82.** $25 \cdot 4/5 \cdot 3/8 =$
83. $100/27 \cdot 3/5 \cdot 9/20 =$ **84.** $125/49 \cdot 36/15 \cdot 3/5 =$
85. $21/20 \cdot 25/49 =$ **86.** $9/121 \cdot 55/28 \cdot 98/3 =$
87. $35/51 \cdot 65/21 \cdot 85/39 =$ **88.** $15/2 \cdot 165/7 \cdot 7/975 =$

Divide.

89. 3/5 by 5/9.   **90.** 2/3 by 8/11.   **91.** 49/72 by 35/36.
92. 25/36 by 35/12.   **93.** 11/17 by 121/34.

Simplify.

94. $\dfrac{3 - \dfrac{2}{5}}{8 - \dfrac{14}{3}} =$ **95.** $\dfrac{\dfrac{6}{5} - \dfrac{3}{7}}{\dfrac{2}{5} + \dfrac{2}{7}} =$ **96.** $\dfrac{\dfrac{9}{11} + \dfrac{1}{\dfrac{4}{7} + \dfrac{8}{3}}}{3} =$

97. $\dfrac{1}{1 - \dfrac{1}{1 - \dfrac{1}{6}}} =$ **98.** $3 - \dfrac{1}{3 - \dfrac{1}{3 - \dfrac{1}{2}}} =$

Evaluate these expressions in longhand.

99. $3^5 =$ **100.** $12^3 =$
101. $14^0 =$ **102.** $10^{-4} =$
103. $5^{-3} =$ **104.** $49^{1/2} =$
105. $64^{3/2} =$ **106.** $25^{-3/2} =$
107. $12^{-2} =$ **108.** $1000^{2/3} =$
109. $5^2 \cdot 5^3 \cdot 5 =$ **110.** $7^3 \cdot 7^{-6} \cdot 7^5 =$
111. $7^5 / 7^2 =$ **112.** $3^6 / 3^9 =$
113. $10^7 \cdot 10^3 / (10^5 \cdot 10^4) =$ **114.** $(5^4)^2 =$
115. $(3 \times 10^4)^{-2} =$ **116.** $(2/5)^3 =$
117. $(2^3 / 3^2)^{-2} =$ **118.** $(10^4 \cdot 10^{-2})^3 =$
119. $(3^2 \cdot 2^3 / 5^2)^{-2} =$

Use your calculator to evaluate these expressions.

120. $2.7^3 =$ **121.** $1.55^5 =$
122. $8.333^0 =$ **123.** $77.2^{-3} =$
124. $123.6^{-2.3} =$ **125.** $(5/13)^2 =$
126. $\pi^3 =$ **127.** $\pi^{-4} =$
128. $\pi^{6/7} =$

Simplify.

129. $10,000^{1/3} =$
130. $49,000^{1/2} =$
131. $2048^{1/4} =$
132. $5400^{1/3} =$
133. $1/1,000^{1/3} =$
134. $1/49,000^{1/2} =$
135. $1/2048^{1/4} =$
136. $1/5400^{1/3} =$

Evaluate.

137. $2 + 8 \cdot 3 - 6 =$
138. $(2 + 34/2)/(23 + 6/2) =$
139. $13 + 6^2 / (23 - 4^2) =$
140. $\{12 - 2^3 /[(25)^{1/2} + 6^2]\}^2 =$
141. $2 + 5^2 \cdot 3 / (34 - 4^{1/2})(34-13) =$

Convert to percent.

142. 0.24
143. 1.54
144. 0.663
145. 3/5
146. 35/79
147. 395/1000
148. 28 is what % of 54?
149. 86 is what % of 105?

Convert to a decimal.

150. 32.6%
151. 0.78%

Convert to a fraction.

152. 71%
153. 38%
154. What is 13% of 358?
155. What is 53% of 6452?
156. If a gallon of gasoline is $1.179 per gallon and it increases to $1.209 per gallon, what is the % increase in cost? (The increase in cost is what % of the original price?)
157. If a technician earning $10.33 per hour gets a promotion and a 5.5% raise, what is the new hourly wage?
158. If a digital multimeter normally sells for $225, what will it cost during a 15% sale?

Calculate the nominal resistance.

159. Grey, red, yellow
160. Orange, orange, brown
161. 3803
162. 17R5

Calculate the upper and lower bounds for these resistors.

163. Orange, white, orange, gold
164. 1672G

165. 26R5F

166. Calculate the expected number of failures in a batch of 10 000 resistors that have a color code of brown, brown, red, gold, and orange.

Unscramble these letters:

M S U
A B E S
D D D N E A
D D D I V E I N
U T E Q N O I T
L I M T L I P R U E
U L I P L I A M T C D N

Now unscramble the circled letters to answer the question.

What's the difference?

_ _ _ _ _ _ _ _ _ _ _ _ _ _ _ _ _ H _ _ _

2
Scientific Notation, Engineering Notation, and Significant Digits

CHAPTER OUTLINE

- **2.1** Decimal Notation
- **2.2** Scientific Notation
- **2.3** Engineering Notation
- **2.4** Significant Digits
- **2.5** Order of Operations

OBJECTIVES

After completing this chapter, you should be able to

1. convert from scientific notation into decimal notation.
2. convert from decimal notation into scientific notation.
3. enter numbers in scientific notation into a scientific calculator.
4. perform arithmetic operations on numbers represented in scientific notation.
5. make measurements to the proper number of significant digits.
6. state the upper and lower bounds of an approximate number.
7. state the accuracy of an approximate number.
8. state the precision of an approximate number.
9. round the result of arithmetic operations to the proper number of significant digits.
10. convert from metric prefix to decimal notation.
11. convert from decimal notation to metric prefix.
12. convert from one metric prefix to another.

NEW TERMS TO WATCH FOR

scientific notation
EE, EEX, or EXP
engineering notation
significant digit
least significant digit
placeholder
metric prefix
exa (E)
peta (P)
tera (T)
giga (G)
mega (M)
kilo (k)

hecto (h)
deka (da)
deci (d)
centi (c)
milli (m)
micro (µ)
nano (n)
pico (p)
femto (f)
atto (a)
accuracy
precision

2.1 ■ DECIMAL NOTATION

As a technician, you will encounter both very large and very small numbers. For example, the frequency of K band microwave ranges from 20 000 000 000 to 40 000 000 000 hertz; the United States of America consumes about 67 000 000 000 000 Btu's of energy each year; and the wavelength of red light from a helium-neon laser is 0.000 000 638 2 meters. A 0.000 010 henry inductor oscillates with a 0.000 000 001 farad capacitor at a frequency of 1 600 000 hertz. These numbers are expressed in decimal notation. Most people are comfortable with this format; but, as the number of decimal places to the left or right of the decimal point increases, it becomes more difficult and tedious to communicate the decimal number to a colleague or to copy it down without losing or adding a zero. There are two methods for handling these extremes conveniently: scientific notation and engineering notation. These two methods are covered in this chapter.

Various sections of this chapter refer to the values of the decimal places in the decimal number system. These values are reviewed in Figure 2-1.

FIGURE 2-1
Decimal Number System

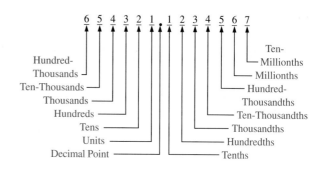

2.2 ■ SCIENTIFIC NOTATION

6.7×10^{16} and 6.382×10^{-7} are two of the numbers used above, but here they have been converted to scientific notation. In a number expressed in **scientific notation**, the decimal point has been shifted so that one digit, not a zero, lies to the left of the decimal point. That number is multiplied by a power of ten. The exponent indicates how many places and in which direction the decimal point must be moved to convert the number back to decimal form.

2.2.1 Converting from Scientific Notation to Decimal Notation

To convert from scientific notation to decimal notation, use the exponent as a guide. If the exponent is positive, move the decimal point that number of places to the right. Fill in zeros as placeholders as needed.

EXAMPLE 2.1 Convert 9.32×10^8 to decimal notation.

Solution The exponent is understood to be positive. Move the decimal point eight places to the right.

$$9.32 \times 10^8 = 932\,000\,000$$

If the exponent is negative, move the decimal point that number of places to the left. Use zeros as placeholders if necessary.

EXAMPLE 2.2 Convert 7.56×10^{-6} to decimal notation.

Solution The exponent is negative. Move the decimal point six places to the left.

$$7.56 \times 10^{-6} = 0.000\,007\,56$$

EXAMPLE 2.3 Convert 10^4 to decimal notation.

Solution 10^4 is the same as 1×10^4. Move the decimal point four places to the right.

$$10^4 = 10\,000$$

EXAMPLE 2.4 Convert 10^{-6} to decimal notation.

Solution 10^{-6} is the same as 1×10^{-6}. Move the decimal point six places to the left.

$$10^{-6} = 0.000\,001$$

EXAMPLE 2.5

Convert 10^0 to decimal notation.

Solution 10^0 is the same as 1×10^0. The exponent zero says not to move the decimal point at all.

$10^0 = 1$ **Check:** $10^3 = 1000$
$10^2 = 100$
$10^1 = 10$
$10^0 = 1$ It fits!
$10^{-1} = .1$
$10^{-2} = .01$

Notice that numbers less than 1 expressed in scientific notation have negative exponents, and numbers greater than 1 have positive exponents.

SECTION 2.2.1 EXERCISES

Convert from scientific notation to decimal notation.

1. 3.2×10^{-4}
2. 1.625×10^{-6}
3. 4.78×10^2
4. 7.89×10^5
5. 10^{-3}
6. 10^{-6}
7. 10^3
8. 10^6

2.2.2 Converting from Decimal Notation to Scientific Notation

A number is expressed in scientific notation by adjusting the decimal point so that one digit, non-zero, exists to the left of the decimal point. That number is multiplied by a power of ten to indicate the number of places and direction the decimal point must be moved to return to decimal notation.

A large number, greater than one, is converted to scientific notation by moving the decimal point to the left. The exponent used is positive to indicate that the decimal point must be moved back to the right to return to decimal notation.

EXAMPLE 2.6

Express 356 000 000 miles in scientific notation.

Solution The decimal point is understood to be on the right. It must be moved eight places to the left so that only the 3 remains to the left of the decimal point: 3.56×10^8. The exponent is positive to indicate that the decimal point must be moved to the right to return to decimal notation.

Check: To check your answer, convert back to decimal notation.
$3.56 \times 10^8 = 356\ 000\ 000.$

A small number, less than 1, is converted to scientific notation by sliding the decimal point to the right so that one digit, non-zero, exists to the left of the decimal point. The exponent used is negative to indicate that the decimal point must be moved back to the left to return to decimal notation.

46 ■ CHAPTER 2 SCIENTIFIC NOTATION, ENGINEERING NOTATION, AND SIGNIFICANT DIGITS

EXAMPLE 2.7 Express 0.000 000 011 7 in scientific notation.

Solution The decimal point must be moved eight places to the right: 1.17×10^{-8}. The exponent used is negative to indicate that the decimal point must be moved back to the left to return to decimal notation.

Check: To check your answer, convert back to decimal notation.
$$1.17 \times 10^{-8} = 0.000\ 000\ 011\ 7$$

EXAMPLE 2.8 Convert 1 000 000 to scientific notation.

Solution Move the decimal point six places to the left: 1.0×10^6 or just 10^6.

EXAMPLE 2.9 Convert 0.000 000 000 01 to scientific notation.

Solution Move the decimal 11 places to the right: 1.0×10^{-11} or just 10^{-11}.

SECTION 2.2.2 EXERCISES PART 1

Convert from decimal notation to scientific notation.

1. 987 000 000
2. 654 300
3. 0.000 211
4. 0.000 039 5
5. 100 000
6. 10 000 000
7. 0.000 001
8. 0.000 000 01

What if a number is expressed in exponential notation, but not in scientific notation? For example, 567×10^4 has more than one digit to the left of the decimal, so it is not in scientific notation. One method to convert it to scientific notation takes two steps. First convert it to decimal notation, and then convert it back into scientific notation.

EXAMPLE 2.10 Express 567×10^4 in scientific notation.

Solution First convert to decimal.
$$567 \times 10^4 = 5\ 670\ 000$$

Now convert to scientific notation.
$$5\ 670\ 000 = 5.67 \times 10^6$$

As you gain confidence, you can take a more direct approach. To convert 567×10^4 to decimal notation, the decimal point must be moved four places to the right. If the decimal point is placed between the 5 and 6, it would have to be moved six places to the right to convert to decimal notation: $567 \times 10^4 = 5.67 \times 10^6$.

2.2 SCIENTIFIC NOTATION

EXAMPLE 2.11 Express 0.0034×10^{-5} in scientific notation.

Solution 1
$$0.0034 \times 10^{-5} = 0.000\,000\,034 = 3.4 \times 10^{-8}$$

Solution 2 To convert $0.003\,4 \times 10^{-5}$ to scientific notation, the decimal point must be moved five places to the left. If the decimal point is placed between the 3 and 4, it would have to be moved eight places to the left to convert to decimal notation.

$$0.003\,4 \times 10^{-5} = 3.4 \times 10^{-8}$$

EXAMPLE 2.12 Express 125×10^{-8} and 0.76×10^{12} in scientific notation.

Solution To convert 125×10^{-8} to scientific notation, begin "using up" the -8 by shifting the decimal point to the left. Stop when the decimal point lies between the 1 and 2.

$$125 \times 10^{-8} = 1.25 \times 10^{-6}$$

To convert 0.76×10^{12} to scientific notation, begin "using up" the 12 by shifting the decimal point to the right. Stop when the decimal point lies between the 7 and 6.

$$0.76 \times 10^{12} = 7.6 \times 10^{11}$$

SECTION 2.2.2 EXERCISES PART 2

Convert to scientific notation.

9. 236×10^3
10. 345×10^{-5}
11. 0.0012×10^4
12. 0.0084×10^{-3}

2.2.3 Using a Calculator

Calculators often display numbers in scientific notation. The digits are displayed, followed by a space, and then the exponent. For example, $1.206\,4$ actually means 1.206×10^4, although the base number 10 is not shown. A common mistake by beginning students is to record this result as 1.206^4, which is quite different.

$$1.206^4 = 1.206 \cdot 1.206 \cdot 1.206 \cdot 1.206 = 2.115$$

whereas
$$1.206 \times 10^4 = 12060$$

+/− Key Many scientific calculators have a +/− key that is used to change (toggle) the sign of a number or exponent. To enter negative numbers, first enter the digits and then press the +/− key. Enter 4.09 and press the +/− key several times. Note that the sign toggles each time. The − sign is used to tell the calculator to subtract; the +/− key is used to toggle the sign of a number or exponent.

EE, EEX, or EXP Key Each scientific calculator has an **EE** or **EEX** (enter exponent) key or an **EXP** (exponent) key that is used to enter an exponent. To enter a number in scientific notation into the calculator, follow these steps.

1. Enter the digits of the decimal portion.
2. Use the +/− key to toggle the sign of the number if needed.
3. Press the EE or EEX (enter exponent) or EXP (exponent) key.
4. Enter the exponent.
5. Use the +/− key to toggle the sign of the exponent as needed.

EXAMPLE 2.13 List the sequence of keystrokes required to enter 5.63×10^{-7} into the calculator.

Solution

5 . 6 3 EE 7 +/−

The +/− can come before or after the 7.
A common error is to enter an extra 10 with this sequence: **5 . 6 3 × 10 EE 7+/−**. The EE key enters the base number 10 automatically. To avoid this error, say to yourself "times ten to the" as you press the EE key.

EXAMPLE 2.14 List the sequence of keystrokes required to enter -8.71×10^{12} into the calculator.

Solution **8 . 7 1 +/− EE 1 2**

SECTION 2.2.3 EXERCISES

Enter these numbers into your calculator.

1. 2.69×10^4
2. -5.796×10^3
3. 1.82×10^{-4}
4. -3.586×10^{-4}

Enter a decimal number less than or equal to 0.001 into your calculator and hit the = key. Scientific calculators will convert your entry into scientific notation.

2.2.4 Using Scientific Notation in Arithmetic Operations

To multiply numbers expressed in scientific notation in longhand, multiply their decimal parts and add the exponents. If the result has more than one digit to the left of the decimal point, adjust the decimal point back into scientific notation.

$$(a \times 10^b)(c \times 10^d) = ac \times 10^{b+d}$$

EXAMPLE 2.15 Multiply $(4 \times 10^{-4})(5.1 \times 10^6)$.

Solution

$(4 \times 10^{-4})(5.1 \times 10^6) = 20.4 \times 10^{-4+6} = 20.4 \times 10^2 = 2.04 \times 10^3$

2.2 SCIENTIFIC NOTATION ■ 49

To work this same problem on your calculator, enter this sequence:

$$4 \text{ EE } 4 +/- \times 5.1 \text{ EE } 6 =$$

To divide numbers expressed in scientific notation in longhand, divide their decimal parts and subtract their exponents. If necessary, adjust the decimal point so that the result is expressed in scientific notation.

$$(a \times 10^b)/(c \times 10^d) = a/c \times 10^{b-d}$$

EXAMPLE 2.16 Divide $1.2 \times 10^7 / 4 \times 10^4$.

Solution
$$1.2 \times 10^7 / 4 \times 10^4 = 0.3 \times 10^{7-4} = 0.3 \times 10^3 = 3 \times 10^2$$

To work this same problem on your calculator, enter this sequence:

$$1.2 \text{ EE } 7 / 4 \text{ EE } 4 =$$

Before adding numbers expressed in scientific notation in longhand, check to see that their exponents are the same. If not, adjust decimal points as needed until they are. Add the mantissas and use the common exponent. Adjust the decimal point so that the result is again in scientific notation.

$$a \times 10^n + b \times 10^n = (a + b) \times 10^n$$

EXAMPLE 2.17 Add $4.56 \times 10^{-4} + 5.89 \times 10^{-2}$.

Solution The exponents are not the same. Rewrite 4.56×10^{-4} as 0.0456×10^{-2}.

$$
\begin{array}{r}
0.0456 \times 10^{-2} \\
\underline{5.89 \times 10^{-2}} \\
5.9356 \times 10^{-2}
\end{array}
$$

The answer is in scientific notation. Later in this chapter we will see how to round off the result. This problem can be worked on your calculator as is by entering this sequence:

$$4.56 \text{ EE } 4 +/- + 5.89 \text{ EE } 2 - =$$

Before subtracting numbers expressed in scientific notation in longhand, check to see that their exponents are the same. If not, adjust decimal points as needed until they are. Subtract the decimal parts and use the common exponent. Adjust the decimal point so that the result is again in scientific notation.

$$a \times 10^n - b \times 10^n = (a - b) \times 10^n$$

50 ■ CHAPTER 2 SCIENTIFIC NOTATION, ENGINEERING NOTATION, AND SIGNIFICANT DIGITS

EXAMPLE 2.18 Subtract $4.23 \times 10^5 - 3.71 \times 10^4$.

Solution $4.23 \times 10^5 - 3.71 \times 10^4 =$ The exponents are not the same.
Rewrite 4.23×10^5 as 42.3×10^4.
The problem is now $42.3 \times 10^4 - 3.71 \times 10^4 =$
$42.3 \times 10^4 - 3.71 \times 10^4 = 38.59 \times 10^4 = 3.859 \times 10^5$

This problem can be worked on your calculator as is by entering this sequence:

$$4.23\,EE\,5 - 3.71\,EE\,4 =$$

SECTION 2.2.4 EXERCISES

Work in longhand and with a calculator.

1. $2.46 \times 10^{-6} \cdot 5.9 \times 10^{-3}$
2. $7.98 \times 10^4 / 5.34 \times 10^{-2}$
3. $1.34 \times 10^3 + 9.45 \times 10^2$
4. $8.35 \times 10^{-5} - 9.36 \times 10^{-6}$
5. $(5.54 \times 10^3 + 4.88 \times 10^2) / 2.3 \times 10^4$

2.3 ■ ENGINEERING NOTATION

Some powers of ten have been given names that are used as prefixes to the units involved. Some examples are *micro*amps, *kilo*grams, *milli*meters, and *mega*hertz. These are called **metric prefixes**. Here is a list of the most commonly seen prefixes, their meanings, and their symbols.

Prefix	Power of Ten	Symbol	Example
mega	10^6	M	Mohm
kilo	10^3	k	kmeter
milli	10^{-3}	m	mliter
micro	10^{-6}	μ	μamps

A more complete list is given at the end of this section. These metric prefixes are used to express a number in engineering notation.

A number expressed in engineering notation consists of a decimal portion that lies between 1 and 1000, followed by a power of 10 that is a multiple of three. For example, 196 Mhertz, 12.6 kmeter, 386 msecond, and 724 μamp are examples of quantities expressed in engineering notation.

2.3.1 Converting from Decimal Notation to Engineering Notation

To express a number in engineering notation, adjust the decimal point so that the decimal portion of the number lies between 1 and 1000 and the corresponding power of 10 is a multiple of three.

EXAMPLE 2.19	Convert 350 000 000 meters into engineering notation.
Solution	Move the decimal six times to the left to get 350×10^6. Replace 10^6 with its symbol, M.

$$350\ 000\ 000 \text{ meters} = 350 \text{ Mmeters}$$

EXAMPLE 2.20	Convert 0.000 452 grams into engineering notation.
Solution	Move the decimal six times to the right to get 452×10^{-6}. Replace 10^{-6} with its symbol, μ. 452 μgrams.

$$0.000452 \text{ grams} = 452 \text{ μgrams}$$

SECTION 2.3.1 EXERCISES

Convert these decimal numbers into engineering notation.

1. 720 000 000 hertz = _____
2. 823 000 meters per sec = _____
3. 0.005 3 grams = _____
4. 0.000 000 019 amps = _____

Some calculators have an engineering notation key, ENG, that places the calculator in the engineering mode. Results to operations are presented in engineering notation. Place your calculator in engineering mode, enter one of the numbers from the last exercise, and press the = sign. The number should be displayed in engineering notation.

2.3.2 Converting from Engineering Notation to Decimal Notation

Removing the metric prefix is easy. Replace the prefix with its equivalent power of ten. Use the exponent as a guide to move the decimal point.

EXAMPLE 2.21	Convert 3.46 Mohms to a decimal number.
Solution	M means 10^6. 3.46×10^6 ohms = 3 460 000 ohms.

EXAMPLE 2.22	Convert 89.3 μmeters to meters.
Solution	μ means 10^{-6}. 89.3×10^{-6} meters = 0.000 089 3 meters.

SECTION 2.3.2 EXERCISES

Convert to decimal notation.

1. 56.2 μamps
2. 120 μhenrys
3. 3 msiemens
4. 12 mmeters
5. 946 knewtons
6. 486 kohms
7. 377 Mhertz
8. 83 Mmeters

Here is a more complete list of metric prefixes:

Prefix	Power of Ten	Symbol
exa	10^{18}	E
peta	10^{15}	P
tera	10^{12}	T
giga	10^{9}	G
mega	10^{6}	M
kilo	10^{3}	k
hecto	10^{2}	h
deka	10^{1}	da
deci	10^{-1}	d
centi	10^{-2}	c
milli	10^{-3}	m
micro	10^{-6}	μ
nano	10^{-9}	n
pico	10^{-12}	p
femto	10^{-15}	f
atto	10^{-18}	a

2.4 ■ SIGNIFICANT DIGITS

Use your calculator to divide 22 by 7. My calculator gives an answer with 12 digits, 3.14285714286. How many of those digits should be used? Should the result be rounded off? These questions are answered with a study of significant digits and the terms *precision* and *accuracy*.

Some numbers we use are exact. For example, there are 12 months in a year and 100 centimeters in a meter. Other numbers are the result of measurements and are approximations or estimates of the actual quantity.

The digits in an approximate number that are known to be accurate enough to use are called **significant digits.** The characteristics of the measuring instrument determines the number of significant digits in the approximation. For example, Figure 2-2 shows a scale that is marked off in tenths of a foot in measuring the length of object A.

FIGURE 2-2
Length of Object A

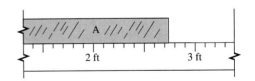

Object A is between 2.7 and 2.8 feet long. The end of object A falls about 0.3 of the way between 2.7 and 2.8. The length of object A is recorded as 2.73 ft. All three of these digits are considered significant, even though the last digit is an estimate.

FIGURE 2-3
Length of Object B

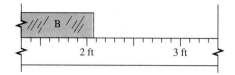

Using the same scale, object B measures between 2.0 and 2.1 ft (Figure 2-3). Its length is about 0.7 of the way between 2.0 and 2.1. The length of object B is recorded as 2.07 ft.

FIGURE 2-4
Length of Object C

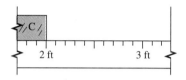

In Figure 2-4, object C falls right on the 2 ft tic mark. Its length is recorded as 2.00 ft to indicate that it was measured with an instrument that can measure to the hundredth of a foot (with the hundredth place an estimate).

FIGURE 2-5
Length of Object D

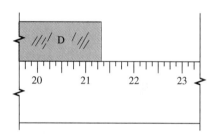

In some applications, measurements are made to 1/2 the smallest increment marked on the measuring device. The rule in Figure 2-5 is marked in 1/8s of an inch. Measurements are often made to the nearest 1/16s of an inch. For example, object D measures between 21 1/4 and 21 3/8, or between 21 4/16 and 21 6/16. Its length is recorded as 21 5/16. How should that be written in decimal form? 1/16" equals 0.0625" or approximately 0.06". Round to the nearest 1/100". 21 5/16" = 21.3125" which rounds to 21.31".

Figure 2-6A shows a B&K Precision Electronic Multimeter, Model 214, that can be used to measure resistance, voltage, or current. The function switch is used to select between voltage, current, or resistance measurements. This is an example of an analog meter. Analog meters have a needle and scale as opposed to the direct readout of a digital meter. The scale of an analog meter is shown in Figure 2-6B. The top scale is used to measure resistance in Ω. Note that it is not linear, that is, the divisions change across the scale. On the right between 0 and 1, there is a large mark at 0.5 and each small mark is worth 0.1. Toward the left, between 20 and 50, there are large marks at 30 and 40 and each small division is worth 5: 20, 25, 30, 35, 40, 45, and 50. Between 50 and 200, there are large marks at 100 and 150, and each small mark is worth 25: 50, 75, 100, 125, 150, 175, and 200. So, before reading the top scale, the divisions must be studied carefully

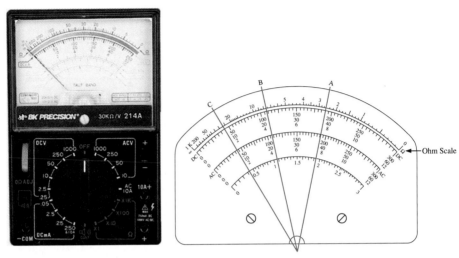

FIGURE 2-6
B&K Precision Electronic Multimeter, Model 214

After connecting a resistor to the leads of the meter, the range switch shown on the lower right in Figure 2-6A is adjusted so that the needle of the meter swings toward the middle of the scale. Readings at the far left or right are avoided.

EXAMPLE 2.23 If the range switch is set to 1k and the needle swings to position A, determine the resistance of the resistor being measured and the number of significant digits in the reading.

Solution The reading falls between 2 and 3, which is unfortunately divided into 6 divisions, making each division worth 1/6 or 0.166. The needle falls between 2.5 and 2.666 kΩ. Let's estimate the reading as 2.53 kΩ. The 2 and the 5 are definite, and the 3 is definitely an estimate. It is counted as a significant digit. This reading has three significant digits.

If the range switch is set to 100 kΩ and the needle swings to position A, the reading is 254 kΩ, still with three significant digits.

EXAMPLE 2.24 **a.** If the range switch is set to 10k Ω and the needle swings to position B, determine the resistance of the resistor being measured and the number of significant digits in the reading.

b. If the range switch is set to 100k Ω and the needle swings to position C, determine the resistance of the resistor being measured and the number of significant digits in the reading.

Solution **a.** The reading is between 7 and 8. Let's estimate it at 7.6. The reading is 7.6 · 10 kΩ or 76 kΩ with 2 significant digits.

b. The reading is between 20 and 30. Let's estimate it at 26. The reading is 26 · 100 kΩ or 2600 kΩ with 2 significant digits.

The scale below the ohm scale is for measuring DC volts or amps. Note that it increases from left to right. To measure a DC voltage, set the function switch to DC volts, and set the range switch to a value larger than that to be measured so that the needle does not peg on the right. Note that the DC scale has three sets of numbers. The setting of the range switch determines which of the three sets to read. If the range switch is set to 300 volts, then the top set of numbers is used and a full-scale reading is 300 volts. If a 3-volt range is selected, the top set of numbers is again used, but this time the full-scale reading is 3 volts and the decimal point has to be adjusted from 300 to 3 by moving the decimal point of the reading two places to the left.

EXAMPLE 2.25

a. Suppose the range switch is set to 1.2 volts and the needle swings to setting C. Determine the voltage being measured and the number of significant digits in the answer.
b. Suppose the range switch is set to 6 volts and the needle swings to setting B. Determine the voltage being measured and the number of significant digits in the answer.
c. Suppose the range switch is set to 3000 volts and the needle swings to setting A. Determine the voltage being measured and the number of significant digits in the answer.

Solution **a.** Read the 12-volt scale but adjust the decimal point one position to the left so that the full-scale reading is 1.2 volts. The reading falls between the 0 and 0.2 marks. The needle is on the 0.18 mark. In this region the scale can be read to 3 significant digits. Call the measurement 0.180 V.

b. The 60 scale will be used, but a full-scale reading is 6.0 V. The needle lies between 20 and 30, which in this case has been scaled down to 2.0 to 3.0 V. The reading is estimated at 2.17 V. The 2 and 1 are definite, the 7 is an estimate. This reading has three significant digits.

c. The 300 scale will be used, but a full scale reading is 3000 V. Adjust the decimal point one place to the right. The needle lies between 1500 and 2000, with each small division worth 50 V. The reading is estimated at 1920 V. The 1 and 9 are definite, the 2 is an estimate. This reading has three significant digits.

Note that the DC scale is linear; the divisions are the same across the full scale. Any DC reading will have three significant digits.

SECTION 2.4 EXERCISES PART 1

In Figures 2-7, 2-8, and 2-9, measure each object to the proper number of significant digits.

1. A	2. B	3. C
4. D	5. E	6. F
7. G	8. H	9. I
10. J	11. K	12. L
13. M	14. N	15. O
16. P	17. Q	18. R
19. S	20. T	21. U
22. V	23. W	24. X
25. Y	26. Z	

FIGURE 2-7
Problems 1–10, Section 2.4 Exercises

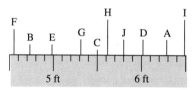

FIGURE 2-8
Problems 11–20, Section 2.4 Exercises

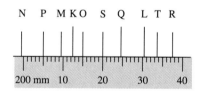

FIGURE 2-9
Problems 21–26, Section 2.4 Exercises

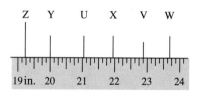

27–31. If the function switch is set to Ω, and the range switch is set to 100 kΩ, determine the amount of resistance being measured for each setting of the needle in Figure 2-10.

27. A	28. B	29. C
30. D	31. E	

32–36. If the function switch is set to DC volts and the range switch is set to 3 volts, determine the voltage being measured for each setting of the needle in Figure 2-10.

2.4 SIGNIFICANT DIGITS ■ 57

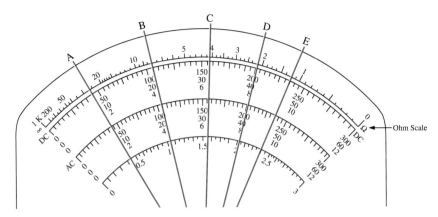

FIGURE 2-10
Problems 27–41, Section 2.4 Exercises

32. A   **33.** B   **34.** C
35. D   **36.** E

37–41. If the function switch is set to DC volts and the range switch is set to 600 volts, determine the voltage being measured for each setting of the needle in Figure 2-10.

37. A   **38.** B   **39.** C
40. D   **41.** E

The right-most significant digit is called the **least significant digit**. It is an estimation. The significant digits to the left of it are known to be exact.

FIGURE 2-11
Upper and Lower Bounds

```
            ┌ 172.5              ┌ 173.7 ┌
      173 ┤                173.7 ┤
            └ 173.5              └ 173.75
      174 ┤                173.8 ┤
            ┌ 174.5              ┌ 173.85
      175 ┤                173.9 ┤
            └                    └ 173.95
            A                    B
```

When we say that an object weighs 174 pounds, we mean that its weight is closer to 174 pounds than it is to 173 or 175 pounds. As shown in Figure 2-11A, 174 pounds actually means any weight between 173.5 and 174.5 pounds. In other words, 174 is the object's weight to the nearest pound. On another more accurate scale, the same object measures 173.8 pounds. Its weight is closer to 173.8 pounds than it is to 173.7 or 173.9 pounds. As shown in Figure 2-11B, 173.8 actually means a weight between 173.75 and 173.85 pounds. 173.8 is the object's weight to the nearest tenth of a pound.

Following the same reasoning, what does a weight of 16 000 pounds mean? If the 1 and 6 are the only significant digits, then the weight is to the nearest 1000 pounds. The object weighs between 15 500 and 16 500 pounds. If the 1, 6, and following 0 are significant, then the weight is to the nearest 100 pounds and the actual weight is between 15 950 and 16 050 pounds. When converted to scientific notation, only the significant digits are retained and the accuracy of the measurement is apparent. 1.600×10^4 indicates that two of the zeros are significant and that the weight was measured to the nearest ten pounds. In that case, the actual weight lies between 15 995 and 16 005 pounds.

EXAMPLE 2.26 If the distance between two points measures 456.3 kilometers, what is the minimum and maximum distance indicated?

Solution The distance was measured to the nearest tenth. The actual distance lies between 456.25 and 456.35 kilometers.

EXAMPLE 2.27 If the width of a trace on a printed circuit board measures 4.50×10^{-2} in, what is its actual width?

Solution Since the zero was included in the scientific notation, it is significant. 4.50×10^{-2} = .0450. The width was measured to the nearest ten-thousandth. The actual width is between 0.04495 and 0.04505 in.

SECTION 2.4 EXERCISES PART 2

For each measurement given, list the minimum and maximum bounds for the actual value.

42. 156 lb
43. 0.0083 in
44. 5.79×10^3 ohms
45. 7.256×10^{-3} meters
46. 3.00×10^4 mph
47. 7.002×10^{-1} amps
48. 1200 miles

When represented in scientific notation, all of the digits of a number are significant digits; none of them are there to hold a place. 4.83×10^5 has three significant digits; $2.000\ 87 \times 10^{-4}$ has six; $4.500\ 0 \times 10^6$ has five. However, in decimal notation some of the zeros can be placeholders. 0.000 085 2 has three significant digits since the four zeros to the right of the decimal point are there to push the significant digits over into the proper place. They are placeholders. Rewritten, it becomes 8.52×10^{-4}, which shows the three significant digits. Here are some rules for determining the number of significant digits in a decimal notation number.

1. All non-zero digits are significant and zeros falling between non-zero digits are significant. 9 006.03 has 6 significant digits.

2. If there are no non-zero digits to the left of a decimal point, the zeros to the right of the decimal point between the decimal point and the first non-zero digit are not significant. They are placeholders. 0.000 87 has 2 significant digits.

3. If there are non-zero digits to the left of the decimal point, all of the digits to the right of the decimal point are significant. 3.600 0 has 5 significant digits.

4. If a number has no digits to the right of the decimal point and the number ends in one or more zeros, then the number of significant digits is often unclear. In 1600 pounds, the 1 and 6 could be significant, or the 1, 6, and first zero could be, or all four digits could be significant, depending on the scale used to determine the weight.

5. When a number is written in scientific notation, all digits in the mantissa are significant.

SECTION 2.4 EXERCISES Part 3

Determine the number of significant digits in each number.

49. 4000.7
50. 2.004
51. 0.00098
52. 1.00098
53. 1.0000
54. 3.700×10^{-5}
55. 186 000

2.4.1 Accuracy

The number of significant digits in a number is called its **accuracy**. For example, 6.05×10^{-6} has three significant digits, so it has an accuracy of three. Accuracy is used to determine the number of significant digits in multiplication and division problems and in raising a number to an exponent.

Multiplication The number of significant digits (accuracy) of the product is limited by the less accurate of the multiplicand and the multiplier. Multiply the numbers as they are on your calculator, then round off to match the less accurate of the multiplicand and the multiplier.

EXAMPLE 2.28 Multiply $3.07 \times 10^{-6} \cdot 6.398 \times 10^{-5}$.
Place the proper number of significant digits in the product.

Solution $$3.07 \times 10^{-6} \cdot 6.398 \times 10^{-5} = 1.964186 \times 10^{-10}$$

3.07×10^{-6} is the less accurate with three significant digits. Round the product to three significant digits. 1.96×10^{-10} is the product.

EXAMPLE 2.29 Multiply $8.94 \times 10^{-4} \cdot 3.0 \times 10^5$.

Solution $$8.94 \times 10^{-4} \cdot 3.0 \times 10^5 = 268.2$$

The multiplier is accurate to two digits. Limit the product to two significant digits. 270 is the answer. The zero is a placeholder and is not a significant digit.

EXAMPLE 2.30 Double 4.678×10^3.

Solution
$$2 \cdot 4.678 \times 10^3 = 9.356 \times 10^3$$

Here, 2 is an exact number and will not limit the accuracy of the answer.

SECTION 2.4.1 EXERCISES PART 1

Multiply. Limit the product to the proper accuracy.

1. $2.67 \cdot 4.768$
2. $38.2 \cdot 900.6$
3. $69.4 \times 10^{-6} \cdot 45.7 \times 10^3$
4. $1.000 \times 10^7 \cdot 2.00 \times 10^{-5}$

Division In division, the accuracy of the quotient is limited by the less accurate of the divisor and the dividend. Divide the numbers as they are using your calculator, then round off to match the less accurate of the divisor and dividend.

EXAMPLE 2.31 Divide 507.3/39.

Solution
$$507.3/39 = 13.007692$$

39 has an accuracy to two. Round the answer to two significant digits. 13 is the answer.

EXAMPLE 2.32 Divide $7.58 \times 10^{-3}/8.5 \times 10^4$.

Solution
$$7.58 \times 10^{-3}/8.5 \times 10^4 = 8.9176471 \times 10^{-8}$$

8.5×10^4 has an accuracy of two. Round the answer to two digits. 8.9×10^{-8} is the answer.

SECTION 2.4.1 EXERCISES PART 2

Divide. Limit the quotient to the proper accuracy.

5. $52.71/42$
6. $7.6/29.31$
7. $3.47 \times 10^{-4}/8.2 \times 10^{-2}$
8. $7.11 \times 10^4/4.005 \times 10^3$

Exponential Expressions When a number is raised to an exponent that is a positive whole number, the operation is actually multiplication. For example, 3^4 is $3 \cdot 3 \cdot 3 \cdot 3$. The accuracy of the base number will determine the accuracy of the result.

EXAMPLE 2.33 $(3.04 \times 10^{-5})^4$

Solution By the calculator, $(3.04 \times 10^{-5})^4 = 8.5407171 \times 10^{-19}$. The sequence used is **3 . 04 EXP 5 +/− INV x^y 4** = 3.04×10^{-5} has an accuracy of 3, as will the result. $(3.04 \times 10^{-5})^4 = 8.54 \times 10^{-19}$.

2.4 SIGNIFICANT DIGITS

When a number is raised to a positive fractional exponent, the process is equivalent to division. The accuracy of the result is determined by the accuracy of the base number.

EXAMPLE 2.34 $(6.004)^{1/2}$

Solution By the calculator, $(6.004)^{1/2} = 2.4503061$. Sequence used: $6.004 \sqrt{}$ 6.004 has an accuracy of 4, so does the result. $(6.004)^{1/2} = 2.450$.

When a number is raised to a negative exponent, it can be rewritten as a reciprocal. $a^{-b} = 1/a^b$ 1 is exact and does not limit the result. The accuracy of the result is once again determined by the accuracy of the base number.

EXAMPLE 2.35 $(5.04)^{-2/3}$

Solution By the calculator, $(5.04)^{-2/3} = 0.3401832$. The sequence used is **5.04 INV** x^y **(2/3+/−)=**. The parentheses are used to group −2/3 together as the exponent. Otherwise, the problem becomes $5.04^2/ - 3$, which yields −8.4672. 5.04 has an accuracy of 3, so does the result. $(5.04)^{-2/3} = 0.340$.

SECTION 2.4.1 EXERCISES Part 3

Limit the result to the proper number of significant digits.

9. 0.0056^3 **10.** $7.00^{1/3}$ **11.** $(5.81 \times 10^3)^{-2}$

2.4.2 Precision

The **precision** of a number is determined by the decimal place of the least significant digit. For example, 7.04 is precise to the hundredth place. 42 is precise to the unit's place.

EXAMPLE 2.36 Determine the precision of 8.15×10^{-3}.

Solution 8.15×10^{-3} is the same as .00815 and is precise to the hundred-thousandth place.

EXAMPLE 2.37 Determine the precision of 6.47×10^5.

Solution $6.47 \times 10^5 = 647000$. 7 is the least significant digit, and it appears in the thousand's place. 6.47×10^5 is precise to the thousand's place.

SECTION 2.4.2 EXERCISES PART 1

Determine the precision of each number.

1. 783
2. 0.00021
3. 2.006
4. 3.9×10^4
5. 3.9×10^{-3}
6. 3.900×10^5

The concept of precision is used to determine the number of significant digits to use in the answers to addition and subtraction problems.

Addition In addition problems, the precision of the sum is limited by the least precise addend. The least precise addend is the addend whose least significant digit is in the furthest place to the left. Calculate the sum on your calculator and then round to the place of least precision. Confused? It's easy.

EXAMPLE 2.38 Add these measurements: 12.12″, 8.0651″, and 9.458″.

Solution

$$\begin{array}{r} 12.1\underline{2} \\ 8.065\underline{1} \\ 9.45\underline{8} \\ \hline 29.6431″ \end{array}$$

The least significant digit of each addend has been underlined. 12.12 is the least precise of the addends, precise to hundredths. The sum can only be precise to hundredths. Round to 29.64″.

EXAMPLE 2.39 Add these measurements: 4.6×10^3, 8.23×10^5, and 2.44×10^2.

Solution

$$\begin{array}{r} 4\underline{6}00. \\ 82\underline{3}000. \\ 24\underline{4}. \\ \hline 827844. \end{array}$$

The least significant digit of each addend has been underlined. 823 000 is the least precise addend because it is precise only to the thousand's place. Round 827 844 to the nearest thousand, 828 000.

SECTION 2.4.2 EXERCISES PART 2

Add. Limit the sum to the proper number of significant digits.

7. $8.2 + 12.3 + 23.45 + 111.888$
8. $154.76 + 486.397 + 233.006$
9. $7.1 \times 10^{-2} + 4.7 \times 10^{-3} + 9.8 \times 10^{-3}$
10. $4.7 \times 10^{-4} + 4.9 \times 10^{-3} + 5.91 \times 10^{-2}$
11. $6.89 \times 10^2 + 2.36 \times 10^3 + 2.444 \times 10^4$
12. $2.144 \times 10^4 + 3.668 \times 10^3 + 8.30 \times 10^3$

Subtraction The precision of the difference in a subtraction problem is limited to the precision of the less precise of the minuend and subtrahend. On your calculator, subtract the numbers as they are and then round to match the less precise of the minuend and subtrahend.

EXAMPLE 2.40 $12.62 - 8.345$.

Solution

$$12.6\underline{2}$$
$$8.34\underline{5}$$
$$\overline{4.275}$$

The least significant digits of the minuend and the subtrahend has been underlined. 12.62 is precise to the hundredth's place and is the least precise of the two. Round the difference to the nearest hundredth. 4.28 is the difference.

EXAMPLE 2.41 $9.01 \times 10^{-2} - 5.7 \times 10^{-4}$.

Solution

$$0.090\underline{1}$$
$$0.0005\underline{7}$$
$$\overline{0.08953}$$

The subtrahend is larger than the minuend. The difference is negative. The least significant digit of the minuend and subtrahend have been underlined. 0.0901 is the least precise. Round the difference to the nearest ten-thousandth. 0.0895 is the difference.

SECTION 2.4.2 EXERCISES PART 3

Express the difference with the proper number of significant digits.

13. $4.56 - 2.314$
14. $300.6 - 0.452$
15. $900.7 - 343.56$
16. $4.7 - 2.7656$
17. $5.6 \times 10^{-1} - 8.5 \times 10^{-3}$
18. $3.9 \times 10^{-1} - 2.56 \times 10^{-8}$
19. $4.98 \times 10^{4} - 3.2 \times 10^{3}$
20. $5.67 \times 10^{4} - 8.9 \times 10^{2}$

Now, how do you determine the significant digits in a problem that contains a variety of operations?

2.5 ■ ORDER OF OPERATIONS

Review the order of operations presented in Chapter 1.

To determine the significant digits of a result from several operations, first find the preliminary answer on the calculator. Then step back through the problem and evaluate each step to determine accuracy or precision as needed.

EXAMPLE 2.42

(386 + 200.5)/29.

Solution By the calculator, (386 + 200.5)/29 = 20.224138.
The sequence entered is: **3 8 6 + 2 0 0 . 5 = 2/9 =**
Note: If the first = sign is not entered, 200.5 will be divided by 29, not the entire numerator.
(386 + 200.5) / 29 =
First the numbers inside the parentheses are added. 386 is precise to the unit's place and 200.5 to the tenth's place. 386 is the least precise and the sum is rounded to the unit's place.

$$587/29$$

We need to know the accuracy of the sum since it is being divided by 29. The sum, 587, has an accuracy of 3. 29 has an accuracy of 2, which limits the quotient to an accuracy of 2. Round the quotient to 2 significant digits.

$$(386 + 200.5) / 29 = 20$$

EXAMPLE 2.43

$(134^2 - 26^2)^{1/2}$.

Solution By the calculator, $(134^2 - 26^2)^{1/2} = 131.45341$. The sequence entered is **1 3 4 INV x^2 − 2 6 INV x^2 = $\sqrt{}$**. First square each number inside the parentheses, find the difference, and then take the square root.

$$(134^2 - 26^2)^{1/2}$$

134 has an accuracy of 3 and so does its square, 18000 (one zero is significant). 26 has an accuracy of 2 and so does its square, 680.

$$(18000 - 680)^{1/2}$$

To subtract the two numbers, their precision must be considered. 18000 is precise to the hundreds; 680 is precise to the tens. Their difference, 17320, is precise only to the hundreds, 17,300.

$$(17300)^{1/2}$$

17300 has an accuracy of 3, and so will its square root. Round the answer to 131.

$$(134^2 - 26^2)^{1/2} = 131$$

EXAMPLE 2.44

Evaluate. $1.06 \times 10^{-3} \cdot 8.0 \times 10^5 + 2.3 \times 10^4 / 8.4 \times 10^2$

Solution By the calculator, the result is 875.38095.
The sequence entered is
1 . 0 6 EXP 3 +/− × 8 . 0 EXP 5 + 2 . 3 EXP 4 /8 . 4EXP 2.

Note: No parentheses are needed here. The calculator multiplies the first two numbers, divides the last two, and adds the two results together. Since the product of the first two numbers will be added to the quotient of the last two, we need to determine the precision of those two intermediate results.

$$1.06 \times 10^{-3} \cdot 8.0 \times 10^5 + 2.3 \times 10^4 / 8.4 \times 10^2$$

1.06×10^{-3} has an accuracy of 3 and 8.0×10^5 has an accuracy of two. Their product is limited to two significant digits, 850. 2.3×10^4 and 8.4×10^2 each have an accuracy of two and so does their quotient, 27. 850 is precise to the ten's place, 27 is precise to the unit's place. Their sum is precise to the ten's place. Round the original result to 880.

$$1.06 \times 10^{-3} \cdot 8.0 \times 10^5 + 2.3 \times 10^4 / 8.4 \times 10^2 = 880$$

SECTION 2.5 EXERCISES

Determine the result to the proper number of significant digits.

1. $581 + 34.3/892.6$
2. $(581 + 34.3)/892.6$
3. $(72^2 + 63^2)^{1/2}$
4. $(4.8 \times 10^1 + 7.1 \times 10^2)^2 + 6.800 \times 10^5$

■ SUMMARY

1. In a number expressed in scientific notation, the decimal point has been shifted so that one digit, not a zero, lies to the left of the decimal point. That number is multiplied by a power of ten.

2. To convert from scientific notation to decimal notation, use the exponent as a guide. If the exponent is positive, move the decimal point that number of places to the right. If the exponent is negative, move the decimal point that number of places to the left. Use zeros as placeholders if necessary.

3. A number is expressed in scientific notation by adjusting the decimal point so that one digit, non-zero, exists to the left of the decimal point. That number is multiplied by a power of ten to indicate the number of places and direction that the decimal point must be moved to return to decimal notation.

4. To multiply numbers expressed in scientific notation in longhand, multiply their decimal parts and add the exponents. If the result has more than one digit to the left of the decimal point, adjust the decimal point back into scientific notation.

$$(a \times 10^b)(c \times 10^d) = ac \times 10^{b+d}$$

5. To divide numbers expressed in scientific notation in longhand, divide their decimal parts and subtract their exponents. If necessary, adjust the decimal point so that the result is expressed in scientific notation.

$$(a \times 10^b)/(c \times 10^d) = a/c \times 10^{b-d}$$

6. Before adding or subtracting numbers expressed in scientific notation in longhand, check to see that their exponents are the same. If not, adjust decimal points as needed

until they are. Add the decimal parts and use the common exponent. Adjust the decimal point so that the result is again in scientific notation.

$$a \times 10^n + b \times 10^n = (a + b) \times 10^n$$
$$a \times 10^n - b \times 10^n = (a - b) \times 10^n$$

7. Some powers of ten have been given names called *metric prefixes* that are used to scale the units being used. Here is a list of the most commonly seen prefixes, their meanings, and their symbols.

Prefix	Power of Ten	Symbol	Example
mega	10^6	M	Mohm
kilo	10^3	k	kmeter
milli	10^{-3}	m	mliter
micro	10^{-6}	μ	μamps

8. To convert from decimal notation to metric prefix, adjust the decimal point until the desired power of ten is present. Replace the power of ten with the symbol.

9. To convert from metric prefix to decimal notation, replace the prefix with its equivalent power of ten. Use the exponent as a guide to adjust the decimal point.

10. The digits in an approximate number that are known to be accurate enough to use in calculations are called *significant digits*.

11. The right-most significant digit is called the *least significant digit*. It is an estimation. The significant digits to the left of it are known to be exact.

12. Here are some rules for determining the number of significant digits in a decimal notation number.

   a. All non-zero digits are significant and zeros falling between non-zero digits are significant. 9006.03 has 6 significant digits.
   b. If there are no non-zero digits to the left of a decimal point, the zeros to the right of the decimal point between the decimal point and the first non-zero digit are not significant. They are placeholders. 0.00087 has 2 significant digits.
   c. If there are non-zero digits to the left of the decimal point, all of the digits to the right of the decimal point are significant. 3.6000 has 5 significant digits.
   d. If a number has no digits to the right of the decimal point and the number ends in one or more zeros, then the number of significant digits is often unclear. In 1600 pounds, the 1 and 6 could be significant, or the 1, 6, and first zero could be, or all four digits could be significant, depending on the scale used to determine the weight.
   e. When a number is written in scientific notation, all digits in the decimal part are significant.

13. The number of significant digits in a number is called its *accuracy*.

14. In a multiplication problem, the accuracy of the product is limited by the less accurate of the multiplicand and the multiplier.

15. In division, the accuracy of the quotient is limited by the less accurate of the divisor and dividend.

16. In raising a number to an exponent that is a whole number or a rational number, the accuracy of the base number will determine the accuracy of the result.
17. The precision of a number is determined by the decimal place of the least significant digit.
18. In addition problems, the precision of the sum is limited by the least precise addend.
19. The precision of the difference in a subtraction problem is limited to the precision of the less precise of the minuend and subtrahend.

SOLUTIONS TO CHAPTER 2 SECTION EXERCISES

Solutions to Section 2.2.1 Exercises

1. 0.000 32
2. 0.000 001 625
3. 478
4. 789 000
5. 0.001
6. 0.000 001
7. 1 000
8. 1 000 000

Solutions to Section 2.2.2 Exercises, Part 1

1. 9.87×10^8
2. 6.543×10^5
3. 2.11×10^{-4}
4. 3.95×10^{-5}
5. 10^5
6. 10^7
7. 10^{-6}
8. 10^{-8}

Solutions to Section 2.2.2 Exercises, Part 2

9. 2.36×10^5
10. 3.45×10^{-3}
11. 1.2×10
12. 8.4×10^{-6}

Solutions to Section 2.2.3 Exercises

1. 2 . 6 9 EE 4
2. 5 . 7 9 6 +/− EE 3
3. 1 . 8 2 EE 4 +/−
4. 3 . 5 8 6 +/− EE +/− 4

Solutions to Section 2.2.4 Exercises

1. 1.5×10^{-8}
2. 1.49×10^6
3. 2.29×10^3
4. 7.41×10^{-5}
5. 2.6×10^{-1}

Solutions to Section 2.3.1 Exercises

1. 720 Mhertz
2. 823 km/sec
3. 5.3 mg
4. 19 namps

Solutions to Section 2.3.2 Exercises

1. 0.000 056 2 amps
2. 0.000 120 henrys
3. 0.003 siemens
4. 0.012 meters
5. 946 000 newtons
6. 486 000 ohms
7. 377 000 000 hertz
8. 83 000 000 meters

68 ■ CHAPTER 2 SCIENTIFIC NOTATION, ENGINEERING NOTATION, AND SIGNIFICANT DIGITS

Solutions to Section 2.4 Exercises, Part 1

1. 6.28 ft
2. 4.73 ft
3. 5.50 ft
4. 6.03 ft
5. 4.98 ft
6. 4.55 ft
7. 5.32 ft
8. 5.64 ft
9. 6.47 ft
10. 5.80 ft
11. 212.8 mm
12. 230.5 mm
13. 209.5 mm
14. 199.7 mm
15. 215.0 mm
16. 205.0 mm
17. 224.8 mm
18. 237.3 mm
19. 220.0 mm
20. 233.8 mm
21. 21 1/16 or 21.06 in
22. 22 7/8 or 22.88 in
23. 23 11/16 or 23.69 in
24. 21 15/16 or 21.94 in
25. 20 or 20.00 in
26. 19 1/4 or 19.25 in
27. 2700 kΩ or 2.7 MΩ
28. 800 kΩ
29. 418 kΩ
30. 217 kΩ
31. 125 kΩ
32. 0.44 V
33. 1.05 V
34. 1.57 V
35. 2.01 V
36. 2.36 V
37. 89 V
38. 210 V
39. 313 V
40. 410 V
41. 470 V

Solutions to Section 2.4 Exercises, Part 2

42. 155.5 − 156.5 lb
43. 0.00825 − 0.00835 in
44. 5.785 × 10³ − 5.795 × 10³ ohms
45. 7.2555 × 10⁻³ − 7.2565 × 10⁻³ meters
46. 2.995 × 10⁴ − 3.005 × 10⁴ mph
47. 0.70015 − 0.70025 amps
48. If the 1 and 2 are the only significant digits, the distance lies between 1150 and 1250 miles. If 1, 2, and the first 0 are significant, the distance lies between 1195 and 1205 miles. If all four digits are significant, the distance lies between 1199.5 and 1200.5 miles.

Solutions to Section 2.4 Exercises, Part 3

49. 5
50. 4
51. 2
52. 6
53. 5
54. 4
55. 3, 4, 5, or 6

Solutions to Section 2.4.1 Exercises, Part 1

1. 12.7
2. 3.44×10^4
3. 3.17
4. 2.00×10^2

Solutions to Section 2.4.1 Exercises, Part 2

5. 1.3
6. 0.26
7. 4.2×10^{-3}
8. 17.8

Solutions to Section 2.4.1 Exercises, Part 3

9. 1.8×10^{-7} **10.** 1.91   **11.** 2.96×10^{-8}

Solutions to Section 2.4.2 Exercises, Part 1

1. units   **2.** hundred-thousandths   **3.** thousandths
4. thousands   **5.** ten-thousandths   **6.** hundreds

Solutions to Section 2.4.2 Exercises, Part 2

7. 155.8   **8.** 874.16   **9.** 8.6×10^{-2}
10. 6.45×10^{-2} **11.** 2.749×10^4 **12.** 3.341×10^4

Solutions to Section 2.4.2 Exercises, Part 3

13. 2.25   **14.** 300.1   **15.** 557.1
16. 1.9   **17.** 0.55   **18.** 0.39
19. 4.66×10^4 **20.** 5.58×10^4

Solutions to Section 2.5 Exercises

1. 581   **2.** 0.689
3. 96   **4.** 1.26×10^6

CHAPTER 2 PROBLEMS

Answer True or False. If false, correct the statement to make it true.

1. In converting from scientific notation to decimal notation, an exponent of −4 means move the decimal point four places to the right.
2. Numbers less than one have a negative exponent when expressed in scientific notation.
3. In scientific notation, the decimal portion is always between 0 and 1.
4. A large number (greater than one) is converted to scientific notation by moving the decimal point to the left.
5. A significant digit is one that is used to hold a decimal place.
6. Accuracy is the decimal place of the least significant digit.
7. Precision is the number of significant digits.
8. In addition and subtraction problems, precision is used to determine the number of significant digits in the answer.
9. In multiplication and division problems, precision is used to determine the number of significant digits in the answer.
10. m means micro or 10^{-6}.
11. k means kilo or 10^6.

12. M means milli or 10^6.
13. μ means micro or 10^{-6}.
14. 0.0067 amps = 6.7 μamps
15. 2.67×10^{-4} meters = 0.267 mmeters

Convert from scientific notation to decimal notation.

16. $2.67 \times 10^{-4} =$
17. $1.08 \times 10^{-3} =$
18. $1.6 \times 10^5 =$
19. $3.89 \times 10^4 =$

Convert from decimal notation to scientific notation.

20. 0.0000529
21. 0.01
22. 3100
23. 24689

For each of these measurements, list the upper and lower bounds.

24. 7.8×10^{-2} meters
25. 3.62×10^{-6} amps
26. 1.50×10^3 sq.ft.
27. 1600 lb (two significant digits)

Add. Limit the sum to the proper number of digits.

28. $7.6 + 2.05 + 3.008 =$
29. $12 + 1310 + 15.6 =$ (1310 has an accuracy of 3)
30. $7.6 \times 10^{-4} + 3.1 \times 10^{-3} + 4.6 \times 10^{-4} =$
31. $8.2 \times 10^2 + 9.13 \times 10^3 + 7.001 \times 10^3 =$

Subtract. Limit the difference to the proper number of digits.

32. $2.08 - 1.6 =$
33. $32.1 - 16.2 =$
34. $2.0 \times 10^{-3} - 3.1 \times 10^{-4} =$
35. $6.1 \times 10^4 - 4.91 \times 10^3 =$

Multiply. Limit the product to the proper number of digits.

36. $1.32 \cdot 0.06 =$
37. $112.6 \cdot 1.06 =$
38. $3 \times 10^{-6} \cdot 4.06 \times 10^3 =$
39. $2.001 \times 10^{-2} \cdot 5 \times 10^2 =$

Divide. Limit the quotient to the proper number of digits.

40. $26/0.0862 =$
41. $4.08/2.1 =$
42. $3.26 \times 10^{-4}/8 =$
43. $8.9 \times 10^{-2}/4 \times 10^2 =$

Convert from decimal to engineering notation.

44. 17 000 ohms = _____ kohms
45. 12 100 000 meter = _____ Mmeter
46. 0.001 2 amps = _____ mamp
47. 0.000 000 6 meters = _____ mmeter

Convert from engineering notation to decimal.

48. 67.4 knewtons = _____ newtons
49. 2.004 µmeter = _____ meters
50. 19 Mohms = _____ ohms
51. 7.9 µfarads = _____ farads

Convert from one metric prefix to another.

52. 0.725 mmeter = _____ µmeter
53. 5.66 Mhertz = _____ khertz
54. 45.5 pfarad = _____ µfarad
55. 46.8 mgram = _____ µgram

Determine the accuracy of these numbers.

56. 123.6
57. 2007
58. 2.300×10^{-2}
59. 0.0006
60. 2060.60
61. 0.000600
62. 100.001
63. 3.50 mA

Determine the precision of these numbers.

64. 123.6
65. 2007
66. 2.300×10^{-2}
67. 0.0006
68. 2060.60
69. 0.000600
70. 100.001
71. 3.50 mA

Unscramble these letters:

A E M G
I O C M R
I O P C
E A R T
E O T M F

Now unscramble the circled letters to solve the puzzle.

U.S. CUSTOMARY   POST   BREAK

_____ _____X

3
Algebra

CHAPTER OUTLINE

- **3.1** Signed Numbers
- **3.2** Addition of Signed Numbers
- **3.3** Subtraction of Signed Numbers
- **3.4** Multiplication of Signed Numbers
- **3.5** Division of Signed Numbers
- **3.6** Addition and Subtraction of Monomials
- **3.7** Multiplication of Monomials
- **3.8** Division of Monomials
- **3.9** Powers and Roots of Monomials
- **3.10** Multiplication of Binomials
- **3.11** Multiplying a Polynomial Times a Polynomial
- **3.12** Factoring
- **3.13** Factoring Trinomials
- **3.14** Factoring Other Trinomials
- **3.15** Reduction of Fractions
- **3.16** Raising Fractions to Higher Terms
- **3.17** Least Common Multiple (LCM)
- **3.18** Adding and Subtracting Fractions
- **3.19** Multiplication of Fractions
- **3.20** Division of Fractions
- **3.21** Complex Fractions

OBJECTIVES

After completing this chapter, you should be able to

1. use $>$ and $<$ symbols correctly.
2. add, subtract, multiply, and divide signed numbers.
3. apply the rules for the order of operations.
4. recognize like terms and combine them.
5. multiply and divide monomials.

6. multiply binomials.
7. multiply a polynomial by a polynomial.
8. factor common terms from a polynomial.
9. factor the difference of two squares.
10. factor a trinomial into two binomials.
11. find the least common multiple of algebraic expressions.
12. add, subtract, multiply, and divide variable fractions.
13. simplify complex fractions.

NEW TERMS TO WATCH FOR

variables
subscripts
signed numbers
magnitude (or absolute value)
sign
greater than >
less than <
like
unlike
term
variable terms
constant terms
coefficients
factor
like terms
unlike terms
polynomial
monomial
binomial
trinomial
difference of two squares
descending order
factoring
literal fractions

Chapter 1 reviewed the basic arithmetic operations. This chapter will present many of the same principles and operations, but the scope will be expanded to include *signed numbers* and *variables*. **Variables** are used to represent quantities being studied. For example, the variable "I" is used to represent the current; the variable "R" is used to represent the resistance of a resistor; and "V" is used to represent the voltage across a resistor. The current flowing through a resistor can be calculated if the voltage across the resistor and the resistance of the resistor are known. How is the calculation done?

$$V = I \cdot R$$

Regardless of the values of I and R, the voltage is the product of the current and the resistance. The variables V, I, and R are used to represent the infinite combinations of situations that can occur.

Many circuits have more than one resistor. The resistors are distinguished from each other by using **subscripts,** R_1 (which is read R sub 1), R_2, R_3, and so on. Each of these designations is a different variable. R_T often represents the total resistance in a circuit. This chapter will study expressions that involve both variables and numbers.

The operations studied so far involve positive numbers. We will encounter many applications in this book that also incorporate negative numbers. Here we begin a study of signed numbers.

3.1 ■ SIGNED NUMBERS

Signed numbers consist of two parts, a **magnitude** (or **absolute value**) and a **sign, +** or −. −23 has a magnitude of 23 and a negative sign. 8 has a magnitude of 8 and its sign is understood to be positive. It is often useful to visualize positive and negative numbers on a number line as shown in Figure 3-1, with positive numbers to the right of zero and negative numbers to the left of zero. The larger the magnitude of the number, the farther it lies away from zero.

FIGURE 3-1
Number Line

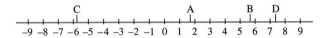

A number that falls to the right of another is said to be more positive or less negative or greater than the other. The symbol for **greater than** is >. In Figure 3-1, 5>2; B>A; −2 > −7. Likewise, a number that falls to the left of another on the number line is said to be less positive, or more negative, or less than the other. The symbol for **less than** is <. For example, 2<5; − 7< − 2; − 6<0. In Figure 3-1, A<B and C<B. Also, D>C, C<A, B<D, and A>C.

SECTION 3.1 EXERCISES

Insert the proper symbol, < or >, between the numbers.

1. 12 7
2. 13 −7
3. −7 12
4. −3 −6
5. −1 −12
6. 0 −6
7. 4 0
8. −7 0
9. −1.6 1.55
10. −2/3 −3/2

3.2 ■ ADDITION OF SIGNED NUMBERS

To add two signed numbers, check their signs and determine whether they are the same **(like)** or different **(unlike)**. To add numbers with like signs, add the magnitudes and affix the common sign.

EXAMPLE 3.1 4 + 8 = .

Solution Both signs are understood to be positive. They are like signs. Add the magnitudes. 4 + 8 = 12, and use the common sign + .

$$4 + 8 = +12.$$

EXAMPLE 3.2 $-4 + -2 =$.

Solution Both signs are negative. They are like. Add the magnitudes, $4 + 2 = 6$. Use the common sign $-$.

$$-4 + -2 = -6$$

To add numbers with unlike signs, take the difference of the magnitudes and affix the sign of the number with the greater magnitude.

EXAMPLE 3.3 $-6 + 4 =$.

Solution The signs are unlike. Take the difference in the magnitudes. $6 - 4 = 2$. Use the sign of the number with the larger magnitude. 6 has a greater magnitude than 4. Since 6 is negative, the answer is negative.

$$-6 + 4 = -2.$$

EXAMPLE 3.4 $8 + -5 =$.

Solution The signs are unlike. Take the difference in the magnitudes. $8 - 5 = 3$. Use the sign of the number with the greater magnitude. 8 has a greater magnitude than 5. The sign of 8 is positive, so the answer is positive.

$$8 + -5 = 3$$

SECTION 3.2 EXERCISES Add these signed numbers.

1. $3 + 5 =$
2. $12 + -6 =$
3. $-6 + -3 =$
4. $-4 + 6 =$
5. $5 + 2 =$
6. $-34 + -3 =$
7. $-6 + 1 =$
8. $9 + -4 =$
9. $-4 + -4 =$
10. $-7 + 3 =$

3.3 ■ SUBTRACTION OF SIGNED NUMBERS

To subtract signed numbers, change the sign of the subtrahend and proceed as in addition.

$$a - b = a + -b$$

EXAMPLE 3.5 $34 - -8 =$.

Solution Change the sign of the subtrahend and proceed as in addition.

$$34 - -8 = 34 + 8 = 42.$$

EXAMPLE 3.6

$-6 - 7 =$.

Solution Change the sign of the subtrahend and proceed as in addition. $-6 - 7 = -6 + -7 =$. The signs are like. Add the magnitudes and use the common sign.

$$-6 - 7 = -13$$

EXAMPLE 3.7

$-8 - -5 =$.

Solution Change the sign of the subtrahend and proceed as in addition. $-8 - -5 = -8 + 5 =$. The signs are unlike. Take the difference in the magnitudes. $8 - 5 = 3$. Use the sign of the number with the larger magnitude. 8 is negative, so the answer is negative.

$$-8 - -5 = -3$$

In the following example, several numbers are included inside parentheses preceded by a $-$ sign. There are two approaches to the solution.

Method 1: Combine all numbers within the parentheses into a single number. Then follow the rules for subtraction of signed numbers.

Method 2: Clear the parentheses by using the distributive rule. This is equivalent to multiplying each number within parentheses by -1. Each sign within parentheses must be changed to remove parentheses.

EXAMPLE: 3.8

$12 + 6 - (18 - 9 - 6) =$.

Solution 1 $12 + 6 - (18 - 9 - 6) = 12 + 6 - 3 = 15$

Solution 2 $12 + 6 - (18 - 9 - 6) = 12 + 6 - 18 + 9 + 6 = 15$

EXAMPLE 3.9

$-12 - (-6 - 18 + 4) =$

Solution 1 $-12 - (-6 - 18 + 4) = -12 - (-20) = -12 + 20 = 8$

Solution 2 $-12 - (-6 - 18 + 4) = -12 + 6 + 18 - 4 = 8$

Many errors are made by subtracting signed numbers incorrectly within complex algebra problems. Work the following exercises until you can complete all 22 problems with 100% accuracy.

SECTION 3.3 EXERCISES

Subtract these signed numbers.

1. $7 - 4 =$
2. $8 - 10 =$
3. $-3 - 6 =$
4. $5 - -8 =$
5. $-6 - -7 =$
6. $-8 - -3 =$

7. $12 - 4 =$
8. $12 - -4 =$
9. $-12 - 4 =$
10. $-12 - -4 =$
11. $7 - 9 =$
12. $-9 - 4 =$
13. $9 - -4 =$
14. $7 - -23 =$
15. $-6 - 3 =$
16. $-5 - 9 =$
17. $-4 - -8 =$
18. $-8 - -4 =$
19. $3 - 9 =$
20. $7 - -5 =$
21. $11 - (-6 + 13 - 5)$
22. $3 - 8 - (12 + 4 - 33)$

3.4 ■ MULTIPLICATION OF SIGNED NUMBERS

To multiply signed numbers, multiply the magnitudes. If the signs are like, affix a positive sign. If the signs are unlike, affix a negative sign.

EXAMPLE 3.10 $-5 \cdot 4 =$.

Solution $5 \cdot 4 = 20$. Signs are unlike, so attach a negative sign.
$$-5 \cdot 4 = -20$$

EXAMPLE 3.11 $-2 \cdot -6 =$.

Solution $2 \cdot 6 = 12$. Signs are like, so attach a positive sign.
$$-2 \cdot -6 = 12$$

EXAMPLE 3.12   Evaluate -4^2 and $(-4)^2$.

Solution
$$-4^2 = -(4 \cdot 4) = -16$$
$$(-4)^2 = -4 \cdot -4 = 16$$

The x^2 function on your calculator will square whatever number is showing in the display. To evaluate $(-4)^2$ on your calculator, enter 4, then $+/-$ to change the sign, and then INV x^2. To evaluate -4^2, enter 4 INV x^2, then $+/-$.

EXAMPLE 3.13 $(-4)^{-2} =$.

Solution $(-4)^{-2} = 1/(-4)^2 = 1/16$

To raise -4 to the -2 power on the calculator, use the x^y key. Follow this sequence: $4+/-$ INV x^y $2+/-=$. The result is in decimal form, 0.0625.

If several signed numbers are being multiplied, determine the sign of the product as follows:

1. If the number of negative factors being multiplied is odd, the product is negative.
2. If the number of negative factors being multiplied is even, the product is positive.

EXAMPLE 3.14 $(5)(-2)(-3)(4)(-1) =$.

Solution   Three of the factors are negative; three is odd; so the product is negative.

$$(5)(-2)(-3)(4)(-1) = -120$$

3.5 ■ DIVISION OF SIGNED NUMBERS

To divide signed numbers, divide the magnitudes. If the signs are like, the quotient is positive. If the signs are unlike, the quotient is negative.

EXAMPLE 3.15 $-8/4 =$.

Solution $8/4 = 2$. The signs are unlike. Attach a negative sign.

$$-8/4 = -2$$

EXAMPLE 3.16 $-12/-6 =$.

Solution $12/6 = 2$. The signs are like. Attach a positive sign.

$$-12/-6 = 2$$

SECTIONS 3.4 AND 3.5 EXERCISES

Perform the indicated operation.

1. $3 \cdot 8 =$
2. $4 \cdot -23 =$
3. $-9 \cdot 3 =$
4. $-8 \cdot -3 =$
5. $-6 \cdot -1 =$
6. $-1 \cdot -1 =$
7. $-6 \cdot 2 \cdot -4 \cdot -3 \cdot 2 =$
8. $-3 \cdot -2 \cdot -1 \cdot 4 \cdot -3 =$
9. $8/3 =$
10. $10/-5 =$
11. $-12/4 =$
12. $-15/-3 =$
13. $-2/-1 =$
14. $-1/-1 =$

3.6 ■ ADDITION AND SUBTRACTION OF MONOMIALS

In the algebraic expression $2a - 3b - 6a + 7c - 12 + 5$, each quantity being added or subtracted is called a **term**. This expression has six terms; 2a, 3b, 6a, 7c, 12, and 5. *a, b,* and *c* are called **variables**. Four of the terms contain variables and are called **variable terms**, *2a, 3b, 6a,* and *7c.* Two of the terms contain no variables and are called **constant terms**, 12 and 5. Each variable term consists of a variable and a multiplier. The multipliers 2, 3, 6, and 7 are called **coefficients.** Each variable and each coefficient is called a **factor** of its term. 2 and *a* are factors of 2a. In the expression above, the first term and third term have the same variable; they are called **like terms** and can be combined: $2a - 6a = -4a$. All constant terms are like terms and can be combined: $-12 + 5 = -7$. The other terms have a different variable; they are called **unlike terms** and cannot be combined with the "*a*" terms or the constant terms. So,

$$2a - 3b - 6a + 7c - 12 + 5 = -4a - 3b + 7c - 7$$

An algebraic expression that contains one or more terms is called a **polynomial.** (To officially be a polynomial, none of the terms can have a variable in the denominator or a radical.) A polynomial that contains one term is called a **monomial;** a polynomial that contains two terms is called a **binomial;** and a polynomial that contains three terms is called a **trinomial.**

The trinomial $4t^2u^2 + 4t^4u^3 - 5t^4u^3$ has three terms, $4t^2u^2, 4t^4u^3,$ and $5t^4u^3$. *t* and *u* are variables and 4, 4, and 5 are coefficients. In the last two terms, the variables raised to their powers are the same, t^4u^3. They are like terms and can be combined.

$$4t^4u^3 - 5t^4u^3 = -1t^4u^3 = -t^4u^3$$

The exponents in the first term differ from those in the second and third terms. $4t^2u^2$ and $4t^4u^3$ are unlike terms and cannot be combined. So,

$$4t^2u^2 + 4t^4u^3 - 5t^4u^3 = 4t^2u^2 - t^4u^3$$

SECTION 3.6 EXERCISES

Combine like terms.

1. $5t^2 + 7t + 4u^2 + 2t - 3t^2 =$
2. $z^3 - 1 - 4z - 7z^2 - 5z^3 - 6 =$
3. $12x^2y^3 + 8x^2y^3 - 5x^2 - 3y^2 =$
4. $7 + 2ab - 4a^2b + 8ab^2 - 13 - 4a^2b - 5ab + 15ab - 4a^2b$
5. $6s^2 - (5s - 16s^2 - 12 + 2t) =$

3.7 ■ MULTIPLICATION OF MONOMIALS

The multiplication of monomials is an application of the rule for the multiplication of exponential quantities. To multiply numbers with like bases, add the exponents and use the common base.

$$(a^m)(a^n) = a^{m+n}$$

80 ■ CHAPTER 3 ALGEBRA

EXAMPLE 3.17 $(10^6)(10^3) =$.

Solution   The bases are like, so this rule can be used to find the product.
$$(10^6)(10^3) = 10^{6+3} = 10^9$$

EXAMPLE 3.18 $(a^3)(a^5)(b^7) =$.

Solution   The two "a" terms can be multiplied by adding exponents.
$$(a^3)(a^5) = a^{3+5} = a^8$$

The "b" term is unlike and cannot be combined with the "a" terms.
$$(a^3)(a^5)(b^7) = a^8 b^7$$

Multiplication of two or more monomials is taken in two steps.

Step 1.   Multiply the coefficients (don't forget their signs).
Step 2.   Multiply variables with like bases by adding exponents.

EXAMPLE 3.19 $(2a)(-7a^4) =$.

Solution   **Step 1.** $(2)(-7) = -14$
Step 2. $(a)(a^4) = a^{1+4} = a^5$
$(2a)(-7a^4) = -14a^5$

EXAMPLE 3.20 $(6u^2v^4)(-12u^8v^2) =$.

Solution   **Step 1.** $(6)(-12) = -72$
Step 2. $(u^2)(u^8) = u^{2+8} = u^{10}$
$(v^4)(v^2) = v^{4+2} = v^6$
$(6u^2v^4)(-12u^8v^2) = -72u^{10}v^6$

EXAMPLE 3.21 $(12X^2Y)(-2X^5Y^6)(2XYZ^3)(-6V^3Y^4Z^5)(-VY) =$

Solution   **Step 1.** $12 \cdot -2 \cdot 2 \cdot -6 \cdot -1 = -288$
Step 2. $X^2 \cdot X^5 \cdot X^1 = X^{2+5+1} = X^8$
$Y^1 \cdot Y^6 \cdot Y^1 \cdot Y^4 \cdot Y = Y^{1+6+1+4+1} = Y^{13}$
$Z^3 \cdot Z^5 = Z^{3+5} = Z^8$
$V^3 \cdot V^1 = V^{3+1} = V^4$
$(12X^2Y)(-2X^5Y^6)(2XYZ^3)(-6V^3Y^4Z^5)(-VY) = -288V^4X^8Y^{13}Z^8$

EXAMPLE 3.22

24 μmeter·2.3 mmeter=.

Solution Replace the metric prefixes with their values.

$$24 \times 10^{-6} \cdot 2.3 \times 10^{-3} =$$

Step 1. $24 \cdot 2.3 = 55.2$

Step 2. $(10^{-6})(10^{-3}) = 10^{-9}$

Replace 10^{-9} with its metric prefix, nano.

meter·meter = meter^2

24 μmeter·2.3 mmeter=55.2 nmeter^2

SECTION 3.7 EXERCISES

Multiply.

1. $(4y^2)(-3y^3) =$
2. $(5d^4)(-6e^4)(-3d^3e^3) =$
3. $(-8a^3b^2c^3)(3a^6b^3c)(-2acb) =$
4. $12r \cdot 3rs^2 \cdot -2rs^3t^4 \cdot -rs^2t^3 =$
5. $4x^3y^3x^3 \cdot -2xy^2z \cdot y$
6. $3mno^2 \cdot 2n^3p^3 \cdot -4n^2o^3 \cdot 2m^3n^3p^3 =$

3.8 ■ DIVISION OF MONOMIALS

The division of monomials is an application of the rule for division of exponential quantities. To divide numbers with like bases, subtract the exponent in the denominator from the exponent in the numerator and use the common base.

$$a^m/a^n = a^{m-n}$$

If some of the exponents are negative, use the rules for subtraction of signed numbers.

EXAMPLE 3.23

$10^5/10^3 =$.

Solution The bases are like. Subtract the exponents.

$$10^5/10^3 = 10^{5-3} = 10^2$$

EXAMPLE 3.24

$3^{-4}/3^{-7} =$.

Solution Bases are like. Subtract the exponents.

$$3^{-4}/3^{-7} = 3^{-4--7} = 3^{-4+7} = 3^3$$

Division of monomials is taken in two steps.

Step 1. Divide the coefficients (don't forget their signs).

Step 2. Divide the variables with like bases by subtracting the exponents.

EXAMPLE 3.25 $8a^3/(-2a^7) =$

Solution
1. $8/-2 = -4$
2. $a^3/a^7 = a^{3-7} = a^{-4}$
   $8a^3/(-2a^7) = -4a^{-4}$ or $-4/a^4$

EXAMPLE 3.26 $-12s^5t^5u^7/(-60s^9t^2u^4v) =$.

Solution
1. $-12/-60 = 1/5$
2. $s^{5-9} = s^{-4}$
   $t^{5-2} = t^3$
   $u^{7-4} = u^3$
   $-12s^5t^5u^7/(-60s^9t^2u^4v) = s^{-4}t^3u^3/(5v)$ or $t^3u^3/(5s^4v)$

EXAMPLE 3.27 $3r^5s^{-3}t \cdot 4r^2st^2/(6rst) =$.

Solution
1. $3 \cdot 4 / 6 = 2$
2. $r^5 \cdot r^2 / r = r^{5+2-1} = r^6$
   $s^{-3} \cdot s^1 / s^1 = s^{-3+1-1} = s^{-3} = 1/s^3$
   $t \cdot t^2 / t = t^{1+2-1} = t^2$
   $3r^5s^{-3}t \cdot 4r^2st^2/(6rst) = 2r^6t^2/s^3$

EXAMPLE 3.28 $-4u^6v^3(3x^2v^3)/(5u^3vx^3) =$.

Solution
1. $-4 \cdot 3 / 5 = -12/5$
2. $u^6 / u^3 = u^3$
   $v^3 \cdot v^3 / v = v^5$
   $x^2 / x^3 = 1/x$
   $-4u^6v^3(3x^2v^3)/(5u^3vx^3) = -12u^3v^5/(5x)$ or $-12/5 \, x^{-1}u^3v^5$

SECTION 3.8 EXERCISES   Simplify.

1. $10^4/10^6 =$
2. $a^7/a^4 =$
3. $4a^5b^3c / -3a^7b^2c^3 =$
4. $35x^4y^3 \cdot 7x^3y^5z / (-21x^5y^5z^5) =$
5. $u^5v^2w^2 \cdot -4uvw^4 / (-12uvw) =$

3.9 ■ POWERS AND ROOTS OF MONOMIALS

When a monomial is raised to a power, each factor of the monomial must be raised to the power. For example, $(3x^2y^3)^4$ is a monomial $3x^2y^3$ being raised to the power 4. 3, x^2, and y^3 must each be raised to the fourth power.

$$(3x^2y^3)^4 = 3^4 \cdot (x^2)^4 \cdot (y^3)^4$$
$$3^4 = 81$$

To raise an exponent to a power, multiply the exponents.

$$(x^2)^4 = x^{2 \cdot 4} = x^8$$
$$(y^3)^4 = y^{3 \cdot 4} = y^{12}$$
$$(3x^2y^3)^4 = 81\, x^8 y^{12}$$

EXAMPLE 3.29 Simplify $(2x^4y^3z^2)^5$.

Solution
$$(2x^4y^3z^2)^5 = 2^5 \cdot x^{4 \cdot 5} \cdot y^{3 \cdot 5} \cdot z^{2 \cdot 5}$$
$$= 32x^{20}y^{15}z^{10}$$

Raising a monomial to the 1/2 power is equivalent to taking the square root of the monomial, and the rules presented in the first part of this section are valid. For example, $(25x^4)^{1/2}$ means take the square root of 25 and the square root of x^4.

$$25^{1/2} \cdot (x^4)^{1/2} = 5\, x^{4 \cdot 1/2} = 5x^2$$

However, if the exponent of a variable in the monomial is not even, then we end up with a fractional exponent like 3/2 or 7/2. A variation of this approach is to rewrite the monomial with as many factors that have even exponents as possible and take the square root of those factors. For example, to take the square root of $45x^7y^3$,

$$(45x^7y^3)^{1/2} = (3 \cdot 3 \cdot 5 \cdot x^6 \cdot x \cdot y^2 \cdot y)^{1/2}$$

The square root of $3 \cdot 3$ is 3; the square root of x^6 is x^3; the square root of y^2 is y.

$$(45x^7y^3)^{1/2} = (3 \cdot 3 \cdot 5 \cdot x^6 \cdot x \cdot y^2 \cdot y)^{1/2} = 3x^3y(5xy)^{1/2}$$

The factors that are not duplicated and the variables with odd exponents remain in the radical.

EXAMPLE 3.30 Simplify $(72x^5y^3z^6)^{1/2}$.

Solution
$$(72x^5y^3z^6)^{1/2} = (2 \cdot 6 \cdot 6 \cdot x \cdot x^4 \cdot y \cdot y^2 \cdot z^6)^{1/2} = 6x^2yz^3\,(2xy)^{1/2}$$

Taking the cube root is the same as raising a quantity to the 1/3 power. Factor the quantity into variables with exponents that are multiples of 3. Factor coefficients into factors that appear three times.

EXAMPLE 3.31 Simplify $(40s^7t^5u^4)^{1/3}$.

Solution $(40s^7t^5u^4)^{1/3} = (5 \cdot 2 \cdot 2 \cdot 2 \cdot s \cdot s^6 \cdot t^2 \cdot t^3 \cdot u \cdot u^3)^{1/3} = 2s^2tu(5st^2u)^{1/3}$

SECTION 3.9 EXERCISES

Simplify.

1. $(2t^5u^3v^2)^6 =$
2. $(-3p^3q^2)^4 =$
3. $(5xy^4z^3)^3 =$
4. $(36x^6y^2z^4)^{1/2} =$
5. $(12t^5u^7v^4)^{1/2} =$
6. $(54a^3b^6c^9)^{1/3} =$
7. $(512m^5n^7o^5)^{1/3} =$
8. $(81r^8s^4t^{12})^{1/4} =$
9. $(80r^3s^6t^7)^{1/4} =$

3.10 ■ MULTIPLICATION OF BINOMIALS

A binomial is an expression that contains two terms, such as $(a + b)$ and $(2s^4t^6 - 8s^3k^7)$. When multiplying one binomial by another, care must be taken to ensure that each term of one binomial gets multiplied by each term of the other. This task can be organized by thinking "FOIL." FOIL is an acronym for **First, Outer, Inner, Last**.

FIGURE 3-2
FOIL

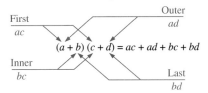

As shown in Figure 3-2, multiply the first terms of each binomial together, then the outer, then the inner, and finally the last terms of each binomial. This procedure ensures that each term gets multiplied.

$$(a + b)(c + d) = ac + ad + bc + bd.$$

Then combine any like terms that result.

EXAMPLE 3.32 $(2x + 3y)(4x - 8y) = $.

Solution
First: $2x \cdot 4x = 8x^2$
Outer: $2x \cdot -8y = -16xy$
Inner: $3y \cdot 4x = 12xy$
Last: $3y \cdot -8y = -24y^2$

$$(2x + 3y)(4x - 8y) = 8x^2 - 16xy + 12xy - 24y^2$$
$$= 8x^2 - 4xy - 24y^2$$

EXAMPLE 3.33 $(3u^4v^3 + 2w^5x^6)(u^2v - 4w^4x^2)$

Solution   **First:** $(3u^4v^3)(u^2v) = 3u^6v^4$
Outer: $(3u^4v^3)(-4w^4x^2) = -12u^4v^3w^4x^2$
Inner: $(2w^5x^6)(u^2v) = 2u^2vw^5x^6$
Last: $(2w^5x^6)(-4w^4x^2) = -8w^9x^8$

There are no like terms to combine.
$$(3u^4v^3 + 2w^5x^6)(u^2v - 4w^4x^2) =$$
$$3u^6v^4 - 12u^4v^3w^4x^2 + 2u^2vw^5x^6 - 8w^9x^8$$

In the next two examples, the binomials to be multiplied differ only in their middle signs. In this situation, the outer and inner products add to zero. The product that results is called a "**difference of two squares.**"

EXAMPLE 3.34 $(a+b)(a-b) =$.

Solution $(a + b)(a - b) = a^2 - ab + ab - b^2 = a^2 - b^2$

Note that the product can be written directly as the square of the first term minus the square of the second term.

EXAMPLE 3.35 $(4r + 3s)(4r - 3s) =$.

Solution   The binomials differ only in their middle signs. The product is the square of the first term minus the square of the second term.
$$(4r + 3s)(4r - 3s) = 16r^2 - 9s^2$$

What if the two binomials to be squared are exactly the same? Squaring a binomial can be accomplished by FOIL. The result, though, is always equal to the first term squared plus twice the product of the first and last term plus the last term squared. Following this special pattern is quicker than resorting to FOIL.

EXAMPLE 3.36 $(a + b)^2 =$.

Solution 1   By FOIL.
$$(a + b)^2 = (a + b)(a + b) = a^2 + ab + ab + b^2 = a^2 + 2ab + b^2$$

Solution 2 By special product.

1. First term squared: a^2
2. Twice the product of the two terms: $2ab$
3. Second term squared: b^2

$$(a + b)^2 = a^2 + 2ab + b^2$$

EXAMPLE 3.37 $(2s^3 - 3t^4)^2 = \underline{}$.

Solution
1. First term squared: $(2s^3)^2 = 4s^6$
2. Twice the product of the two terms: $2(2s^3)(-3t^4) = -12s^3t^4$
3. Second term squared: $(-3t^4)^2 = 9t^8$

$$(2s^3 - 3t^4)^2 = 4s^6 - 12s^3t^4 + 9t^8$$

SECTION 3.10 EXERCISES

Multiply.

1. $(k + 1)(k - 1) =$
2. $(y + 2)(y - 6) =$
3. $(r + 2)(r - 2) =$
4. $(2y - 3z)(3y + 2z) =$
5. $(2w - 5)(3w + 3t) =$
6. $(r - 4)^2 =$
7. $(3rt^2 - 4w^3)^2 =$

3.11 ■ MULTIPLYING A POLYNOMIAL TIMES A POLYNOMIAL

A trinomial is an expression that contains three terms, such as $(a + b - c)$ and $(12r^5s^2t^4 + 73.8t^3 - 22^2rst)$. To multiply a trinomial by a binomial, simply multiply each term of the trinomial by each term of the binomial and then combine like terms.

EXAMPLE 3.38 $(3a^2 - 4b^3 + 5c^2)(2a^3 - 3bc) = \underline{}$.

Solution
$$(3a^2 - 4b^3 + 5c^2)(2a^3 - 3bc) =$$
$$6a^5 - 8a^3b^3 + 10a^3c^2 - 9a^2bc + 12b^4c - 15bc^3$$

There are no like terms to combine.

EXAMPLE 3.39 $(7b^2 - 8ab + 4a^2)(a - b) = \underline{}$.

Solution Multiply each term of the trinomial by each term of the binomial.

$$7ab^2 - 8a^2b + 4a^3 - 7b^3 + 8ab^2 - 4a^2b$$

Combine like terms; there are two ab^2 terms and two a^2b terms.

$$15ab^2 - 12a^2b + 4a^3 - 7b^3$$
$$(7b^2 - 8ab + 4a^2)(a - b) = 4a^3 - 12a^2b + 15ab^2 - 7b^3$$

The terms in the product have been arranged in **descending order** according to the variable a. That is, the highest order term a^3 is listed first, followed by a^2, a^1, and a^0.

As the number of terms in the factors increase it becomes more difficult to identify the like terms. You may prefer to list the like terms that result in vertical columns, as shown in the next example.

EXAMPLE 3.40 $(3x^2 + 6xy + 2y^2)(2x^2 - 3xy - y^3) = $.

Solution Multiply each term of the second trinomial by each term of the first trinomial. When like terms result, list them in their own column.

$$\begin{array}{cccccccc} 6x^4 & -9x^3y & -3x^2y^3 & -18x^2y^2 & -6xy^4 & -6xy^3 & -2y^5 \\ & +12x^3y & & +4x^2y^2 & & & \\ \hline 6x^4 & +3x^3y & -3x^2y^3 & -14x^2y^2 & -6xy^4 & -6xy^3 & -2y^5 \end{array}$$

Arrange the terms of the product in descending order according to x and ascending order according to y when possible.

$$(3x^2 + 6xy + 2y^2)(2x^2 - 3xy - y^3)$$
$$= 6x^4 + 3x^3y - 14x^2y^2 - 3x^2y^3 - 6xy^3 - 6xy^4 - 2y^5$$

SECTION 3.11 EXERCISES

Multiply.

1. $(4r - 3t)(2r + 3s + 4t) =$
2. $(2r^2s^2 - 3rs^3)(r^3 - s^2 - rs^3) =$
3. $(2a + 3b)(3a - 4b - 6c) =$
4. $(3w^2y^4 - 2wy^2)(3a - 4wy + 3aw^2y^3) =$
5. $(2t^2 + tu + 3u^2)(4t^2 - 3tu - 2u^2) =$

3.12 ■ FACTORING

Now that we've mastered the process of multiplication, let's attack the opposite process, factoring. **Factoring** means dividing an expression into its factors. One type of factoring is an application of the distributive rule.

$$ab + ac = a(b + c)$$

a is common to each of the terms and is factored out, or written outside parentheses. To factor an algebraic expression, first look for factors that are common to both the coefficients and the constant terms and factor them out.

CHAPTER 3 ALGEBRA

EXAMPLE 3.41 Factor $12a^4 - 16b^6 + 8$.

Solution
$$12 = 2 \cdot 2 \cdot 3$$
$$16 = 2 \cdot 2 \cdot 2 \cdot 2$$
$$8 = 2 \cdot 2 \cdot 2$$

2·2 is common to each of these numbers. Factor out 4.
$$4(3a^4 - 4b^6 + 2)$$
$$12a^4 - 16b^6 + 8 = 4(3a^4 - 4b^6 + 2)$$

To check your work, use the distributive rule to clear the parentheses of the result. The product should be the same as the original expression.
$$4(3a^4 - 4b^6 + 2) = 12a^4 - 16b^6 + 8 \text{ It checks!}$$

Watch for variables that appear in each term of the expression to be factored. If a variable is common to each term, factor out that variable raised to the highest exponent that is common to each term. For example, in an expression with three terms, if one term contains x^3, one contains x^5, and one contains x^7, factor out x^3.

EXAMPLE 3.42 Factor $36s^3y^7 - 12sy^5$.

Solution Each term contains 12; each term contains s^1; and each term contains y^5. Factor $12sy^5$ out of each term. $12sy^5(3s^2y^2 - 1)$
$$36s^3y^7 - 12sy^5 = 12sy^5(3s^2y^2 - 1)$$

To check the result, clear the parentheses and compare to the original expression.
$$12sy^5(3s^2y^2 - 1) = 36s^3y^7 - 12sy^5 \text{ It checks!}$$

Watch for special patterns such as the difference of two squares. The difference of two squares factors into two binomials that differ only in their middle sign. The terms of the binomial are the square roots of two squares.

EXAMPLE 3.43 Factor $a^2 - b^2$.

Solution This expression is of the difference of two squares form. It factors into $(a+b)(a-b)$.
$$a^2 - b^2 = (a + b)(a - b)$$

We should check this one.

$$(a + b)(a - b) = a^2 - ab + ab - b^2$$

The middle terms add to zero.

$$(a + b)(a - b) = a^2 - b^2 \text{ It checks!}$$

EXAMPLE 3.44   Factor $25t^2 - 81y^2$.

Solution   This expression can be rewritten as $(5t)^2 - (9y)^2$ to emphasize the fact that it is the difference of two squares. It factors into

$$(5t)^2 - (9y)^2 = (5t + 9y)(5t - 9y)$$

The *sum of two squares* like $a^2 + b^2$ cannot be factored.

SECTION 3.12 EXERCISES   Factor.

1. $4X^2Y - 2XY^2 =$
2. $24r^3s^4 + 8rst =$
3. $2x^2 - 2y^2 =$
4. $2x^2 + 2y^2 =$
5. $16R^2 - 25T^2 =$
6. $40X^2 - 90Y^2 =$

3.13 ■ FACTORING TRINOMIALS

When two binomials of the form $(ax + b)(cx + d)$ where a and c are coefficients, x is a variable, and b and d are constants, are multiplied together using FOIL, the product is

$$acx^2 + adx + bcx + bd$$

The middle two terms are like terms and combine into $(ad + bc)x$. So,

$$(ax + b)(cx + d) = acx^2 + (ad + bc)x + bd$$

The resulting trinomial has an x^2 term, an x term, and a constant term. Examples of trinomials of this type are $12x^2 - 8x + 2$ and $1.6x^2 + 3.5$ (the coefficient of the x term is 0).

To factor trinomials of the form $acx^2 + (ad + bc)x + bd$, follow these steps.

1. Factor out any common coefficient to reduce the size of the numbers involved and the amount of work involved. This step is often overlooked.
2. Factor the x^2 term into ax and cx. $(ax\ )(cx\ )$
3. Choose signs for b and d as follows:
   a. If bd is positive, the signs will both be the same as the sign of $(ad + bc)$. $(ax +)(cx +)$ or $(ax -)(cx -)$.
   b. If bd is negative, the signs will be different and you will have to decide which goes where. $(ax +)(cx -)$ or $(ax -)(cx +)$.

4. Factor the constant into b and d. $(ax + b)(cx + d)$.
5. Use FOIL to see if the two binomials yield the original trinomial.
6. If not, try swapping b and d. Try different factors of the constant or the x^2 term until the correct combination is found. It gets easier with practice.

EXAMPLE 3.45 Factor $x^2 + 4x - 12$.

Solution $x^2 + 4x - 12 = (\ )(\ )$

1. There are no common coefficients to factor out.
2. Factor x^2 into $(x\ )(x\ )$.
3. The minus sign in front of the 12 indicates that the signs in the two binomials are different. $(x +)(x -)$
4. The product of the inner terms and outer terms must add to the middle term $+ 4x$. Factor 12 into $(6)(2)$ to produce the difference of 4. The $+$ sign goes with the 6 and the $-$ with the 2, to make the sum of $+ 4$. $(x + 6)(x - 2)$
5. Check the result by multiplying the binomials.

$$(x + 6)(x - 2) = x^2 - 2x + 6x - 12 = x^2 + 4x - 12 \text{ It checks!}$$
$$x^2 + 4x - 12 = (x + 6)(x - 2)$$

EXAMPLE 3.46 Factor $6r^2 - 19r + 10$.

Solution
1. 19 is prime and cannot be factored. There are no common factors of the coefficients and constant to be factored.
2. Factor the r^2 term into $(3r\ )(2r\ )$. This is an initial guess. The other possibility is $(6x\ )(x\ )$.
3. Determine the signs. The constant term is positive, so the signs in the binomials are the same, and, since the middle term $- 19r$ is negative, they are both negative.

$$(3r -)(2r -)$$

4. Factor the constant into $(2)(5)$. $(3r - 2)(2r - 5)$. The product of the outer terms yields $- 15r$ and the product of the inner terms yields $- 4r$. These add to $- 19r$, the middle term.

$$6r^2 - 19r + 10 = (2r - 5)(3r - 2)$$

SECTION 3.13 EXERCISES

Factor.

1. $t^2 - 8t - 48 =$
2. $y^2 - 5y - 6 =$
3. $2r^2 - 19r + 24 =$
4. $5r^2 - 2r - 24 =$
5. $12x^2 + 79x + 25 =$

3.14 ■ FACTORING OTHER TRINOMIALS

Trinomials of the form $ax^2 + bxy + cy^2$ can sometimes be factored using the process described in the last section.

EXAMPLE 3.47 Factor $6x^2 - xy - y^2$.

Solution $6x^2 - xy - y^2 = (3x + y)(2x - y)$

EXAMPLE 3.48 Factor $25s^2 - 248st - 20t^2$.

Solution $25s^2 - 248st - 20t^2 = (25s + 2t)(s - 10t)$

SECTION 3.14 EXERCISES

Factor.

1. $8u^2 - 13uv - 6v^2 =$
2. $6s^2 - 21st + 18t^2 =$ Hint: Look for common factors first.
3. $56 m^2 - 5 mn - 6 n^2 =$
4. $-15 t^2 + 29tu - 12 u^2 =$
5. $18 s^2 - 42 su + 20 u^2 =$

The following sections address **literal fractions**, fractions that contain both coefficients and variables. All the rules and procedures presented in the first chapter are valid here. The rules are summarized as they are used. Return to the first chapter as needed to master this material.

3.15 ■ REDUCTION OF FRACTIONS

A procedure for reducing fractions that consist of a monomial divided by a monomial was presented in Section 3.9 Division of Monomials. Let's concentrate here on fractions that contain binomials and trinomials. A common error made by beginning students is to cancel terms that appear in both the numerator and the denominator. For example, in the fraction

$$\frac{3x - 6y}{3x - 12y}$$

can $3x$ be canceled? NO, IT CANNOT! However, a 3 can be factored from the numerator, and a 3 can be factored from the denominator.

$$\frac{3x - 6y}{3x - 12y} = \frac{\cancel{3}(x - 2y)}{\cancel{3}(x - 4y)}$$

Now 3 appears as a factor in the numerator and the denominator and can be canceled.

$$\frac{3x - 6y}{3x - 12y} = \frac{3(x - 2y)}{3(x - 4y)} = \frac{x - 2y}{x - 4y}$$

Factors can be canceled but terms cannot.

EXAMPLE 3.49 Reduce.

$$\frac{40x^2 - 120x + 20x^4}{10xy^2 + 100x^2y^3 - 40x^5}$$

Solution The numerator factors into $2 \cdot 10x(2x - 6 + x^3)$.
The denominator factors into $10x(y^2 + 10xy^3 - 4x^4)$.
$10x$ appears as a factor in the numerator and in the denominator and will cancel.

$$\frac{40x^2 - 120x + 20x^4}{10xy^2 + 100x^2y^3 - 40x^5} = \frac{2(2x - 6 + x^3)}{y^2 + 10xy^3 - 4x^4}$$

EXAMPLE 3.50 Reduce.

$$\frac{36t^2 + 132tu + 105u^2}{360t^2 - 490u^2}$$

The numerator factors into

$$3(12t^2 + 44tu + 35u^2) = 3(6t + 7u)(2t + 5u)$$

The denominator factors into $10(36t^2 - 49u^2)$. $(36t^2 - 49u^2)$ is the difference of two squares.

$$10(36t^2 - 49u^2) = 10(6t - 7u)(6t + 7u)$$

$(6t + 7u)$ is a factor in both the numerator and the denominator and will cancel.

$$\frac{36t^2 + 132tu + 105u^2}{360t^2 - 490u^2} = \frac{3(2t + 5u)}{10(6t - 7u)}$$

SECTION 3.15 EXERCISES Reduce these fractions.

1. $(2x - 2y + 2z)/2$
2. $(16x^2 + 4x^2y - 8x^4y^4)/(120x^2y - 200x^2y^2 - 40x^4)$
3. $(8c^2 + 2cd - 15d^2)/(16c^2 - 25d^2)$
4. $(27t^2 - 36tu + 12\,u^2)/(60t^2 + 8tu - 32u^2)$

3.16 ■ RAISING FRACTIONS TO HIGHER TERMS

Raising fractions to higher terms is the opposite process from reducing a fraction to lowest terms. This process is used in finding a least common denominator. To raise a fraction to higher terms, determine what factor the original denominator must be multiplied by to become the new denominator. Multiply both the numerator and the denominator by this factor so that the value of the original fraction will not be altered.

EXAMPLE 3.51 Express $5a^3/4b^4$ as a fraction with a denominator of $12\,a^4b^7$.

Solution $4b^4$ must be multiplied by $3a^4b^3$ to produce the new denominator, $12a^4b^7$. Multiply both numerator and denominator by $3a^4b^3$.

$$\frac{(5a^3)(3a^4b^3)}{(4b^4)(3a^4b^3)} = \frac{15a^7b^3}{12a^4b^7}$$

EXAMPLE 3.52 Express $-13xyz / 120b^5x^4z^3$ as a fraction with a denominator of $480b^6x^6y^6z^6$. $120b^5x^4z^3$ must be multiplied by $4bx^2y^6z^3$ to produce the new denominator. Multiply both numerator and denominator by $4bx^2y^6z^3$.

$$\frac{-13xyz}{120\,b^5x^4z^3} \times \frac{4bx^2y^6z^3}{4bx^2y^6z^3} = \frac{-52bx^3y^7z^4}{480b^6x^6y^6z^6}$$

SECTION 3.16 EXERCISES

1. Express $-23m^2/n^3$ as a fraction with a denominator of $2n^5m^6$.
2. Express $14xyz^4/3ax^2z$ as a fraction with a denominator of $21a^4b^2x^2z$.

3.17 ■ LEAST COMMON MULTIPLE (LCM)

To find the least common multiple (LCM) of two or more numbers, break each number into its prime factors. Each prime factor must appear in the LCM the maximum number of times that it appears in any of the numbers. For example, let's find the LCM of $12a^4b^3c$ and $72a^2b^2c^2$.

$$12a^4b^3c = 2 \cdot 2 \cdot 3 \cdot a \cdot a \cdot a \cdot a \cdot b \cdot b \cdot b \cdot c$$
$$72a^2b^2c^2 = 2 \cdot 2 \cdot 2 \cdot 3 \cdot 3 \cdot a \cdot a \cdot b \cdot b \cdot c \cdot c$$

The LCM of $12a^4b^3c$ and $72a^2b^2c^2$ contains as factors three 2s, two 3s, four a's, three b's, and two c's.

$$\text{LCM} = 2 \cdot 2 \cdot 2 \cdot 3 \cdot 3 \cdot a^4 \cdot b^3 \cdot c^2 = 72a^4b^3c^2$$

Note that it is not necessary to write the variables out as prime factors as shown above. Instead, for each variable choose the largest exponent that appears.

EXAMPLE 3.53 Determine the LCM for $36g^3$, $27ghi^3$, $9h^2i^2$, and $12ghi$.

Solution
$$36g^3 = 2 \cdot 2 \cdot 3 \cdot 3 \cdot g^3$$
$$27ghi^3 = 3 \cdot 3 \cdot 3 \cdot g \cdot h \cdot i^3$$
$$9h^2i^2 = 3 \cdot 3 \cdot h^2 \cdot i^2$$
$$12ghi = 3 \cdot 2 \cdot 2 \cdot g \cdot h \cdot i$$
$$\text{LCM} = 2 \cdot 2 \cdot 3 \cdot 3 \cdot 3 \cdot g^3 \cdot h^2 \cdot i^3 = 108g^3h^2i^3$$

SECTION 3.17 EXERCISES

Find the LCM for each set of numbers.

1. $-18p^4q^3r^6$, $27p^5$, $3pq^2$
2. $4Rt^3u^4$, $5R^4$, $9t^3$, $6u^8$

3.18 ■ ADDING AND SUBTRACTING FRACTIONS

Determine whether the fractions being combined are like or unlike.

1. **Like fractions:** Add or subtract the numerators and use the like denominator as the denominator of the result.
2. **Unlike fractions:** Find the LCD (least common denominator, the LCM of the denominators involved), and raise each fraction to higher terms with the LCD as the denominator.

EXAMPLE 3.54 $(5a - 6b)/(2a^4bc^2) - (7a - 3b)/(2a^4bc^2) = $.

Solution Each fraction has $2a^4bc^2$ as a denominator. The fractions are like. Combine the numerators.

$$\frac{5a - 6b}{2a^4bc^2} - \frac{7a - 3b}{2a^4bc^2} = \frac{5a - 6b - (7a - 3b)}{2a^4bc^2}$$

Remove the parentheses by multiplying each term within parentheses by -1 (change each sign).

$$= \frac{5a - 6b - 7a + 3b}{2a^4bc^2}$$

Combine like terms.

$$= \frac{-2a - 3b}{2a^4bc^2}$$

Basically we are finished; but, since both terms in the numerator are negative, factor out -1.

$$= \frac{-1(2a + 3b)}{2a^4bc^2}$$

Multiplying the numerator by -1 is the same as multiplying the entire fraction by -1.

$$\frac{5a - 6b}{2a^4bc^2} - \frac{7a - 3b}{2a^4bc^2} = -\frac{2a + 3b}{2a^4bc^2}$$

EXAMPLE 3.55 Find the LCD. $1/(24x^3y^7) + 3/(5x^4y^7) - 5/(12x^2y^2z^2)$.

Solution $24 = 2 \cdot 2 \cdot 2 \cdot 3$
5 is prime.
$12 = 2 \cdot 2 \cdot 3$

2 must appear 3 times, 3 must appear once, and 5 must appear once.

$$2 \cdot 2 \cdot 2 \cdot 3 \cdot 5 = 120$$

120 is the coefficient in the LCD. The highest power of x that appears is 4; the highest power of y is 7; and the highest power of z is 2. x^4, y^7, and z^2 must be factors in the LCD.

$$\text{LCD} = 120x^4y^7z^2$$

EXAMPLE 3.56 Combine the fractions in the last example into one fraction. $1/(24x^3y^7) + 3/(5x^4y^7) - 5/(12x^2y^2z^2) = $.

Solution The LCD was determined in the last example.

$$\text{LCD} = 2^3 \cdot 3 \cdot 5 \cdot x^4 \cdot y^7 \cdot z^2$$

Raise each fraction to higher terms with the LCD as the denominator.

$$\frac{(5xz^2)}{24x^3y^7(5xz^2)} + \frac{3(24z^2)}{5x^4y^7(24z^2)} - \frac{5(10x^2y^5)}{12x^2y^2z^2(10x^2y^5)}$$
$$= \frac{5xz^2 + 72z^2 - 50x^2y^5}{120x^4y^7z^2}$$

EXAMPLE 3.57 $2rs^3/(3ab^2) - 3r^2s/(4a^2b^3) = $.

Solution The fractions are unlike. Find the LCD.

$$3ab^2 = 3 \cdot a \cdot b^2$$
$$4a^2b^3 = 2 \cdot 2 \cdot a^2 \cdot b^3$$

Form a least common denominator by using each prime factor the maximum number of times that it appears in one of the denominators.

$$\text{LCD} = 3 \cdot 2 \cdot 2 \cdot a^2 \cdot b^3 = 12a^2b^3$$

Raise each fraction to higher terms with the LCD as the denominator.

$$\frac{2rs^3(4ab)}{3ab^2(4ab)} - \frac{3r^2s(3)}{4a^2b^3(3)} = \frac{8abrs^3 - 9r^2s}{12a^2b^3}$$

SECTION 3.18 EXERCISES

Add or subtract these fractions.

1. $2a/(3b^2) + 4a/(3b^2) =$
2. $2xy/(u^2v^3) - 4xy/(u^2v^3) =$
3. $4/u - 5/v + 6/w =$
4. $4v/(3u^2) + 3u/(4v^2) =$
5. $12tu/(5v^3w^3) + 8tu/(5vw^2) =$

3.19 ■ MULTIPLICATION OF FRACTIONS

To multiply fractions, follow these steps.

1. Reduce the fractions to lowest terms. Remember that the reduction can take place between the numerator of one fraction and the denominator of another. That is, a factor in the numerator of one fraction can cancel the same factor in the denominator of another fraction. Factor the expressions as necessary to discover all the common factors.

2. Multiply numerator times numerator to determine the numerator of the result, and denominator times denominator to determine the denominator of the result.

$$\frac{a}{b} \times \frac{c}{d} = \frac{a \times c}{b \times d}$$

EXAMPLE 3.58 $12a^7b^2/(x^5y) \cdot x^3y/(3a^4b^3)$

Solution

$$\frac{12a^7b^2}{x^5y} \times \frac{x^3y}{3a^4b^3} = \frac{(4)(3)(a^4)(a^3)(b^2)}{(x^3)(x^2)(y)} \times \frac{x^3y}{3a^4(b^2)(b)} = \frac{4a^3}{bx^2}$$

EXAMPLE 3.59 $(y^2 - 4y - 12)/(y^2 - 1) \cdot (y + 1)/(y + 2) = .$

Solution   The numerator of the first fraction is a trinomial that can be factored, and the denominator is the difference of two squares.

$$\frac{y^2 - 4y - 12}{y^2 - 1} \times \frac{y+1}{y+2} = \frac{(y+2)(y-6)}{(y+1)(y-1)} \times \frac{y+1}{y+2} = \frac{y-6}{y-1}$$

EXAMPLE 3.60 $(t - s)/(25s - 100t) \cdot (5s^2 - 5st - 60t^2)/(7s^2 - 7t^2) = .$

Solution Check for common factors and factor them out.

$$\frac{t - s}{25(s - 4t)} \times \frac{5(s^2 - st - 12t^2)}{7(s^2 - t^2)} =$$

Reduce to lowest terms.

$$\frac{t - s}{(5)(5)(s - 4t)} \times \frac{5(s + 3t)(s - 4t)}{7(s - t)(s + t)} =$$

It appears that all the common factors have been canceled, but the numerator contains a factor $(t - s)$ and the denominator contains a factor $(s - t)$. One is the negative of the other. Factor -1 out of the numerator.

$$\frac{-1(s - t)}{5} \times \frac{s + 3t}{7(s - t)(s + t)} =$$

Now cancel the factors $(s - t)$ and multiply numerator times numerator and denominator times denominator.

$$= \frac{-(s + 3t)}{35(s + t)} = -\frac{s + 3t}{35(s + t)}$$

SECTION 3.19 EXERCISES

Multiply.

1. $3b^3/(4c^4) \cdot 6c^2/(5b^4) =$
2. $(x^2 + 2x - 24)/(3xy^3) \cdot (21x^2y^3)/(x - 4) =$
3. $(4x^2 - 25)/(x^2 + 12x + 36) \cdot (x + 6)/(2x + 5) =$
4. $(9x^2 - 1)/(x^2 - 3x - 18) \cdot (x + 3)/(1 - 3x) =$

3.20 ■ DIVISION OF FRACTIONS

To divide by a fraction, multiply by its reciprocal.

$$\frac{\frac{a}{b}}{\frac{c}{d}} = \frac{a}{b} \times \frac{d}{c} = \frac{a \times d}{b \times c}$$

EXAMPLE 3.61 Divide $26u^3v^4/st^5$ by $10uv/s^3t^3$.

Solution

$$\frac{\frac{26u^3v^4}{st^5}}{\frac{10uv}{s^3t^3}} = \frac{26u^3v^4}{st^5} \times \frac{s^3t^3}{10uv} = \frac{(2)(13)(u)(u^2)(v)(v^3)}{(s)(t^3)(t^2)} \times \frac{(s)(s^2)(t^3)}{(2)(5)(u)(v)} = \frac{13s^2u^2v^3}{5t^2}$$

EXAMPLE 3.62 Divide $(2x^2 + 5x - 12)/(x^2 + 2x - 8)$ by $(4x^2 - 9)/(x^2 - 4)$.

Solution

$$\frac{\frac{2x^2 + 5x - 12}{x^2 + 2x - 8}}{\frac{4x^2 - 9}{x^2 - 4}} = \frac{2x^2 + 5x - 12}{x^2 + 2x - 8} \times \frac{x^2 - 4}{4x^2 - 9}$$

$$= \frac{(x + 4)(2x - 3)}{(x + 4)(x - 2)} \times \frac{(x - 2)(x + 2)}{(2x - 3)(2x + 3)} = \frac{x + 2}{2x + 3}$$

SECTION 3.20 EXERCISES

1. Divide $44m^5n^7/p^4q^8$ by $84mn/p^3q^6$
2. Divide $(2a + 3b)/(5a - 7b)$ by $(8a + 3b)/(5a - 7b)$
3. Divide $(14r^2 - 22rs - 12s^2)/(6r^2 + 9rs - 15s^2)$ by $(4r^2 + 16rs + 16s^2)/(4r^2 + 20rs + 25s^2)$.

3.21 ■ COMPLEX FRACTIONS

Complex fractions have more than one division line, for example $(4y^3/b - 1/b^2)/(7by)$ and $(1 - 2a^3/3b)/(1 + 5a/4b^3)$. Simplifying complex fractions means writing an equivalent expression that contains only one division line. This involves most of the skills mastered so far. These exercises are an excellent way to sharpen your skills in algebra and gain confidence in your work. Attack!

EXAMPLE 3.63 Simplify $(4y^3/b - 1/b^2)/(7by)$.

Solution Combine the two fractions in the overall numerator into one. The LCD of those two fractions is b^2. Raise $4y^3/b$ to higher terms with b^2 as the denominator.

$$\frac{\frac{4y^3}{b} - \frac{1}{b^2}}{7by} = \frac{\frac{4y^3(b)}{b(b)} - \frac{1}{b^2}}{7by} = \frac{\frac{4by^3 - 1}{b^2}}{7by} =$$

$7by$ can be written as a fraction $7by/1$. To divide by a fraction, invert and multiply.

$$= \frac{\frac{4by^3 - 1}{b^2}}{7by} = \frac{\frac{4by^3 - 1}{b^2}}{\frac{7by}{1}} = \frac{4by^3 - 1}{b^2} \times \frac{1}{7by} = \frac{4by^3 - 1}{7b^3y}$$

3.21 COMPLEX FRACTIONS

EXAMPLE 3.64 Simplify $(1 - 2a^3/3b)/(1 + 5a/4b^3)$.

Solution Combine the numerator into one fraction and the denominator into one fraction. The LCD of the fractions in the numerator is $3b$; and the LCD of the fractions in the denominator is $4b^3$.

$$\frac{1 - \dfrac{2a^3}{3b}}{1 + \dfrac{5a}{4b^3}} = \frac{\dfrac{3b}{3b} - \dfrac{2a^3}{3b}}{\dfrac{4b^3}{4b^3} + \dfrac{5a}{4b^3}} = \frac{\dfrac{3b - 2a^3}{3b}}{\dfrac{4b^3 + 5a}{4b^3}} =$$

To divide by a fraction, invert and multiply. b is the only common factor.

$$= \frac{3b - 2a^3}{3b} \times \frac{4b^3}{4b^3 + 5a} = \frac{3b - 2a^3}{3\not{b}} \times \frac{4b^2(\not{b})}{4b^3 + 5a} = \frac{4b^2(3b - 2a^3)}{3(4b^3 + 5a)}$$

EXAMPLE 3.65 Simplify $(1/a^2 - b^2)/(a + 1/b)$.

Solution

$$\frac{\dfrac{1}{a^2} - b^2}{a + \dfrac{1}{b}} = \frac{\dfrac{1}{a^2} - \dfrac{b^2(a^2)}{(a^2)}}{\dfrac{a(b)}{(b)} + \dfrac{1}{b}} = \frac{\dfrac{1 - a^2b^2}{a^2}}{\dfrac{ab + 1}{b}} =$$

$1 - a^2b^2$ can be written as the difference of two squares and factored.

$$= \frac{\dfrac{1 - (ab)^2}{a^2}}{\dfrac{ab + 1}{b}} = \frac{1 - (ab)^2}{a^2} \times \frac{b}{ab + 1} = \frac{(1 - ab)(1 + ab)}{a^2} \times \frac{b}{1 + ab}$$

$$= \frac{b(1 - ab)}{a^2}$$

SECTION 3.21 EXERCISES

Simplify these complex fractions.

1. $\dfrac{1 - \dfrac{1}{s^2}}{1 - \dfrac{1}{s}} =$

2. $\dfrac{\dfrac{t}{2} - 1}{\dfrac{4}{t} - t} =$

3. $$\dfrac{\dfrac{1}{u}-\dfrac{1}{u^2}}{\dfrac{1}{u}-\dfrac{1}{u^2}}=$$

4. $$b-\dfrac{a}{a-\dfrac{3}{b}}=$$

5. $$\dfrac{1}{a}-\dfrac{a}{a-\dfrac{3}{1/a}}=$$

■ SUMMARY

1. Signed numbers consist of two parts, a magnitude (or absolute value) and a sign, + or −.
2. A number that falls to the right of another on the number line is said to be more positive, or less negative, or greater than the other. The symbol for "greater than" is >.
3. A number that falls to the left of another on the number line is said to be less positive, or more negative, or less than the other. The symbol for "less than" is <.
4. To add numbers with like signs, add the magnitudes and affix the common sign.
5. To add numbers with unlike signs, take the difference of the magnitudes and affix the sign of the number with the greater magnitude.
6. To subtract signed numbers, change the sign of the subtrahend and proceed as in addition.
$$a - b = a + -b$$
7. To multiply signed numbers, multiply the magnitudes. If the signs are like, affix a positive sign. If the signs are unlike, affix a negative sign.
8. If several signed numbers are being multiplied, determine the sign of the product as follows:
   a. If the number of negative factors being multiplied is odd, the product is negative.
   b. If the number of negative factors being multiplied is even, the product is positive.
9. To divide signed numbers, divide the magnitudes. If the signs are like, the quotient is positive. If the signs are unlike, the quotient is negative.
10. An algebraic expression that contains one or more terms is called a *polynomial*.
11. A polynomial that contains one term is called a *monomial;* a polynomial that contains two terms is called a *binomial;* and a polynomial that contains three terms is called a *trinomial*.
12. Multiplication of two or more monomials is taken in two steps.
    Step 1.   Multiply the coefficients (don't forget their signs).
    Step 2.   Multiply variables with like bases by adding exponents.

13. Division of monomials is taken in two steps.

    Step 1. Divide the coefficients (don't forget their signs).

    Step 2. Divide variables with like bases by subtracting exponents.

14. When a monomial is raised to a power, each factor of the monomial must be raised to the power.

    $$(axy)^n = a^n x^n y^n$$

15. When multiplying one binomial by another, care must be taken to ensure that each term of one binomial gets multiplied by each term of the other.

    $$(a + b)(c + d) = ac + ad + bc + bd$$

16. If binomials to be multiplied differ only in their middle signs, the outer and inner products add to zero. The product that results is called a "difference of two squares."

    $$(a + b)(a - b) = a^2 - b^2$$

17. To square a binomial, square the first term, add twice the product of the first and last term, and add the last term squared.

    $$(a + b)^2 = a^2 + 2ab + b^2$$

18. To multiply a trinomial by a binomial, simply multiply each term of the trinomial by each term of the binomial and then combine like terms.

19. To factor an algebraic expression, first apply the distributive rule if possible.

    $$ab + ac = a(b + c)$$

20. To factor trinomials of the form $acx^2 + (ad + bc)x + bd$, follow these steps.
    a. Factor out any common coefficient to reduce the size of the numbers involved and the amount of work involved. This step is often overlooked.
    b. Factor the x^2 term into ax and cx. $(ax\ )(cx\ )$
    c. Choose signs for b and d as follows:
        1. If bd is positive, the signs will both be the same as the sign of $(ad + bc)$. $(ax + \ )(cx + \ )$ or $(ax - \ )(cx - \ )$.
        2. If bd is negative, the signs will be different and you will have to decide which goes where. $(ax + \ )(cx - \ )$ or $(ax - \ )(cx + \ )$.
    d. Factor the constant into b and d. $(ax + b)(cx + d)$
    e. Use FOIL to see if the two binomials yield the original trinomial.
    f. If not, try swapping b and d. Try different factors of the constant or the x^2 term until the correct combination is found.

21. *Literal fractions* are fractions that contain both coefficients and variables.

22. To raise a fraction to higher terms, determine what factor the original denominator must be multiplied by to become the new denominator. Multiply both the numerator and the denominator by this factor so that the value of the original fraction will not be altered.

23. To find the least common multiple (LCM) of two or more numbers, break each number into its prime factors. Each prime factor must appear in the LCM the maximum number of times that it appears in any of the numbers.
24. To add or subtract literal fractions, determine whether the fractions being combined are like or unlike.
    a. Like fractions: Add or subtract the numerators and use the like denominator as the denominator of the result.
    b. Unlike fractions: Find the LCD (least common denominator, the LCM of the denominators involved), and raise each fraction to higher terms with the LCD as the denominator.
25. To multiply fractions, follow these steps.
    a. Reduce the fractions to lowest terms. Remember that the reduction can take place between the numerator of one fraction and the denominator of another. That is, a factor in the numerator of one fraction can cancel the same factor in the denominator of another fraction. Factor the expressions as necessary to discover all the common factors.
    b. Multiply numerator times numerator to determine the numerator of the result and denominator times denominator to determine the denominator of the result.

$$\frac{a}{b} \times \frac{c}{d} = \frac{a \times c}{b \times d}$$

26. To divide by a fraction, multiply by its reciprocal.

$$\frac{\frac{a}{b}}{\frac{c}{d}} = \frac{a}{b} \times \frac{d}{c} = \frac{a \times d}{b \times c}$$

■ SOLUTIONS TO CHAPTER 3 EXERCISES

Solutions to Section 3.1 Exercises

1. $12 > 7$
2. $13 > -7$
3. $-7 < 12$
4. $-3 > -6$
5. $-1 > -12$
6. $0 > -6$
7. $4 > 0$
8. $-7 < 0$
9. $-1.6 < -1.55$
10. $-2/3 > -3/2$

Solutions to Section 3.2 Exercises

1. 8
2. 6
3. -9
4. 2
5. 7
6. -37
7. -5
8. 5
9. -8
10. -4

Solutions to Section 3.3 Exercises

1. 3
2. −2
3. −9
4. 13
5. 1
6. −5
7. 8
8. 16
9. −16
10. −8
11. −2
12. −13
13. 13
14. 30
15. −9
16. −14
17. 4
18. −4
19. −6
20. 12
21. 9
22. 12

Solutions to Sections 3.4 and 3.5 Exercises

1. 24
2. −92
3. −27
4. 24
5. 6
6. 1
7. −288
8. 72
9. 2 2/3
10. −2
11. −3
12. 5
13. 2
14. 1

Solutions to Section 3.6 Exercises

1. $2t^2 + 9t + 4u^2$
2. $-4z^3 - 7z^2 - 4z - 7$
3. $20x^2y^3 - 5x^2 - 3y^2$
4. $-12a^2b + 12ab + 8ab^2 - 6$
5. $22s^2 - 5s + 12 - 2t$

Solutions to Section 3.7 Exercises

1. $-12y^5$
2. $90d^7e^7$
3. $48a^{10}b^6c^5$
4. $72r^4s^7t^7$
5. $-8x^4y^6z^4$
6. $-48m^4n^9o^5p^6$

Solutions to Section 3.8 Exercises

1. 10^{-2}
2. a^3
3. $-4/3a^{-2}bc^{-2}$ or $-4b/(3a^2c^2)$
4. $-35/3x^2y^3z^{-4}$ or $-35x^2y^3/(3z^4)$
5. $u^5v^2w^5/3$

Solutions to Section 3.9 Exercises

1. $64t^{30}u^{18}v^{12}$
2. $81p^{12}q^8$
3. $125x^3y^{12}z^9$
4. $6x^3yz^2$
5. $2t^2u^3v^2(3tu)^{1/2}$
6. $3ab^2c^3(2)^{1/3}$
7. $8mn^2o(m^2no^2)^{1/3}$
8. $3r^2st^3$
9. $2st(5r^3s^2t^3)^{1/4}$

Solutions to Section 3.10 Exercises

1. $k^2 - 1$
2. $y^2 - 4y - 12$
3. $r^2 - 4$
4. $6y^2 - 5yz - 6z^2$
5. $6w^2 - 15w + 6wt - 15t$
6. $r^2 - 8r + 16$
7. $9r^2t^4 - 24rt^2w^3 + 16w^6$

Solutions to Section 3.11 Exercises

1. $8r^2 + 12rs + 10rt - 9st - 12t^2$
2. $2r^5s^2 - 3r^4s^3 - 2r^3s^5 - 2r^2s^4 + 3r^2s^6 + 3rs^5$
3. $6a^2 + ab - 12ac - 12b^2 - 18bc$
4. $9aw^4y^7 - 6aw^3y^5 - 12w^3y^5 + 9aw^2y^4 + 8w^2y^3 - 6awy^2$
5. $8t^4 - 2t^3u + 5t^2u^2 - 11tu^3 - 6u^4$

Solutions to Section 3.12 Exercises

1. $2XY(2X - Y)$
2. $8rs(3r^2s^3 + t)$
3. $2(x + y)(x - y)$
4. $2(x^2 + y^2)$
5. $(4R + 5T)(4R - 5T)$
6. $10(2X + 3Y)(2X - 3Y)$

Solutions to Section 3.13 Exercises

1. $(t + 4)(t - 12)$
2. $(y - 6)(y + 1)$
3. $(2r - 3)(r - 8)$
4. $(5r - 12)(r + 2)$
5. $(4x + 25)(3x + 1)$

Solutions to Section 3.14 Exercises

1. $(8u + 3v)(u - 2v)$
2. $3(s - 2t)(2s - 3t)$
3. $(8m - 3n)(7m + 2n)$
4. $-(5t - 3u)(3t - 4u)$
5. $2(3s - 5u)(3s - 2u)$

Solutions to Section 3.15 Exercises

1. $x - y + z$
2. $(4 + y - 2x^2y^4)/[10(3y - 5y^2 - x^2)]$
3. $(2c + 3d)/(4c + 5d)$
4. $[3(3t - 2u)]/[4(5t + 4u)]$

Solutions to Section 3.16 Exercises

1. $\dfrac{-46n^2m^8}{2n^5m^6}$
2. $\dfrac{28a^3b^2xyz^4}{21a^4b^2x^2z}$

Solutions to Section 3.17 Exercises

1. $-54p^5q^3r^6$

2. $180R^4t^3u^8$

Solutions to Section 3.18 Exercises

1. $2a/b^2$

2. $-2xy/(u^2v^3)$

3. $\dfrac{4vw-5uw+6uv}{uvw}$

4. $\dfrac{16v^3+9u^3}{12u^2v^2}$

5. $\dfrac{12tu+8tuv^2w}{5v^3w^3}$

Solutions to Section 3.19 Exercises

1. $\dfrac{9}{10bc^2}$

2. $7x(x+6)$

3. $\dfrac{2x-5}{x+6}$

4. $-\dfrac{3x-1}{x-6}$

Solutions to Section 3.20 Exercises

1. $11m^4n^6/(21pq^2)$
2. $(2a+3b)/(8a+3b)$
3. $(14r^2+29rs-15s^2)/[(3)(2r^2+2rs-4s^2)]$

Solutions to Section 3.21 Exercises

1. $\dfrac{s+1}{s}$

2. $-\dfrac{t}{2(2+t)}$

3. $\dfrac{u-1}{u+1}$

4. $\dfrac{b(ab-a-3)}{ab-3}$

5. $\dfrac{a+2}{2a}$

CHAPTER 3 PROBLEMS

Fill in the blanks with the proper symbol, < or >.

1. 5 6
2. 13 2
3. −5 5
4. −9 −3
5. −3 −5
6. −2 0
7. 0 9
8. −7 −10
9. -5.88 5.55
10. −5/8 −5/9

Add these signed numbers.

11. 7 + 4 =
12. 71 + −34 =
13. −4 + −7 =
14. −9 + 5 =
15. 85 + 2 =
16. −73 + −53 =
17. −66 + 10 =
18. 12 + −73 =
19. −74 + −34 =

20. $-5 + 33 =$
21. $88 + 88 =$
22. $88 + -88 =$
23. $-88 + 88 =$
24. $-88 + -88 =$

Subtract these signed numbers.

25. $3 - 14 =$
26. $18 - 46 =$
27. $-5 - 6 =$
28. $7 - -4 =$
29. $-9 - -22 =$
30. $-4 - -9 =$
31. $82 - 44 =$
32. $27 - -8 =$
33. $-35 - 82 =$
34. $-49 - -9 =$
35. $4 - 8 =$
36. $-21 - 31 =$
37. $6 - -8 =$
38. $53 - -23 =$
39. $-8 - 7 =$
40. $-6 - 19 =$
41. $-77 - -9 =$
42. $-3 - -15 =$
43. $8 - 11 =$
44. $10 - -10 =$
45. $34 - (-8 + 3 - 7) =$
46. $6 - 3 - (44 + 7 - 47) =$
47. $88 - 88 =$
48. $88 - -88 =$
49. $-88 - -88 =$
50. $-88 - 88 =$

Multiply these signed numbers.

51. $7 \cdot 4 =$
52. $6 \cdot -3 =$
53. $-10 \cdot 3 =$
54. $-7 \cdot -11 =$
55. $-4 \cdot -8 =$
56. $-7 \cdot -9 =$
57. $-4 \cdot 7 \cdot -3 \cdot -5 \cdot 3 =$
58. $-2 \cdot -20 \cdot -1 \cdot 3 \cdot -2 =$
59. $-1 \cdot -1 \cdot -1 \cdot -1 =$
60. $-2 \cdot -2 \cdot 2 \cdot -2 =$

Divide these signed numbers.

61. $9/3 =$
62. $20/-4 =$
63. $-24/6 =$
64. $-66/-11 =$
65. $-51/-3 =$
66. $1/-1 =$
67. $125/25 =$
68. $-76/19 =$
69. $81/-27 =$
70. $-1/1 =$

Work in longhand then confirm on your calculator.

71. $8 \cdot 4 + 7 =$
72. $4 \cdot 6 + 21/7 - 10 =$
73. $9 + 5(12 + 6/3) =$
74. $7 + 4^2 =$
75. $(6 + 4)^2 =$

76. $-5^2 =$
77. $(-5)^2 =$
78. $-7^{-2} =$
79. $(-7)^{-2} =$
80. $(-7)^2 + 9 - (33 - 16 - 7)/(2 - 7) + 8 =$
81. $\{4 - 4[4 - 4(4 - -4)]\} =$
82. $(3 - -2)^2/(4 - -3)^2 + (2 - 7)/(5 - -2)^2 =$
83. $(5^2 + 12^2)^{1/2} =$
84. $(60^2 - 48^2)^{1/2} =$
85. $(2 + 2^2 - -2^2)^2 =$
86. $(4 + -8)(3 - -4)(-2 - -2)(-2) =$
87. $[3^2/21 - (-7 + 10)/(6 - -1)]^2 =$
88. $-(4 - 8)/5^2 + (-2 - -6)/(125/-5) =$
89. $(-4 - -5 - 3 - -2)/8 =$
90. $(1 - (1 - (1 - (1 - (1 - 8))))) =$

Combine like terms.

91. $8u^2 + 5v + 3u^2 + 8v - 7t^2 =$
92. $16s^3 - s - 6s - 6s^2 - 6s^3 - 6s =$
93. $12m^2n^3 + 8m^2n^3 - 5m^2 - 7n^2 + 2m^2 =$
94. $6 + 3cd - 5c^2d + 7cd^2 - 33c - 8c^2d - 3cd + 11cd - 8c^2d =$
95. $6t^2 - (8t - 16t^2 - 12t + 2t) =$
96. $7\,xyz + 5\,xy^2z - 3\,x^2yz + 6\,xy^2z - 12\,xyz - 3xy^2z + 5\,xyz^2$
97. $-2(3x - 4y) + 5(-4x + 3y) - 3(2x + 5y) =$
98. $-4(5t - 6u + 5v) + 3(6t - 4u - 7v) - 8(6t - 5u + 4v) =$
99. $2x^3y^2z^2 - 6x^3y^3z^3 - 8x^3y^2z^2 - 6x^3y^3z^3 =$
100. $2x - -7y + 3x - -6y + 4z + 8z - -7z =$

Multiply.

101. $(5t^2)(-4t^3) =$
102. $(2k^5)(4j^7)(-5k^2j^5) =$
103. $(-4t^4u^7v^2)(2t^4u^6v)(-3tuv) =$
104. $34st \cdot 5st^2 \cdot -7st^3u^4 \cdot -8st^2u^3 =$
105. $-6x^7y^2z^4 \cdot -5x^3y^4z^5 \cdot 4y^2z =$
106. $8m^2n^4o^3 \cdot -3n^5p^4 \cdot -4n^5o^2 \cdot 3m^2n^4p^5 =$
107. $2v \cdot 3v \cdot -4v \cdot -2v^3 =$

108. $3t^4uv^3 \cdot -5t^3u^3v \cdot 10t^3uv \cdot -4t^2u^2v^2 =$
109. $3a^3b^2c^4d \cdot -2a^6b^2cd^4 \cdot -3ac^4d^3 \cdot 3b^3d^5 =$
110. $12p^2q^4 \cdot 2p^3q^2 \cdot -3pq \cdot -pq^3 =$

Divide.

111. $10^7/10^3 =$
112. $s^5/s^3 =$
113. $10^6/10^7 =$
114. $a^2/a^5 =$
115. $-9a^7b^4c / (-3a^2b^6c^7) =$
116. $2u^3v^4w^5 \cdot -4u^2vw^4 / (-12uvw) =$
117. $5u^{-3}v^2w^4 / (15u^{-3}v^4w^{-2}) =$
118. $4x^{-4}y^3 \cdot -3x^{-2}y^4z^5 / (-21x^{-3}y^{-2}z^{-7}) =$
119. $25c^3d^4e^{-3} \cdot 2c^4d^{-3}e^2 / (5c^{-2}d^4e) =$
120. $3x^{-2}y^4z^{-8} / (51x^{-2}yz^{-2}) =$

Simplify.

121. $(2a^3b^4c^5)^3 =$
122. $(3u^3v^2w^5)^4 =$
123. $(-4p^2q^3r^4)^4 =$
124. $(-2x^3y^5z^7)^3 =$
125. $(400x^8y^4z^{10})^{1/2} =$
126. $(169s^9t^{12}u^5)^{1/2} =$
127. $(250m^{12}n^{15}o^6)^{1/3} =$
128. $(2000x^2y^{25}z^{11})^{1/3} =$
129. $(4r^{16}s^{44}t^{24})^{1/4} =$
130. $(80f^5g^9h^{17})^{1/4} =$

Multiply.

131. $(t + 1)(t - 1) =$
132. $(u + 4)(u - 5) =$
133. $(t - 2)^2 =$
134. $(x + 5)^2 =$
135. $(3s - 6)(2s + 7) =$
136. $(7t + 5)(4t - 3) =$
137. $(4xy + 6)(2xy - 5) =$
138. $(2st - 1)(3st - 8) =$
139. $(4r - 3s)(2r + 2s) =$
140. $(7t + 2u)(9t - 4u) =$
141. $(4u - 5v)^2 =$
142. $(3t + 4u)^2 =$
143. $(4x - 5y)(4x + 5y) =$
144. $(2m - 5n)(4m + 2n) =$
145. $(4x^2y^3 - 6)^2 =$
146. $(3rt^2 - 4)^2 =$
147. $(5x^3y^4 + 2)(4x^3y^4 - 3) =$
148. $(2a^2 - 5b^3)(2a^2 + 5b^3) =$
149. $(4g^4 - 5h^3)^2 =$
150. $(2w^3 + 2v^2)^2 =$

Multiply these polynomials.

151. $(5a - 2b)(4a + 4b - 3c) =$
152. $(3r^3s^4 - 3r^3s^2)(r^4 - 2s^3 - 3r^2s^3) =$

153. $(6a - 2b)(5a + 4b - 7c) =$
154. $(4x^3y^3 - 3xy^2)(3x - 5xy - 3xy^2z^4) =$
155. $(2t^3 + 5tu + 4u^2)(3t^3 - 6tu^2 - 2u^2) =$
156. $(u + v - w + y - z)(u - v + w - y + z) =$

Factor.

157. $5X^3Y^2 - 15XY^2 =$
158. $51r^3s^2 + 3rst =$
159. $13r^2 - 13s^2 =$
160. $15x^2y^2 - 15s^2t^2 =$
161. $5y^2 + 15z^2 =$
162. $4a^2b^2 + 16 =$
163. $49r^2 - 64t^2 =$
164. $100x^2 - 81y^2 =$

Factor these trinomials.

165. $x^2 + x - 6 =$
166. $t^2 - t - 20 =$
167. $u^2 + 4u - 21 =$
168. $v^2 - 3v - 10 =$
169. $w^2 + 10w + 21 =$
170. $x^2 + 11x + 30 =$
171. $y^2 - 7y + 12 =$
172. $z^2 - 10z + 24 =$
173. $6t^2 - t - 15 =$
174. $21u^2 + 32u - 5 =$
175. $18v^2 - 15v - 12 =$
176. $6w^2 - 23w + 20 =$
177. $15x^2 + 9x - 6 =$
178. $12t^2 + 50t + 8 =$
179. $10s^2 + 9st - 9t^2 =$
180. $24u^2 - 26uv + 6v^2 =$
181. $28w^2 - 23wx - 15x^2 =$
182. $15y^2 - 22yz + 8z^2 =$

Simplify.

183. $(4a - 4b + 4c)/4 =$
184. $(x^2 + 2x - 24)/(x^2 - x - 12) =$
185. $(x^2 - y^2)/(2x^2 - xy - 3y^2) =$
186. $(12t^2 - 27tu + 6u^2)/(16t^2 - 24tu + 5u^2) =$
187. Express $3x^2/y^4$ as a fraction with a denominator of $21x^5y^6$.
188. Express $15rst/4r^2st$ as a fraction with a denominator of $48r^4s^2t^2u^3$.

Find the LCM for each set of numbers.

189. $-25w^3y^4z^2$, $15w^5$, $5wy^2z^4$
190. $12ab^3c^5$, $18b^4c$, $3a^3c$, $6b^8$

Combine these fractions:

191. $14s/(5t^2) + 33s/(5t^2) =$
192. $3pq/(p^2q^5) - 7pq/(p^2q^5) =$
193. $5/a - 5/b + 5/c =$
194. $9t^2/(uv) - 8t/(v^2) + 3t/(v^2) =$

195. $3m/(7n^2) + 3n/(4m^2) =$
196. $100t^3u^4/(3r^3s^3) + 25t^2u^2/(3r^3s^2) =$
197. $12p^4/(5q^4) \cdot 25q^2/(144p^7) =$
198. $(6s^2 + 7st - 20t^2)/(5s^2t^3) \cdot (105s^3t^4)/(93s^2 - st - 4t^2) =$
199. $(t^2 - 12t + 36)/(stu) \cdot (12s^2t^3u^4)/(3t^2 - 15t - 18) =$
200. $(25m^2 - 1)/(9m^2 - 4) \cdot (3m + 2)/(5m - 1) =$

Divide.

201. $55p^4q^2/r^3s^3$ by $105pq^3/2rs^6$
202. $(4d + 5e)/(3d - 7e)$ by $(4d + 5e)/(4d - 7e)$
203. $(4u^2 - 9v^2)/(2u^2 + 2v^2)$ by $(4u^2 - 12uv + 9v^2)/(5u^2 + 5v^2)$
204. $(10x^2 + 4xy - 6y^2)/(25x^2 - 20xy + 4y^2)$ by $(25x^2 - 20xy + 4y^2)/(10x^2 + xy - 2y^2)$

Simplify these complex fractions.

205. $\dfrac{1 - \dfrac{1}{x}}{1 - \dfrac{1}{x^2}} =$

206. $\dfrac{1 - \dfrac{x}{5}}{\dfrac{25}{x} - x} =$

207. $\dfrac{\dfrac{1}{a} + \dfrac{1}{a^2}}{\dfrac{1}{a} - \dfrac{1}{a^2}} =$

208. $\dfrac{1}{s} + \dfrac{t}{s + \dfrac{3}{t}} =$

209. $\dfrac{1}{x} + 4\dfrac{x}{x - \dfrac{5}{1/x}} =$

4
Solving Literal Equations

CHAPTER OUTLINE

- **4.1** Literal Equations
- **4.2** Removing a Term from One Side of the Equation
- **4.3** Removing a Factor from One Side of the Equation
- **4.4** Variable within Parentheses
- **4.5** Raising Both Sides to a Power
- **4.6** Collecting Common Terms
- **4.7** Inverting Both Sides

OBJECTIVES

After completing this chapter, you should be able to use the following techniques to solve literal equations for the unknown variable:

1. add an additive inverse to both sides of the equation.
2. multiply both sides by a reciprocal.
3. isolate parentheses that contain the unknown.
4. raise both sides of the equation to the same power.
5. invert both sides of the equation.
6. collect all terms that contain the unknown on one side of the equation.

NEW TERMS TO WATCH FOR

literal equation or formula
solving the equation for the unknown variable
term
additive inverse
addition property of equality
transpose
factor
multiplication property of equality
reciprocal
radical
radicand
index
radical expression
exponential expression
distributive property
common factor

4.1 ■ LITERAL EQUATIONS

A **literal equation** or **formula** is an equation that contains more than one variable. You will encounter many literal equations in your technical studies. $P = I^2 \cdot R$ is a literal equation with three variables, P, I, and R.

One step in the problem-solving process is to rearrange a literal equation so that the variable whose value is being sought is isolated on one side of the equation with all other variables and constants on the other side. This task is called **solving the equation for the unknown variable**. For example, the equation shown above is solved for P, power. P is isolated on one side of the equation. If current I were the quantity being sought, the equation would have to be rearranged to isolate I in order to get I equal to some expression.

The two sides of the original equation are equal. In solving the equation for the unknown variable, each step performed must maintain that equality. **Whatever is done to one side must also be done to the other.** For example, if

$2x = 5$, then

$4(2x) = 4(5)$ Both sides are multiplied by 4; the equality is maintained.

This chapter will concentrate on solving an equation for an unknown, and a following chapter will show where this step fits in the overall scheme of solving problems. All the literal equations used in this chapter appear in electronics texts. Each is followed by a statement that begins with "where." The "where's" define each of the variables in the equation and provide pertinent information about the units to be used. Chances are good that you will encounter these equations in the future. Do not worry about the meaning of the variables or the equations. That knowledge will come as you continue through this book and from your electronics classes. Concentrate instead on the tools available for solving literal equations.

In the first examples of each section, the procedure being studied is applied to an equation with a single unknown. The remaining examples deal with literal equations. Each example is checked by substituting the result into the original equation. These checks provide an excellent opportunity to improve your skills in algebra.

4.2 ■ REMOVING A TERM FROM ONE SIDE OF THE EQUATION

Review of Terminology

1. In general, plus and minus signs separate an algebraic expression into **terms**. For example, $b^2 - 4ac$ has two terms, b^2 and $4ac$. However, an algebraic expression raised to a power is a single term. $(a^2 - b^2)^3 + c^{1/2} - 6$ has three terms, $(a^2 - b^2)^3, c^{1/2}$, and 6. The minus sign in $(a^2 - b^2)$ does not separate a^2 and b^2 into two separate terms because the expression $a^2 - b^2$ is raised to a power other than 1.

2. The sum of the **additive inverse** of a number and the number itself is zero. The additive inverse of 7 is -7; the additive inverse of $-y$ is y; and the additive inverse of $-8x^3y^6z$ is $8x^3y^6z$.

3. The **addition property of equality** states that a term can be added to each side of an equation.

$$\text{If } a = b, \text{ then } a + c = b + c.$$

Use the addition property of equality to **transpose** a term from one side of an equation to the other by adding the additive inverse of that term to each side of the equation. In the following examples, the term being added to each side is shown in bold print.

EXAMPLE 4.1   Solve $x + 12 = 0$ for x.

Solution   The additive inverse of 12 is -12. -12 must be added to each side of the equation to isolate x.

$$x + 12 + \mathbf{-12} = 0 + \mathbf{-12}$$
$$x = -12$$

Check: $x + 12 = 0$
$-12 + 12 = 0$
$0 = 0$

EXAMPLE 4.2   Solve $y - 15 = 30$ for y.

Solution   15 is being subtracted from y. This is the same as adding -15 to y.

$$y + -15 = 30$$

The additive inverse of -15 is 15. To isolate y, 15 must be added to each side of the equation.

$$y + -15 + \mathbf{15} = 30 + \mathbf{15}$$
$$y = 45$$

Check: $y - 15 = 30$
$45 - 15 = 30$
$30 = 30$

EXAMPLE 4.3   Solve $17/3 - t = 4$ for t.

Solution t is being subtracted from 17/3. The equation can be rewritten

$$17/3 + -t = 4$$

Add t to each side to eliminate the $-$ sign associated with t.

$$17/3 + -t + t = 4 + t$$
$$17/3 = t + 4$$

To isolate t, add -4 to each side.

$$17/3 + -4 = t + 4 + -4$$
$$17/3 - 4 = t$$
$$17/3 - 12/3 = t$$
$$5/3 = t$$

Swap sides so that the variable is on the left.

$$t = 5/3$$

Check: $17/3 - t = 4$
$17/3 - 5/3 = 4$
$12/3 = 4$
$4 = 4$

You might have noticed in the first three examples that when the additive inverse of a term is added to both sides of an equation, the term moves to the other side of the equation and its sign is toggled (changed). In Example 4.3, $17/3 - t = 4$ becomes $17/3 = 4 + t$. t is **transposed** from the left side of the equation to the right by changing its sign from $-$ to $+$. This shortcut will be used in the remainder of the chapter.

EXAMPLE 4.4 $R_1 + R_2 = R_t$ where R_1 and R_2 are resistors in series and R_t is the total resistance. Solve for R_1.

Solution   To isolate R_1, R_2 must be removed from the left side of the equation. Transpose R_2 by changing its sign from $+$ to $-$.

$$R_1 = R_t - R_2$$

Check: $R_1 + R_2 = R_t$
$R_t - R_2 + R_2 = R_t$
$R_t = R_t$

4.3 REMOVING A FACTOR FROM ONE SIDE OF THE EQUATION ■ 115

EXAMPLE 4.5 $f_r/Q = f_2 - f_1$, where f_r is resonant frequency, and f_2 and f_1 are cutoff frequencies, all measured in hertz, and Q is the quality factor. Solve for f_1.

Solution Transpose f_1 to remove its − sign.

$$f_r/Q + f_1 = f_2$$

To isolate f_1, transpose f_r/Q.

$$f_1 = f_2 - f_r/Q$$

Check: $f_r/Q = f_2 - f_1$
$f_r/Q = f_2 - (f_2 - f_r/Q)$
$f_r/Q = f_2 - f_2 + f_r/Q$
$f_r/Q = f_r/Q$

SECTION 4.2 EXERCISES

1. $v - 81 = 13$ Solve for v.
2. $3 - w = 9$ Solve for w.
3. $4 - 7 = x + 12$ Solve for x.
4. $C_1 + C_2 + C_3 = C_T$ where C_1, C_2, and C_3 are capacitors connected in parallel and C_T is the total capacitance, all measured in farads (F). Solve for C_2.
5. $X_L - X_C = X$ where X_L is inductive reactance, X_C is capacitive reactance, and X is total reactance, all measured in ohms (Ω). Solve for X_L.
6. $X_L - X_C = X$ where X_L is inductive reactance, X_C is capacitive reactance, and X is total reactance, all measured in ohms (Ω). Solve for X_C.
7. $V_B = V_E + V_{BE}$, where V_B is the voltage on the base of a transistor, V_E is the voltage on the emitter, and V_{BE} is the voltage from base to emitter, all measured in volts (V). Solve for V_E.
8. $V_{CE} = V_{CC} - I_C R_C$, where V_{CE} is the voltage, measured in volts (V), from collector to emitter in a common-emitter transistor circuit; V_{CC} is the DC supply voltage measured in volts; I_C is the collector current measured in amps (A), and R_C is the collector resistor measured in ohms (Ω). Solve for V_{CC}.

4.3 ■ REMOVING A FACTOR FROM ONE SIDE OF THE EQUATION

Review of Terminology

1. The **factors** of a term are those quantities that are multiplied together to form that term. For example, in the term $2ax^2y$, the factors are 2, a, x, x, and y.
2. The **multiplication property of equality** states that both sides of an equation can be multiplied by the same factor.

$$\text{If } a = b, \text{ then } a \cdot c = b \cdot c$$

3. The **reciprocal** of a number times the number equals one. The reciprocal of 7 is 1/7; the reciprocal of $1/(7y^3)$ is $7y^3$; and the reciprocal of $-4y^2z$ is $1/(-4y^2z)$.

Use the multiplication property of equality to remove a factor from one side of an equation by multiplying both sides of the equation by its reciprocal. In the following examples, the reciprocals being multiplied are shown in bold type.

EXAMPLE 4.6 $12x/5 = 3$. Solve for x.

Solution x is being multiplied by 12/5. The reciprocal of 12/5 is 5/12. Multiply each side by 5/12.

$$12x/5 \cdot \mathbf{5/12} = 3 \cdot \mathbf{5/12}$$
$$x = 5/4$$

Check: $12x/5 = 3$
$12(5/4)/5 = 3$
$12/4 = 3$
$3 = 3$

EXAMPLE 4.7 $y/5 = 3$. Solve for y.

Solution $y/5 = 3$ can be rewritten as

$$y \cdot 1/5 = 3$$

y is multiplied by 1/5. The reciprocal of 1/5 is 5. Multiply each side by 5 to isolate y.

$$y \cdot 1/5 \cdot \mathbf{5} = 3 \cdot \mathbf{5}$$
$$y = 15$$

Check: $y/5 = 3$
$15/5 = 3$
$3 = 3$

EXAMPLE 4.8 $V = I \cdot R$, where V is the voltage in volts (V) across resistor R and I is the current in amps (A) through resistor R. R is measured in ohms (Ω). (Ohm's Law). Solve for R.

Solution To isolate R, I must be removed from the right side of the equation. The reciprocal of I is $1/I$. Multiply both sides of the equation by $1/I$.

$$V = I \cdot R$$
$$V \cdot \mathbf{1/I} = I \cdot R \cdot \mathbf{1/I}$$
$$V/I = R$$

Swap sides to get R on the right.

$$R = V/I$$

Check: $V = I \cdot R$
$V = I \cdot V/I$
$V = V$

You might have noticed in the last three examples that when both sides of an equation are multiplied by the reciprocal of a factor, the factor moves from the numerator on one side to the denominator on the other, or from the denominator on one side to the numerator on the other. In Example 4.8, $V = I \cdot R$ becomes $V/I = R$. I is **transposed** from the numerator on the right to the denominator on the left. In Example 4.7, $y/5 = 3$ becomes $y = 3 \cdot 5$. 5 is **transposed** from the denominator on the left to the numerator on the right. This shortcut will be used in the remainder of the chapter.

EXAMPLE 4.9 $\lambda = c/f$, where λ is the wavelength of an electromagnetic wave in meters, c is the speed of light in meters per second, and f is the frequency of the electromagnetic wave in hertz. Solve for c.

Solution Transpose f.

$$\lambda \cdot f = c$$
$$c = \lambda \cdot f$$

Check: $\lambda = c/f$
$\lambda = \lambda \cdot f/f$
$\lambda = \lambda$

The multiplication property of equality gives us an alternate solution to Example 4.5.

EXAMPLE 4.10 $f_r/Q = f_2 - f_1$, where f_r is resonant frequency, and f_2 and f_1 are cutoff frequencies, all in hertz, and Q is the quality factor (dimensionless). Solve for f_1.

Solution Transpose f_2.

$$f_r/Q - f_2 = -f_1$$

Multiply both sides of the equation by -1.

$$-1(f_r/Q - f_2) = -1(-f_1)$$
$$-f_r/Q + f_2 = f_1$$

Rearrange the left side of the equation so that it does not lead off with a negative sign. (Even though subtraction is not commutative, $-b + a$ can be rearranged as $a - b$.)

$$f_2 - f_r/Q = f_1$$

Swap sides so that f_1 is on the left.

$$f_1 = f_2 - f_r/Q$$

118 ■ CHAPTER 4 SOLVING LITERAL EQUATIONS

This solution agrees with Example 4.5.

If the unknown is in the denominator by itself, eliminate the fraction by transposing the denominator.

EXAMPLE 4.11 $5/3 = 3/V$. Solve for V.

Solution Transpose V.

$$5/3 \cdot V = 3$$

Transpose $5/3$.

$$V = 9/5$$

Check: $5/3 = 3/V$
$5/3 = 3/(9/5)$
$5/3 = 3 \cdot 5/9$
$5/3 = 5/3$

EXAMPLE 4.12 $A_v = R_L/R_E$, where A_v is the voltage amplification of an unbypassed common-emitter transistor amplifier and is dimensionless (no unit of measure); R_L is the load resistance in ohms; and R_E is the emitter resistor in ohms. Solve for R_E.

Solution Transpose R_E to eliminate the fraction.

$$A_v \cdot R_E = R_L$$

Now transpose A_v.

$$R_E = R_L/A_v$$

Check: $A_v = R_L/R_E$
$A_v = R_L/(R_L/A_v)$
$A_v = R_L \cdot (A_v/R_L)$
$A_v = A_v$

If the unknown is in the denominator with other factors, transpose the unknown only, not the complete denominator.

EXAMPLE 4.13 $3 = 1/(14z)$. Solve for z.

Solution Since there is no addition or subtraction in the denominator but only multiplication, transpose z only.

$$3z = 1/14$$

To isolate z, transpose 3.

$$z = 1/42$$

Check: $3 = 1/(14z)$
$3 = 1/(14 \cdot 1/42)$
$3 = 1/(14/42)$
$3 = 42/14$
$3 = 3$

EXAMPLE 4.14 $X_C = 1/(2 \cdot \pi \cdot f \cdot C)$, where X_C is capacitive reactance in ohms (Ω), f is the frequency of the applied signal in hertz (Hz), and C is the capacitance in farads (F). Solve for f.

Solution f is one of four factors in the denominator. Transpose f.

$$X_C \cdot f = 1/(2 \cdot \pi \cdot C)$$

Isolate f by transposing X_C.

$$f = 1/(2 \cdot \pi \cdot C \cdot X_C)$$

Check: $X_C = 1/(2 \cdot \pi \cdot f \cdot C)$
$X_C = 1/\{2 \cdot \pi \cdot C \cdot [1/(2 \cdot \pi \cdot C \cdot X_C)]\}$
$X_C = 2 \cdot \pi \cdot C \cdot X_C/(2 \cdot \pi \cdot C)$
$X_C = X_C$

SECTION 4.3 EXERCISES

Solve each equation for the indicated unknown.

1. $2t/7 = 6$. Solve for t.
2. $7/3 = 1/r$. Solve for r.
3. $5 = 1/(6w)$. Solve for w.
4. $P = I^2 \cdot R$, where P is power in watts (W), I is current in amps (A), and R is resistance in ohms (Ω). Solve for R.
5. $P = V^2/R$, where P is power in watts (W), V is voltage in volts (V), and R is resistance in ohms (Ω). Solve for R.
6. $BW = f_r/Q$, where BW is bandwidth in hertz (Hz), f_r is resonant frequency in hertz (Hz), and Q is dimensionless (no units). Solve for f_r.
7. $X_C = 1/(2 \cdot \pi \cdot f \cdot C)$, where X_C is capacitive reactance in Ohms (Ω), f is the frequency of the applied signal in hertz, and C is the capacitance in farads, (F). Solve for C.
8. $V_2/V_1 = N_2/N_1$, where V_2 is the voltage across the secondary of a transformer, V_1 is the voltage across the primary of a transformer, N_2 is the number of turns in the secondary winding of a transformer, and N_1 is the number of turns in the primary winding of a transformer. Solve for V_1.

4.4 ■ VARIABLE WITHIN PARENTHESES

If the unknown variable is contained within parentheses, isolate the parentheses first. If the quantity within parentheses is not being raised to a power other than 1, the parentheses will no longer be needed and the unknown can be isolated.

EXAMPLE 4.15 $3/(X - 8) = 4$. Solve for X.

Solution Transpose $(X - 8)$, then transpose 4.

$$3 = 4(X - 8)$$
$$3/4 = (X - 8)$$

The parentheses are no longer needed.

$$3/4 = X - 8$$

To isolate X, transpose 8.

$$8 + 3/4 = X$$
$$X = 8\ 3/4$$

Check: $3/(X - 8) = 4$
$3/(35/4 - 8) = 4$
$3/(35/4 - 32/4) = 4$
$3/(3/4) = 4$
$4 = 4$

EXAMPLE 4.16 $5(2x - 7)/6 = 4$. Solve for x.

Solution $(2x - 7)$ is multiplied by 5/6. To isolate the parentheses, transpose 5/6.

$$(2x - 7) = 24/5$$

The parentheses are no longer needed.

$$2x - 7 = 24/5$$

Solve for x.

$$2x = 24/5 + 7$$
$$2x = 24/5 + 35/5$$
$$2x = 59/5$$
$$x = 59/10 = 5.9$$

Check: $5(2x - 7)/6 = 4$
$5(2 \cdot 5.9 - 7)/6 = 4$
$5(11.8 - 7)/6 = 4$
$5 \cdot 4.8/6 = 4$
$24/6 = 4$
$4 = 4$

EXAMPLE 4.17 $Q(f_2 - f_1) = f_r$, where f_r is resonant frequency, and f_2 and f_1 are cutoff frequencies, all measured in hertz (Hz), and Q is the quality factor (dimensionless). Solve for f_2.

Solution f_2 is contained within parentheses. To isolate the parentheses, transpose Q.

$$(f_2 - f_1) = f_r/Q$$

The parentheses are no longer needed.

$$f_2 - f_1 = f_r/Q$$

Transpose f_1.

$$f_2 = f_r/Q + f_1$$

Check: $Q(f_2 - f_1) = f_r$
$Q(f_r/Q + f_1 - f_1) = f_r$
$Q(f_r/Q) = f_r$
$f_r = f_r$

EXAMPLE 4.18
$$R_1 = \frac{R_A R_C}{R_A + R_B + R_C}$$

where R_A, R_B, and R_C are resistors in delta configuration and R_1 is an equivalent resistor in Y configuration, all measured in ohms (Ω). Solve for R_B.

Solution In a sense, R_B is contained within parentheses because the numerator is being divided by the sum of the three terms, R_A, R_B, and R_C. $R_A + R_B + R_C$ must be enclosed in parentheses when this equation is written in single-line format.

$$R_1 = R_A \cdot R_C/(R_A + R_B + R_C)$$

Transpose $(R_A + R_B + R_C)$ to get the unknown out of the denominator and to eliminate the fraction.

$$R_1(R_A + R_B + R_C) = R_A \cdot R_C$$

Transpose R_1 to isolate the parentheses.
$$(R_A + R_B + R_C) = R_A R_C/R_1$$
The parentheses are no longer needed.
$$R_A + R_B + R_C = R_A R_C/R_1$$
Isolate R_B by transposing R_A and R_C.
$$R_B = R_A R_C/R_1 - R_A - R_C$$

Check: $R_1 = R_A \cdot R_C/(R_A + R_B + R_C)$
$R_1 = R_A \cdot R_C/(R_A + R_A R_C/R_1 - R_A - R_C + R_C)$
$R_1 = R_A \cdot R_C/(R_A R_C/R_1)$
$R_1 = R_A \cdot R_C \cdot R_1/(R_A R_C)$
$R_1 = R_1$

EXAMPLE 4.19 $V_1 = V_{CC} \cdot R_1/(R_1 + R_2)$ where resistors R_1 and R_2 form a voltage divider, V_1 is the voltage across R_1 in volts (V), and V_{CC} is the applied voltage in volts (V). Solve for R_2.

Solution Get the unknown out of the denominator by transposing $(R_1 + R_2)$.
$$V_1 \cdot (R_1 + R_2) = V_{CC} \cdot R_1$$
Transpose V_1 to isolate the parentheses.
$$(R_1 + R_2) = V_{CC} \cdot R_1/V_1$$
The parentheses are no longer needed.
$$R_1 + R_2 = V_{CC} \cdot R_1/V_1$$
Now isolate R_2 by transposing R_1.
$$R_2 = V_{CC} \cdot R_1/V_1 - R_1$$

Check: $V_1 = V_{CC} \cdot R_1/(R_1 + R_2)$
$V_1 = V_{CC} \cdot R_1/(R_1 + V_{CC} \cdot R_1/V_1 - R_1)$
$V_1 = V_{CC} \cdot R_1/(V_{CC} \cdot R_1/V_1)$
$V_1 = V_{CC}/(V_{CC}/V_1)$
$V_1 = V_{CC} \cdot (V_1/V_{CC})$
$V_1 = V_1$

SECTION 4.4 EXERCISES

Solve for the indicated unknown.

1. $4 = 3(t - 2)$. Solve for t.
2. $8/(3 + w) = 7$. Solve for w.

3. $8 - (y + 9) = 4$. Solve for y.
4. $6(x/7 - 7)/5 = 4$. Solve for x.
5. $Q = f_r/(f_2 - f_1)$, where f_r is resonant frequency; and f_2 and f_1 are cutoff frequencies, all measured in hertz (Hz); and Q is the quality factor (dimensionless). Solve for f_1.
6. $V_3 = V_{CC} \cdot R_3/(R_1 + R_2 + R_3)$, where resistors R_1, R_2, and R_3, measured in ohms (Ω), form a voltage divider; V_3 is the voltage across R_3 in volts (V); and V_{CC} is the applied voltage in volts, (V). Solve for R_2.
7. $X_C = 1/[2 \cdot \pi \cdot f \cdot (C_1 + C_2)]$, where X_C is the capacitive reactance in ohms (Ω) of two capacitors in parallel C_1 and C_2, both in farads (F), and f is the frequency of the applied signal in hertz (Hz). Solve for C_2.
8. $R_2 = R_B R_C/(R_A + R_B + R_C)$, where R_A, R_B, and R_C are resistors in delta configuration and R_2 is an equivalent resistor in Y configuration, all measured in ohms (Ω). Solve for R_A.
9. $I_B = (V_{CC} - V_{BE})/R_B$, where I_B is the current flowing into the base of a transistor in amps, V_{CC} is the DC supply voltage in volts (V), V_{BE} is the voltage from base to emitter in volts, and R_B is the resistor connected from V_{CC} to the base lead. Solve for V_{CC}.

4.5 ■ RAISING BOTH SIDES TO A POWER

Review of Terminology

1. When raising a power to a power, multiply the exponents. $(a^b)^c = a^{b \cdot c}$. For example,

$$(y^4)^3 = y^{4 \cdot 3} = y^{12}$$
$$(y^3)^{1/2} = y^{3 \cdot 1/2} = y^{3/2}$$
$$(y^2)^{1/2} = y^{2 \cdot 1/2} = y^1 = y$$

2. In the expression $\sqrt[n]{a}$, $\sqrt{}$ is the **radical**, a is the **radicand**, and n is the **index**. If the index is 2, then the operation is square root and the 2 is not written, $\sqrt{a}$. If the index is 3, then the operation is cube root, $\sqrt[3]{a}$. If the index is 5, then the operation is read "5th root of a", $\sqrt[5]{a}$. An expression in this form is called a **radical expression**.

3. A radical expression can be rewritten as an **exponential expression** as follows.

$$\sqrt[n]{a} = a^{1/n}$$

For example, $8^{1/3} = \sqrt[3]{8} = 2$.
Likewise, $a^{m/n} = (a^m)^{1/n} = \sqrt[n]{a^m} = (a^{1/n})^m = (\sqrt[n]{a})^m$
For example, $8^{2/3} = (8^2)^{1/3} = (8^{1/3})^2 = 2^2 = 4$.

4. If $a^2 = b$, then raising both sides of the equation to the 1/2 power gives $(a^2)^{1/2} = (b)^{1/2}$ or $a = b^{1/2}$.

Use this information to solve for a variable that is raised to a power n. First isolate the variable, then raise each side of the equation to the $1/n$ power. For example, if the unknown quantity is being raised to the second power (squared), isolate the squared quantity and raise each side to the 1/2 power (take the square root of each side of the equation). In the following examples, the power that each side is being raised to is shown in bold print.

EXAMPLE 4.20 $y^2 = 7$. Solve for y.

Solution $(y^2)^{\mathbf{1/2}} = 7^{\mathbf{1/2}}$

$$y = 7^{1/2}$$

Check: $y^2 = 7$
$(7^{1/2})^2 = 7$
$7 = 7$

EXAMPLE 4.21 $t^{1/3} = 3$. Solve for t.

Solution t is raised to the 1/3 power. The reciprocal of 1/3 is 3. Cube each side.

$$(t^{1/3})^{\mathbf{3}} = 3^{\mathbf{3}}$$
$$t = 27$$

Check: $t^{1/3} = 3$
$27^{1/3} = 3$
$3 = 3$

EXAMPLE 4.22 $70 - V^2/4 = 10$. Solve for V.

Solution Isolate V^2.

$$70 = 10 + V^2/4$$
$$60 = V^2/4$$
$$240 = V^2$$

The reciprocal of 2 is 1/2. Take the square root of each side.

$$240^{\mathbf{1/2}} = (V^2)^{\mathbf{1/2}}$$
$$240^{1/2} = V$$

Simplify.

$$V = 240^{1/2}$$
$$V = (2 \cdot 2 \cdot 2 \cdot 2 \cdot 3 \cdot 5)^{1/2}$$
$$V = 4(15)^{1/2}$$

4.5 RAISING BOTH SIDES TO A POWER ■ 125

Check:
$$70 - V^2/4 = 10$$
$$70 - (4(15)^{1/2})^2/4 = 10$$
$$70 - 4^2(15)/4 = 10$$
$$70 - 4(15) = 10$$
$$70 - 60 = 10$$
$$10 = 10$$

EXAMPLE 4.23 $Z^2 = R^2 + X^2$, where Z is impedance, R is resistance, and X is reactance, all measured ohms (Ω). Solve for Z.

Solution Do not jump to conclusions and say that $Z = R + X$. Follow the rules. To find Z, raise each side to the 1/2 power (take the square root of each side).

$$(Z^2)^{1/2} = (R^2 + X^2)^{1/2}$$
$$Z = (R^2 + X^2)^{1/2}$$

Note that Z **does not equal** $R + X$.

EXAMPLE 4.24 $Z^2 = R^2 + X^2$, where Z is impedance, R is resistance and X is reactance, all measured in ohms (Ω). Solve for R.

Solution Isolate R^2 by transposing X^2.

$$Z^2 - X^2 - V^2 = R^2$$

Take the square root of each side.

$$(Z^2 - X^2)^{1/2} = (R^2)^{1/2}$$
$$(Z^2 - X^2)^{1/2} = R$$
$$R = (Z^2 - X^2)^{1/2}$$

Check:
$$Z^2 = R^2 + X^2$$
$$Z^2 = [(Z^2 - X^2)^{1/2}]^2 + X^2$$
$$Z^2 = Z^2 - X^2 + X^2$$
$$Z^2 = Z^2$$

EXAMPLE 4.25 $P = V^2/R$, where P is power dissipated in watts (W), V is voltage in volts (V), and R is resistance in ohms (Ω). Solve for V.

Solution First isolate V^2 by transposing R.

$$P \cdot R = V^2$$

Now raise each side to the 1/2 power.

$$(P \cdot R)^{1/2} = (V^2)^{1/2}$$
$$(P \cdot R)^{1/2} = V$$

Swap sides.

$$V = (P \cdot R)^{1/2}$$

Check: $P = V^2/R$
$P = [(P \cdot R)^{1/2}]^2/R$
$P = P \cdot R/R$
$P = P$

EXAMPLE 4.26 $L_M = k(L_1 L_2)^{1/2}$, where L_M is mutual inductance in henries (H), k is the coefficient of coupling (dimensionless) between two inductors L_1 and L_2, measured in henries (H). Solve for L_2.

Solution L_2 is contained within parentheses. To isolate the parentheses, transpose.

$$L_M/k = (L_1 L_2)^{1/2}$$

Square each side.

$$(L_M/k)^2 = [(L_1 L_2)^{1/2}]^2$$
$$(L_M/k)^2 = (L_1 L_2)$$

The parentheses containing L_2 are no longer needed.

$$(L_M/k)^2 = L_1 L_2$$

Isolate L_2 by transposing L_1.

$$(L_M/k)^2/L_1 = L_2$$
$$L_2 = (L_M/k)^2/L_1$$

Check: $L_M = k(L_1 L_2)^{1/2}$
$L_M = k[L_1 (L_M/k)^2/L_1]^{1/2}$
$L_M = k[(L_M/k)^2]^{1/2}$
$L_M = k(L_M/k)$
$L_M = L_M$

The following example requires raising each side of the equation to a power twice.

EXAMPLE 4.27 $Z = (R^2 + X^2)^{1/2}$, where Z is impedance, R is resistance, and X is reactance, all measured in ohms (Ω). Solve for R.

Solution Square both sides of the equation to release R^2 from the parentheses.
$$(Z)^2 = [(R^2 + X^2)^{1/2}]^2$$
$$Z^2 = (R^2 + X^2)$$

The parentheses are no longer needed.
$$Z^2 = R^2 + X^2$$

Isolate R^2 by transposing X^2.
$$Z^2 - X^2 = R^2$$

Take the square root of each side.
$$(Z^2 - X^2)^{1/2} = (R^2)^{1/2}$$
$$(Z^2 - X^2)^{1/2} = R$$
$$R = (Z^2 - X^2)^{1/2}$$

Check:
$$Z = (R^2 + X^2)^{1/2}$$
$$Z = \{[(Z^2 - X^2)^{1/2}]^2 + X^2\}^{1/2}$$
$$Z = [Z^2 - X^2 + X^2]^{1/2}$$
$$Z = [Z^2]^{1/2}$$
$$Z = Z$$

In the next example, the unknown is in the denominator and is under a radical.

EXAMPLE 4.28 $f = 1/[2 \cdot \pi \cdot (LC)^{1/2}]$, where f is the resonant frequency in hertz (Hz) of an inductor L measured in henries (H) and a capacitor C measured in farads (F). Solve for C.

Solution To begin isolating $(LC)^{1/2}$, transpose $(LC)^{1/2}$.
$$f \cdot (LC)^{1/2} = 1/(2 \cdot \pi)$$

Transpose f.
$$(LC)^{1/2} = 1/(2 \cdot \pi \cdot f)$$

To eliminate the need for parentheses on the left side of the equation, square both sides.
$$[(LC)^{1/2}]^2 = [1/(2 \cdot \pi \cdot f)^2]^2$$
$$(LC) = 1/(2 \cdot \pi \cdot f)^2$$
$$LC = 1/(2^2 \cdot \pi^2 \cdot f^2)$$

Transpose L.
$$C = 1/(4 \cdot \pi^2 \cdot f^2 \cdot L)$$

Check:
$$f = 1/(2 \cdot \pi \cdot (LC)^{1/2})$$
$$f = 1/\{2 \cdot \pi \cdot [L/(4 \cdot \pi^2 \cdot f^2 \cdot L)]^{1/2}\}$$

$$f = 1/\{2 \cdot \pi \cdot [1/(4 \cdot \pi^2 \cdot f^2)]^{1/2}\}$$
$$f = 1/\{2 \cdot \pi \cdot [1/(2 \cdot \pi \cdot f)]\}$$
$$f = 1/(1/f)$$
$$f = f$$

SECTION 4.5 EXERCISES

Solve for the indicated unknown.

1. $w^3 = 125$. Solve for w.
2. $6v^{1/4} = 3$. Solve for v.
3. $(1 - k)^{1/3}/5 = 2$. Solve for k.
4. $(z^2 - 4^2)^3 = 343$. Solve for z.
5. $P = I^2 \cdot R$, where P is power in watts (W), I is current in amps (A), and R is resistance in ohms (Ω). Solve for I.
6. $L_M = k(L_1 L_2)^{1/2}$, where L_M is mutual inductance in henries (H), k is the coefficient of coupling (dimensionless) between two inductors L_1 and L_2, measured in henries (H). Solve for L_1.
7. $I_D = I_{DSS}(1 - V_{GS}/V_{GS(off)})^2$, where I_D is drain current measured in amps (A), V_{GS} is voltage from gate to source measured in volts (V), and $V_{GS(off)}$ is the cutoff voltage in a JFET measured in volts (V). Solve for $V_{GS(off)}$.
8. $P_L = 0.5 V_{CEQ}^2/R_L$, where P_L is power in watts delivered to the load resistor R_L, and V_{CEQ} is the quiescent voltage from collector to emitter. Solve for V_{CEQ}.

4.6 ■ COLLECTING COMMON TERMS

Review of Terminology

1. The **distributive property** states that $a(b + c) = ab + ac$.
2. A **common factor** is a factor that occurs in each term of an expression. The common factor can be factored out. This is actually an application of the distributive rule. In the expression $2xy + 3xz - x$, each term contains the variable x. x can be factored out.

$$2xy + 3xz - x = x(2y + 3z - 1)$$

If the unknown being solved for appears in more than one place in the equation, follow this procedure.

1. Collect all the terms that contain the unknown on one side of the equation and all other terms on the other.
2. The unknown will be a common factor; factor it out.
3. Multiply both sides by the reciprocal of all factors of the unknown.

4.6 COLLECTING COMMON TERMS ■ 129

EXAMPLE 4.29 $3x - 12 = 7x + 14$. Solve for x.

Solution   Terms containing x appear on both sides of the equation. Collect all the terms containing x on the right side and all other terms on the left. (The right side was chosen for the x terms so that the sign associated with the x term will be positive.)

$$-12 = 7x + 14 - -3x$$

The two x terms are like terms and can be combined.

$$-12 = 4x + 14$$

Transpose 14.

$$-26 = 4x$$

Transpose 4.

$$-13/2 = x$$
$$x = -13/2$$

Check: $3x - 12 = 7x + 14$
$3(-13/2) - 12 = 7(-13/2) + 14$
$-39/2 - 12 = -91/2 + 14$
$-39 - 24 = -91 + 28$
$-63 = -63$

EXAMPLE 4.30 $(z + 7)/(z - 6) = 5$. Solve for z.

Solution   Transpose $(z - 6)$ to get rid of the fraction.

$$(z + 7) = 5(z - 6)$$

The parentheses on the left are not needed. Use the distributive property on the right side.

$$z + 7 = 5z - 30$$

Collect the z terms on the right and the constants on the left.

$$7 = 5z - 30 - z$$

Combine the z terms and transpose 30.

$$37 = 4z$$

Transpose 4.

$$37/4 = z$$
$$z = 37/4$$

Check: $(z + 7)/(z - 6) = 5$
$(37/4 + 7)/(37/4 - 6) = 5$
$(37/4 + 28/4)/(37/4 - 24/4) = 5$

$$65/4 \,/\, 13/4 = 5$$
$$65/4 \cdot 4/13 = 5$$
$$5 = 5$$

EXAMPLE 4.31 $R_{EQ} = R_1 \cdot R_2/(R_1 + R_2)$, where R_1 and R_2 are resistors connected in parallel, and R_{EQ} is the equivalent resistance. Solve for R_1.

Solution Get rid of the fraction by transposing $(R_1 + R_2)$.

$$R_{EQ} \cdot (R_1 + R_2) = R_1 \cdot R_2$$

Use the distributive rule to eliminate the parentheses on the left side of the equation.

$$R_{EQ} \cdot R_1 + R_{EQ} \cdot R_2 = R_1 \cdot R_2$$

Collect all terms containing R_1 on one side of the equation and all other terms on the other.

$$R_{EQ} \cdot R_2 = R_1 \cdot R_2 - R_{EQ} \cdot R_1$$

All the terms on the right contain R_1. Factor out R_1.

$$R_{EQ} \cdot R_2 = R_1 (R_2 - R_{EQ})$$

Transpose $(R_2 - R_{EQ})$.

$$R_{EQ} \cdot R_2 / (R_2 - R_{EQ}) = R_1$$

Swap sides.

$$R_1 = R_{EQ} \cdot R_2 / (R_2 - R_{EQ})$$

Check: $R_{EQ} = R_1 \cdot R_2/(R_1 + R_2)$

$$R_{EQ} = \dfrac{\dfrac{R_{EQ} \times R_2 \times R_2}{R_2 - R_{EQ}}}{\dfrac{R_{EQ} \times R_2}{R_2 - R_{EQ}} + R_2}$$

$$R_{EQ} = \dfrac{\dfrac{R_{EQ} \times R_2 \times R_2}{R_2 - R_{EQ}}}{\dfrac{R_{EQ} \times R_2 + R_2 \times R_2 \times R_{EQ} \times R_2}{R_2 - R_{EQ}}}$$

$$R_{EQ} = \dfrac{R_{EQ} \times R_2 \times R_2}{R_2 \times R_2}$$

EXAMPLE 4.32 $R_A = (R_1R_2 + R_1R_3 + R_2R_3)/R_2$, where $R_1, R_2,$ and R_3 are resistors connected in Y configuration and R_A is an equivalent delta resistor, all measured in ohms (Ω). Solve for R_2.

4.6 COLLECTING COMMON TERMS ■ 131

Solution Our unknown variable, R_2, appears in the original equation three times. Eliminate the fraction and the parentheses will not be needed. Transpose R_2.

$$R_A R_2 = (R_1 R_2 + R_1 R_3 + R_2 R_3)$$
$$R_A R_2 = R_1 R_2 + R_1 R_3 + R_2 R_3$$

Collect all terms that contain R_2 on one side and all other terms on the other.

$$0 = R_1 R_2 + R_1 R_3 + R_2 R_3 - R_A R_2$$

Transpose $R_1 R_3$.

$$-R_1 R_3 = R_1 R_2 + R_2 R_3 - R_A R_2$$

All the terms on the right contain R_2. Factor it out.

$$-R_1 R_3 = \mathbf{R_2}(R_1 + R_3 - R_A)$$

Now, isolate R_2 by transposing $(R_1 + R_3 - R_A)$.

$$-R_1 R_3 / (R_1 + R_3 - R_A) = R_2$$
$$R_2 = -R_1 R_3 / (R_1 + R_3 - R_A)$$

To remove the leading − sign, factor −1 out of the expression within parentheses.

$$R_2 = -R_1 R_3 / [(-1)(-R_1 - R_3 + R_A)]$$
$$R_2 = R_1 R_3 / (-R_1 - R_3 + R_A)$$

Rearrange the terms in the denominator to lead off with a positive sign.

$$R_2 = R_1 R_3 / (R_A - R_1 - R_3)$$

Check: $R_A = (R_1 R_2 + R_1 R_3 + R_2 R_3)/R_2$

$$R_A = \frac{\dfrac{R_1 R_1 R_3}{R_A - R_1 - R_3} + R_1 R_3 = \dfrac{R_1 R_3 R_3}{R_A - R_1 - R_3}}{\dfrac{R_1 R_3}{R_A - R_1 - R_3}}$$

$$R_A = \frac{\dfrac{R_1 R_1 R_3}{R_A - R_1 - R_3} + \dfrac{R_1 R_3 (R_A - R_1 - R_3)}{R_A - R_1 - R_3} + \dfrac{R_1 R_3 R_3}{R_A - R_1 - R_3}}{\dfrac{R_1 R_3}{R_A - R_1 - R_3}}$$

$$R_A = \frac{R_1 R_1 R_3 + R_1 R_3 (R_A - R_1 - R_3) + R_1 R_3 R_3}{R_1 R_3}$$

Each term in the numerator contains $R_1 R_3$. Factor out $R_1 R_3$ and cancel.

$$R_A = R_1 + (R_A - R_1 - R_3) + R_3$$

The parentheses are not needed.

$$R_A = R_1 + R_A - R_1 - R_3 + R_3$$
$$R_A = R_A$$

SECTION 4.6 EXERCISES

Solve for the indicated unknown.

1. $3y - 6y = 12y - 6$. Solve for y.
2. $(z - 7)/(z + 5) = 3$. Solve for z.
3. $R_{EQ} = R_1 \cdot R_2/(R_1 + R_2)$, where R_1 and R_2 are resistors connected in parallel, and R_{EQ} is the equivalent resistance, all measured in ohms (Ω). Solve for R_2.
4. $R_A = (R_1R_2 + R_1R_3 + R_2R_3)/R_2$, where $R_1, R_2,$ and R_3 are resistors connected in Y configuration and R_A is an equivalent delta resistor, all measured in ohms (Ω). Solve for R_1.
5. Solve $R_2 = R_B \cdot R_C/(R_A + R_B + R_C)$ for R_C, where $R_A, R_B,$ and R_C are resistors in delta configuration and R_2 is an equivalent resistor in Y configuration, all measured in ohms (Ω).
6. $f_0 = [(R_1 + R_3)/(R_1R_2R_3)]^{1/2}/(2\pi C)$, where f_0 is the center frequency of a multiple-feedback, band-pass filter in hertz (Hz), and $R_1, R_2, R_3,$ and C are components in the feedback network. Solve for R_3.

4.7 ■ INVERTING BOTH SIDES

If one fraction is equal to another, then their reciprocals are equal.

$$\text{If } a/b = c/d, \text{ then } b/a = d/c.$$

If the unknown is in the denominator of a fraction, one approach is to invert both sides of the equation.

EXAMPLE 4.33 $3/4 = 6/Y$. Solve for Y.

Solution Y is in the denominator, so invert both sides.

$$4/3 = Y/6$$

To isolate Y, transpose 6.

$$24/3 = Y$$
$$Y = 24/3 = 8$$

Check: $3/4 = 6/Y$
$3/4 = 6/8$
$3/4 = 3/4$

EXAMPLE 4.34 $3/y = 5$. Solve for y.

Solution   Think of 5 as 5/1.

$$3/y = 5/1$$

Invert both sides.

$$y/3 = 1/5$$

Transpose 3.

$$y = 3/5$$

Check: $3/y = 5$
$3/(3/5) = 5$
$5 = 5$

EXAMPLE 4.35 $12 = 1/(1/x - 1/5)$.

Solution Think of the left side as 12/1. Invert both sides.

$$1/12 = (1/x - 1/5)$$

The parentheses are not needed.

$$1/12 = 1/x - 1/5$$

Isolate $1/x$ by transposing 1/5.

$$1/12 + \mathbf{1/5} = 1/x$$
$$5/60 + 12/60 = 1/x$$
$$17/60 = 1/x$$

Invert both sides.

$$60/17 = x/1$$
$$x = 60/17$$

Check: $12 = 1/(1/x - 1/5)$
$12 = 1/(1/(60/17) - 1/5)$
$12 = 1/(17/60 - 1/5)$
$12 = 1/(17/60 - 12/60)$
$12 = 1/(5/60) = 1/(1/12)$
$12 = 12$

EXAMPLE 4.36 $V_2/V_1 = N_2/N_1$, where V_2 is the voltage across the secondary of a transformer, V_1 is the voltage across the primary of a transformer, N_2 is the number of turns in the secondary winding of a transformer, and N_1 is the number of turns in the primary winding of a transformer. Solve for V_1.

Solution $V_2/V_1 = N_2/N_1$

The unknown is in the denominator. Invert both sides of the equation.

$$V_1/V_2 = N_1/N_2$$

Transpose V_2.

$$V_1 = V_2 N_1/N_2$$

This equation was solved in an earlier section using a different procedure.

EXAMPLE 4.37 $R_T = 1/(1/R_1 + 1/R_2 + 1/R_3)$, where R_1, R_2, and R_3 are resistors connected in parallel, and R_T is their equivalent or total resistance, all measured in ohms (Ω). Solve for R_1.

Solution Think of R_T as the fraction $1/R_T$.

$$\frac{R_T}{1} = \frac{1}{\frac{1}{R_1} + \frac{1}{R_2} + \frac{1}{R_3}}$$

Invert both sides.

$$\frac{1}{R_T} = \frac{1}{R_1} + \frac{1}{R_2} + \frac{1}{R_3}$$

Isolate $1/R_1$, by transposing $1/R_2$ and $1/R_3$.

$$\frac{1}{R_T} - \frac{1}{R_2} - \frac{1}{R_3} = \frac{1}{R_1}$$

Think of each side as a fraction.

$$\frac{\frac{1}{R_T} - \frac{1}{R_2} - \frac{1}{R_3}}{1} = \frac{1}{R_1}$$

Invert each side.

$$\frac{1}{\frac{1}{R_T} - \frac{1}{R_2} - \frac{1}{R_3}} = R_1$$

Transpose.

$$R_1 = \frac{1}{\frac{1}{R_T} - \frac{1}{R_2} - \frac{1}{R_3}}$$

This equation is written in this form because it is easy to process on the calculator. This is a nice solution because it leaves the result in the same format as the original.

Check: $R_T = 1/(1/R_1 + 1/R_2 + 1/R_3)$

$$R_T = \cfrac{1}{\cfrac{1}{\cfrac{1}{\cfrac{1}{R_T} - \cfrac{1}{R_2} - \cfrac{1}{R_3}}} + \cfrac{1}{R_2} + \cfrac{1}{R_3}}$$

$$R_T = \cfrac{1}{\cfrac{1}{R_T} - \cfrac{1}{R_2} - \cfrac{1}{R_3} + \cfrac{1}{R_2} + \cfrac{1}{R_3}}$$

$$R_2 = \cfrac{1}{\cfrac{1}{R_T}}$$

$$R_T = R_T$$

SECTION 4.7 EXERCISES

Solve for the indicated unknown by inverting each side of the equation.

1. $3/t = 7/8$. Solve for t.
2. $7/x = 5$. Solve for x.
3. $1/(1/w - 1/8) = 5$. Solve for w.
4. $R_T = 1/(1/R_1 + 1/R_2 + 1/R_3)$, where R_1, R_2, and R_3 are resistors connected in parallel, and R_T is their equivalent or total resistance, all measured in ohms (Ω). Solve for R_3.
5. Solve $f_r/Q = f_2 - f_1$ for Q, where f_r is resonant frequency, and f_2 and f_1 are cutoff frequencies, all measured in hertz (Hz), and Q is the quality factor (dimensionless).
6. Solve $\lambda = c/f$ for f, where λ is the wavelength of an electromagnetic wave in meters, c is the speed of light in meters per second, and f is the frequency of the electromagnetic wave in hertz (Hz).
7. $I_B = (V_{CC} - V_{BE})/R_B$, where I_B is the current flowing into the base of a transistor in amps, V_{CC} is the DC supply voltage in volts (V), V_{BE} is the voltage from base to emitter in volts, and R_B is the resistor connected from V_{CC} to the base lead. Solve for R_B.

SUMMARY

1. A literal equation or formula is an equation that contains more than one variable.
2. In solving an equation for the unknown variable, whatever is done to one side must also be done to the other.
3. To transpose a term from one side of an equation to the other, add the additive inverse of that term to each side of the equation. Shortcut: Move the term to the other side and toggle its sign.
4. To transpose a factor from one side of an equation to the other, multiply both sides of the equation by its reciprocal. Shortcut: Move the factor from the numerator on one side to the denominator on the other or vice versa.

5. If the unknown variable is contained within parentheses, isolate the parentheses first. If the quantity within parentheses is not being raised to a power other than 1, the parentheses will no longer be needed and the unknown can be isolated.

6. To solve for a variable that is raised to a power n, first isolate the variable, then raise each side of the equation to the $1/n$ power.

7. If the unknown being solved for appears in more than one place in the equation, follow this procedure.
   a. Collect all the terms that contain the unknown on one side of the equation and all other terms on the other.
   b. The unknown will be a common factor; factor it out.
   c. Multiply both sides by the reciprocal of all factors of the unknown.

8. If the unknown is in the denominator of a fraction, one approach is to invert both sides of the equation.

SOLUTIONS TO CHAPTER 4 EXERCISES

Solutions to Section 4.2 Exercises

1. 94
2. −6
3. −15
4. $C_T - C_1 - C_3$
5. $X + X_C$
6. $X_L - X$
7. $V_B - V_{BE}$
8. $V_{CE} + I_C R_C$

Solutions to Section 4.3 Exercises

1. 21
2. 3/7
3. 1/30
4. P/I^2
5. V^2/P
6. $BW \cdot Q$
7. $1/(2 \cdot \pi \cdot f \cdot X_C)$
8. $V_2 N_1 / N_2$

Solutions to Section 4.4 Exercises

1. 10/3
2. −13/7
3. −5
4. 217/3
5. $f_2 - f_r/Q$ or $(Qf_2 - f_r)/Q$
6. $(V_{CC} R_3 - V_3 R_1 - V_3 R_3)/V_3$
7. $[1/(2\pi f) - X_C C_1]/X_C$ or $(1 - 2\pi f X_C C_1)/(2\pi f X_C)$
8. $(R_B R_C - R_2 R_B - R_2 R_C)/R_2$
9. $I_B R_B + V_{BE}$

Solutions to Section 4.5 Exercises

1. 5
2. 1/16
3. −999
4. $23^{1/2}$
5. $(P/R)^{1/2}$
6. $L_M^2/(k^2 L_2)$
7. $V_{GS}[1 - (I_D/I_{DSS})^{1/2}]$
8. $(2P_L R_L)^{1/2}$

Solutions to Sections 4.6 Exercises

1. 2/5
2. −11
3. $R_{EQ}R_1/(R_1 - R_{EQ})$
4. $(R_A \cdot R_2 - R_2 R_3)/(R_2 + R_3)$
5. $(R_2 R_A + R_2 R_B)/(R_B - R_2)$
6. $R_1/[R_1 R_2 (2\pi f_0 C)^2 - 1]$

Solutions to Section 4.7 Exercises

1. 24/7
2. 7/5
3. 40/13
4. $1/(1/R_T - 1/R_1 - 1/R_2)$
5. $f_r/(f_2 - f_1)$
6. c/λ
7. $(V_{CC} - V_{BE})/I_B$

■ CHAPTER 4 PROBLEMS

It is sometimes more difficult to determine what someone else has done than to work a problem on your own. Here are some solved problems from the section exercises. Decide what operation has been performed on each line of each of the following solutions.

1. $X_L - X_C = X$. Solve for X_C.
   $X_L = X + X_C$
   $X_L - X = X_C$
   $X_C = X_L - X$

2. $X_C = 1/(2 \cdot \pi \cdot f \cdot C)$. Solve for C.
   $X_C \cdot C = 1/(2 \cdot \pi \cdot f)$
   $C = 1/(2 \cdot \pi \cdot f \cdot X_C)$

3. $C = A\epsilon/d$. Solve for A.
   $Cd/\epsilon = A$
   $A = Cd/\epsilon$

4. $P = V^2/R$. Solve for R.
   $P \cdot R = V^2$
   $R = V^2/P$

5. $P = I^2 \cdot R$. Solve for R.
   $P/I^2 = R$
   $R = P/I^2$

6. $V_2/V_1 = N_2/N_1$. Solve for V_1.
   $V_2 = N_2/N_1 \cdot V_1$
   $V_2 \cdot N_1/N_2 = V_1$
   $V_1 = V_2 N_1/N_2$

7. $Q = f_r/(f_2 - f_1)$. Solve for f_1.
   $Q \cdot (f_2 - f_1) = f_r$
   $Qf_2 - Qf_1 = f_r$
   $Qf_2 = f_r + Qf_1$
   $Qf_2 - f_r = Qf_1$
   $(Qf_2 - f_r)/Q = f_1$
   $f_1 = (Qf_2 - f_r)/Q$

8. $X_C = 1/[2 \cdot \pi \cdot f \cdot (C_1 + C_2)]$. Solve for C_2.
   $X_C(C_1 + C_2) = 1/(2\pi f)$
   $X_C C_1 + X_C C_2 = 1/(2\pi f)$
   $X_C C_2 = 1/(2\pi f) - X_C C_1$
   $C_2 = [1/(2\pi f) - X_C C_1]/X_C$

9. $L_M = k(L_1 L_2)^{1/2}$. Solve for L_1.
   $(L_M)^2 = [k(L_1 L_2)^{1/2}]^2$
   $L_M^2 = k^2 L_1 L_2$
   $L_M^2/(k^2 L_2) = L_1$
   $L_1 = L_M^2/(K^2 L_2)$

10. $f_o = [(R_1 + R_3)/(R_1 R_2 R_3)]^{1/2}/(2\pi C)$. Solve for R_2.
    $2\pi f_o C = [(R_1 + R_3)/(R_1 R_2 R_3)]^{1/2}$
    $[2\pi f_o C]^2 = \{[(R_1 + R_3)/(R_1 R_2 R_3)]^{1/2}\}^2$
    $[2\pi f_o C]^2 = (R_1 + R_3)/(R_1 R_2 R_3)$
    $[2\pi f_o C]^2 \cdot R_2 = (R_1 + R_3)/(R_1 R_3)$
    $R_2 = (R_1 + R_3)/[R_1 R_3 (2\pi f_o C)^2]$

11. $R_A = (R_1 R_2 + R_1 R_3 + R_2 R_3)/R_2$. Solve for R_1.
    $R_A R_2 = R_1 R_2 + R_1 R_3 + R_2 R_3$
    $R_A R_2 - R_2 R_3 = R_1 R_2 + R_1 R_3$
    $R_A R_2 - R_2 R_3 = R_1(R_2 + R_3)$
    $(R_A R_2 - R_2 R_3)/(R_2 + R_3) = R_1$
    $R_1 = (R_A R_2 - R_2 R_3)/(R_2 + R_3)$

12. Solve $A_v = g_m R_d r_{ds} / (R_d + r_{ds})$ for R_d.
    $A_v \cdot (R_d + r_{ds}) = g_m R_d r_{ds}$
    $A_v R_d + A_v r_{ds} = g_m R_d r_{ds}$
    $A_v r_{ds} = g_m R_d r_{ds} - A_v R_d$
    $A_v r_{ds} = R_d(g_m r_{ds} - A_v)$
    $A_v r_{ds} / (g_m r_{ds} - A_v) = R_d$
    $R_d = A_v r_{ds}/(g_m r_{ds} - A_v)$

13. $\lambda = c/f$. Solve for f.
    $\lambda/1 = c/f$
    $1/\lambda = f/c$
    $c/\lambda = f$
    $f = c/\lambda$

14. $R_T = 1/(1/R_1 + 1/R_2 + 1/R_3)$. Solve for R_3.
    $R_T/1 = 1/(1/R_1 + 1/R_2 + 1/R_3)$
    $1/R_T = 1/R_1 + 1/R_2 + 1/R_3$
    $1/R_T - 1/R_1 - 1/R_2 = 1/R_3$
    $(1/R_T - 1/R_1 - 1/R_2)/1 = 1/R_3$
    $1/(1/R_T - 1/R_1 - 1/R_2) = R_3$
    $R_3 = 1/(1/R_T - 1/R_1 - 1/R_2)$

Solve for the indicated unknown.

15. $u + 66 = 44$. Solve for u.
16. $6 + v = 12$. Solve for v.
17. $1 - 9 = y + 17$. Solve for y.
18. $9t/11 = 22$. Solve for t.
19. $5/8 = 1/y$. Solve for y.
20. $7 = 3/(5x)$. Solve for x.
21. $3 = 7(k - 4)$. Solve for k.
22. $6/(9 - w) = 4$. Solve for w.
23. $11 + (y - 8) = 3$. Solve for y.
24. $5(x/4 + 8)/3 = 8$. Solve for x.
25. $y^4 = 81$. Solve for y.
26. $5w^{1/3} = 3$. Solve for w.
27. $(3 + s)^{1/3}/8 = 4$. Solve for s.
28. $6y + 5y = 10y - 4$. Solve for y.
29. $(u + 4)/(u - 3) = 5$. Solve for u.
30. $5/t = 9/11$. Solve for t.
31. $67/y = 2$. Solve for y.
32. $1/(1/z + 1/5) = 8$. Solve for z.
33. $5t = 3 - 5(5 - 2t)$. Solve for t.
34. $0.7(y + 3) = 0.2$. Solve for y.
35. $3\ 1/4 - 3/2\ x = 7$. Solve for x.
36. $(5v/7 - 4/3)^2 = 25/9$. Solve for v.
37. $2/3\ x - 5/7\ x = 3x + 5$. Solve for x.
38. $1/(3z + 6)^{1/2} = 4$. Solve for z.
39. $3.3x + 4.4x = 5.5$. Solve for x.
40. $R_1 + R_2 = R_t$, where R_1 and R_2 are resistors in series and R_t is the total resistance, all measured in Ω. Solve for R_2.
41. $C_1 + C_2 + C_3 = C_T$, where C_1, C_2, and C_3 are capacitors connected in parallel and C_T is the total capacitance, all measured in farads (F). Solve for C_3.
42. Solve $f_r/Q = f_2 - f_1$ for f_2, where f_r is resonant frequency, and f_2 and f_1 are cutoff frequencies, all measured in hertz (Hz), and Q is the quality factor (dimensionless).
43. $V_B = V_E + V_{BE}$, where V_B is the voltage on the base of a transistor, V_E is the voltage on the emitter, and V_{BE} is the voltage from base to emitter. Solve for V_{BE}.
44. $I_E = I_B + I_C$, where I_E is the emitter current in a transistor, I_B is the base current, and I_C is the collector current. Solve for I_C.

45. Solve $f_r/Q = f_2 - f_1$ for Q, where f_r is resonant frequency, and f_2 and f_1 are cutoff frequencies, all measured in hertz (Hz), and Q is the quality factor (dimensionless).
46. Solve $V = I \cdot R$ for I, where V is the voltage in volts (V) across resistor R and I is the current in amps (A) through resistor R. R is measured in ohms (Ω).
47. Solve $\lambda = c/f$ for f, where λ is the wavelength of an electromagnetic wave in meters, c is the speed of light in meters per second, and f is the frequency of the electromagnetic wave in hertz (Hz).
48. Solve $BW = f_r/Q$ for Q, where BW is bandwidth in hertz (Hz), f_r is resonant frequency in hertz (Hz), and Q is dimensionless—no units.
49. Solve $Q(f_2 - f_1) = f_r$ for Q, where f_r is resonant frequency, and f_2 and f_1 are cutoff frequencies, all measured in hertz (Hz), and Q is the quality factor (dimensionless).
50. Solve $A_v = R_L/R_E$ for R_L, where A_v is the voltage amplification of an unbypassed common-emitter transistor amplifier and is dimensionless (no unit of measure), R_L is the load resistance in Ω, and R_E is the emitter resistor in Ω.
51. Solve $L_M = k(L_1 L_2)^{1/2}$ for k, where L_M is mutual inductance in henries, k is the coefficient of coupling (dimensionless) between two inductors L_1 and L_2, measured henries.
52. $V_2/V_1 = N_2/N_1$ where V_2 is the voltage across the secondary of a transformer, V_1 is the voltage across the primary of a transformer, N_2 is the number of turns in the secondary winding of a transformer, and N_1 is the number of turns in the primary winding of a transformer. Solve for N_2.
53. $V_2/V_1 = N_2/N_1$ where V_2 is the voltage across the secondary of a transformer, V_1 is the voltage across the primary of a transformer, N_2 is the number of turns in the secondary winding of a transformer, and N_1 is the number of turns in the primary winding of a transformer. Solve for N_1.
54. Solve $V_3 = V_{CC} \cdot R_3/(R_1 + R_2 + R_3)$ for V_{CC}, where resistors R_1, R_2, and R_3 form a voltage divider, V_3 is the voltage across R_3 in volts, and V_{CC} is the applied voltage in volts.
55. $I_B = (V_{CC} - V_{BE})/R_B$, where I_B is the current flowing into the base of a transistor in amps, V_{CC} is the DC supply voltage in volts (V), V_{BE} is the voltage from base to emitter in volts, and R_B is the resistor connected from V_{CC} to the base lead. Solve for V_{BE}.
56. $f = 1.44/[(R_1 + 2R_2)C_{ext}]$, where f is the frequency of oscillation of a 555 timer in hertz (Hz), R_1 and R_2 are external timing resistors in ohms (Ω), and C_{ext} is an external timing capacitor in farads (F). Solve for R_2.
57. Solve $V_3 = V_{CC} \cdot R_3/(R_1 + R_2 + R_3)$ for R_1, where resistors R_1, R_2, and R_3 form a voltage divider, V_3 is the voltage across R_3 in volts, and V_{CC} is the applied voltage in volts.
58. $I_E = (V_{EE} - V_{BE})/(R_E + R_B/\beta)$, where I_E is the current flowing in the emitter lead of an emitter-biased transistor amplifier, V_{EE} is the DC voltage on the emitter lead, V_{BE} is the voltage from base to emitter, R_E and R_B are bias resistors, and β is current gain of the transistor (dimensionless). Solve for R_E.

59. $I_E = (V_{EE} - V_{BE})/(R_E + R_B/\beta)$, where I_E is the current flowing in the emitter lead of an emitter-biased transistor amplifier, V_{EE} is the DC voltage on the emitter lead, V_{BE} is the voltage from base to emitter, R_E and R_B are bias resistors, and β is current gain of the transistor (dimensionless). Solve for R_B.

60. Solve $Z = [R^2 + (X_L - X_C)^2]^{1/2}$ for X_C, where Z is impedance, R is resistance, X_C is capacitive reactance, and X_L is inductive reactance, all measured in Ω.

61. Solve $Z = [R^2 + (X_L - X_C)^2]^{1/2}$ for R, where Z is impedance, R is resistance, X_C is capacitive reactance, and X_L is inductive reactance, all measured in Ω.

62. Solve $f = 1/(2 \cdot \pi \cdot (LC)^{1/2})$ for L, where f is the resonant frequency in hertz of an inductor L measured in henries and a capacitor C measured in farads.

63. $f_c = 1/[2\pi (R_A R_B C_A C_B)^{1/2}]$, where f_c is the critical frequency of a Sallen-Fey low-pass filter, and R_A, R_B, C_A, and C_B are components in the network. Solve for R_B.

64. $f_o = (f_{c1} f_{c2})^{1/2}$, where f_o is the center frequency and f_{c1} and f_{c2} are critical frequencies, all measured in hertz (Hz). Solve for f_{c2}.

65. Solve $f_r = [Q^2/(Q^2 + 1)]^{1/2}/[2\pi(LC)^{1/2}]$ for L, where f_r is the parallel resonant frequency of inductor L and capacitor C including the effect of the circuit quality Q.

66. Solve $V_1 = V_{CC} \cdot R_1/(R_1 + R_2)$ for R_1, where resistors R_1 and R_2 form a voltage divider, V_1 is the voltage across R_1 in volts, and V_{CC} is the applied voltage in volts.

67. Solve $R_2 = R_B \cdot R_C/(R_A + R_B + R_C)$ for R_B, where R_A, R_B, and R_C are resistors in delta configuration and R_2 is an equivalent resistor in Y configuration, all measured in Ω.

68. Solve $V_3 = V_{CC} \cdot R_3/(R_1 + R_2 + R_3)$ for R_3, where resistors R_1, R_2, and R_3 form a voltage divider, V_3 is the voltage across R_3 in volts, and V_{CC} is the applied voltage in volts.

69. Solve $R_A = (R_1 R_2 + R_1 R_3 + R_2 R_3)/R_2$ for R_3, where R_1, R_2, and R_3 are resistors connected in Y configuration, and R_A is an equivalent delta resistor, all measured in Ω.

70. $V_{DC(out)} = R_L V_{DC(in)}/(R_w + R_L)$, where $V_{DC(out)}$ is the DC voltage out of an LC filter in volts; $V_{DC(in)}$ is the DC voltage in, in volts; R_w is the winding resistance of the inductor in ohms; and R_L is the load resistor in ohms. Solve for R_L.

71. Solve $R_T = 1/(1/R_1 + 1/R_2 + 1/R_3)$ for R_2, where R_1, R_2, and R_3 are resistors connected in parallel, and R_T is their equivalent or total resistance, all measured in Ω.

72. Solve $C_T = 1/(1/C_1 + 1/C_2 + 1/C_3)$ for C_3, where C_1, C_2, and C_3 are resistors connected in series, and C_T is their equivalent or total capacitance, all measured in farads.

73. Rework Problem 57 by inverting both sides of the equation.

74. Rework Problem 58 by inverting both sides of the equation.

75. Rework Problem 59 by inverting both sides of the equation.

5
Units of Measure

CHAPTER OUTLINE

- **5.1** Systems of Units of Measure
- **5.2** Quantity—Length, Symbol—s
- **5.3** Quantity—Area, Symbol—A
- **5.4** Quantity—Volume, Symbol—V
- **5.5** Quantity—Time, Symbol—t
- **5.6** Quantity—Velocity, Symbol—v
- **5.7** Quantity—Acceleration, Symbol—a
- **5.8** Quantity—Mass, Symbol—m
- **5.9** Quantity—Force, Symbol—F
- **5.10** Quantity—Density, Symbol—ρ
- **5.11** Quantity—Pressure, Symbol—P
- **5.12** Quantity—Temperature, Symbol—T
- **5.13** Quantity—Charge, Symbol—Q
- **5.14** Quantity—Current, Symbol—I or i
- **5.15** Quantity—Voltage or Electromotive Force (EMF), Symbol—V_s or E
- **5.16** Quantity—Resistance, Symbol—R
- **5.17** Quantity—Conductance, Symbol—G
- **5.18** Quantity—Capacitance, Symbol—C
- **5.19** Quantity—Inductance, Symbol—L
- **5.20** Quantity—Frequency, Symbol—f
- **5.21** Quantity—Energy, Symbol—W
- **5.22** Quantity—Power, Symbol—P

OBJECTIVES

After completing this chapter, you should be able to

1. use dimensional analysis to reduce units within the US system of units.
2. use dimensional analysis to reduce units within the SI system of units.

3. use dimensional analysis to convert units from US to SI and vice versa.

4. recognize and use SI and US units for these quantities: length, area, volume, time, velocity, acceleration, mass, force, density, pressure, temperature, charge, current, electromotive force, resistance, conductance, capacitance, inductance, frequency, energy, and power.

NEW TERMS TO WATCH FOR

International System of Units (SI)
United States Customary Units (US)
meter (m)
micron (μm)
angstrom (A)
dimensional analysis
reduction
conversion
liters
velocity
acceleration
mass
kilogram
slug
forces
newton (N)
dyne
kilopound (kip)
density
pressure
pascal (Pa)
atmosphere
inches of mercury (inHg)
millibar (mb)
millimeters of mercury (mm-Hg)
degrees Kelvin (K)
degrees Celsius (°C)
degrees Fahrenheit (°F)
charge

Coulomb (C)
current
amp (A)
electromotive force (EMF)
volts (V)
voltage drop
potential difference
resistance
ohms (Ω)
conductance
siemens (S)
capacitor
capacitance
farad (F)
inductor
inductance
henrys (H)
cycle
frequency
Hertz (Hz)
energy
joule (J)
newton-meter
foot-pounds
British thermal unit (Btu)
quad
calorie
power
watt (w)

5.1 ■ SYSTEMS OF UNITS OF MEASURE

In your work as a technician, you will encounter a variety of quantities, each with its own symbol. For many quantities, you will have a choice of units of measure. For example, in interfacing a pressure switch to a process controller, you will encounter a quantity called *pressure* whose symbol is P. What is pressure measured in? What are its units? Pounds per square foot, pounds per square inch, newtons per square meter (pascals) are just a few. In this chapter, a variety of quantities will be presented and their units of measure will be discussed.

The metric system of measure has been in existence since 1790. By 1960 it had evolved into the **Système International d'unités (SI)** or **International System of units.** Virtually every country except the United States of America and Yemen has adopted this system. The United States still uses the British System of units that became the **United States Customary Units (US)** when England converted to the SI system. US and SI units will be compared in this chapter.

Length is one of the most familiar quantities. Let's begin with it.

5.2 ■ QUANTITY—LENGTH, SYMBOL—s

In SI units, the basic unit of length is the **meter (m)**. Any divisions or multiples of the meter are accomplished by using the metric prefixes. Examples are centimeter (cm), millimeter (mm), micrometer or **micron (μm)**, nanometer (nm), kilometer (km), and **angstrom (A)**.

$$1 \text{ angstrom} = 1 \times 10^{-10} \text{ meters}$$
$$1 \text{ micron} = 1 \text{ micrometer} = 1 \text{ }\mu\text{m} = 1 \times 10^{-6} \text{ m}$$

In US units, the standard unit for length is the **foot (ft or ')**. Other units include **inch (in or ")**, **yard (yd), mile (mi), nautical mile, furlong,** and **fathom**. Here are some of the relationships between these units.

$$12 \text{ inches} = 1 \text{ foot}$$
$$3 \text{ feet} = 1 \text{ yard}$$
$$5280 \text{ feet} = 1 \text{ mile}$$
$$6076 \text{ feet} = 1 \text{ nautical mile}$$
$$660 \text{ feet} = 1 \text{ furlong}$$
$$6 \text{ feet} = 1 \text{ fathom}$$

One conversion factor between the two systems is easy to remember to six significant digits.

$$1 \text{ in} = 2.54000 \text{ cm}$$

Here are several examples showing a technique called **dimensional analysis** that organizes the work of **reduction,** changing from one unit to another within the same system of units, and **conversion,** changing from a unit in one system to a unit in another.

In this method, the units tell you whether to multiply or divide by a factor. Never again will your brains be scrambled trying to make that decision. Comments about each step taken follow the step. This gives you a chance to decipher the action on your own, before reading the comment.

EXAMPLE 5.1   Convert 4.2 miles to inches.

Solution

$$4.2 \text{ mi} \times \frac{\text{ft}}{\text{mi}}$$

Unless you know how many inches there are in one mile, work the reduction in two steps, first to feet and then to inches. To reduce miles to feet, multiply 4.2 mi by a fraction that has miles in the denominator so that it will cancel the original unit, miles. Put feet in the numerator.

$$4.2 \text{ mi} \times \frac{5280 \text{ ft}}{1 \text{ mi}}$$

Fill in the numbers from the proper reduction fact. 5280 ft = 1 mi. This makes the fraction equivalent to 1; multiplication by 1 does not change the original quantity. The mile's cancel, leaving feet in the numerator.

$$4.2 \text{ mi} \times \frac{5280 \text{ ft}}{1 \text{ mi}} \times \frac{\text{in}}{\text{ft}}$$

Now multiply by a fraction that has feet in the denominator so that feet will cancel, and inches in the numerator. Every unit cancels except inches, which is what we wanted.

$$4.2 \text{ mi} \times \frac{5280 \text{ ft}}{1 \text{ mi}} \times \frac{12 \text{ in}}{1 \text{ ft}}$$

Fill in the reduction fact. 12 in = 1 ft.

$$4.2 \cdot 5280 \cdot 12 / 1 / 1 = 266\ 112 \text{ in}$$

Multiply the numbers in the numerator and divide by the numbers in the denominator.

$$4.2 \text{ mi} = 270\ 000 \text{ in}$$

Significant digits: 5280 and 12 are exact, but 4.2 has an accuracy of two. Limit the result to two significant digits.

EXAMPLE 5.2   Convert 34.6 meters to miles.

Solution

$$34.6 \text{ m} \times \frac{\text{cm}}{\text{m}} \times \frac{\text{in}}{\text{cm}} \times \frac{\text{ft}}{\text{in}} \times \frac{\text{mi}}{\text{ft}}$$

Set up a string of fractions that guides the units from meters to miles. Notice that whatever is in the numerator of one fraction is in the denominator of the next so that those units will cancel. The only unit left is miles.

$$34.6 \text{ m} \times \frac{100 \text{ cm}}{1 \text{ m}} \times \frac{1 \text{ in}}{2.54000 \text{ cm}} \times \frac{1 \text{ ft}}{12 \text{ in}} \times \frac{1 \text{ mi}}{5280 \text{ ft}}$$

Fill in the reduction and conversion facts.

$$34.6 \cdot 100 / 2.54000 / 12 / 5280 = 0.0214994$$

If the number is in the numerator, hit the multiply key. If it is in the denominator, hit the divide key.

$$34.6 \text{ meters} = 0.0215 \text{ miles}$$

34.6 has an accuracy of three. Round the answer to three significant digits.

In the following exercises, 1 mi, 1 km, and so on mean exactly 1, 1.000 . . ., unlimited accuracy. The units of conversion will determine the number of significant digits in the result.

SECTION 5.2 EXERCISES

Reduce or convert. Use the appropriate number of significant digits.

1. 1 mi = _____ km
2. 1 km = _____ mi
3. 1 yd = _____ cm
4. 1 ft = _____ cm
5. 59.5 mi = _____ m
6. 1 μm = _____ in

5.3 ■ QUANTITY—AREA, SYMBOL—A

Dimensionally, area is equivalent to a length squared. In SI units, the standard unit of measure of area is square meters (m²). In US units, the standard unit is square feet (ft²). Other choices are square inches (in²), square miles (mi²), and acres.

$$640 \text{ acres} = 1 \text{ square mile}$$

EXAMPLE 5.3 How many ft² are in 1 acre?

Solution

$$1 \text{ acre} \times \frac{1 \text{ mi}^2}{640 \text{ acre}} \times \frac{5280 \text{ ft}}{1 \text{ mi}} \times \frac{5280 \text{ ft}}{1 \text{ mi}} = 43\ 560 \text{ ft}^2$$

mi² is the same as mi · mi. Use the conversion factor twice. The numbers in these factors are exact and do not limit the accuracy of the result. The problem asked how many ft² in exactly 1 acre so 1 does not limit the accuracy of the result either. The answer can be left as it is.

$$1 \text{ acre} = 43\ 560 \text{ ft}^2$$

EXAMPLE 5.4

Convert 1 square inch to square centimeters.

$$1 \text{ in}^2 \times \frac{2.54 \text{ cm}}{1 \text{ in}} \times \frac{2.54 \text{ cm}}{1 \text{ in}} = 6.451\ 60 \text{ cm}^2$$

Unitwise, in² is the same as in · in. Use the conversion factor twice to cancel both inches. That leaves cm · cm or cm².

$$1 \text{ in}^2 = 6.451\ 60 \text{ cm}^2$$

Significant digits: 2.540 00 has an accuracy of six. Round the result to six significant digits. The following approach is a little more elegant.

$$1 \text{ in}^2 \times \left[\frac{2.54 \text{ cm}}{1 \text{ in}}\right]^2 = 6.451\ 60 \text{ cm}^2$$

SECTION 5.3 EXERCISES

1. How many square feet are in one square yard?
2. How many acres are in one square kilometer?
3. How many square inches are in one square foot?
4. How many square meters are in one acre?
5. How many acres are in one square mile?
6. How many square inches are in one square yard?

5.4 ■ QUANTITY—VOLUME, SYMBOL—V

The volume of a rectangular prism (box) is computed by multiplying length by width by height. So, dimensionally, volume is equivalent to length cubed. In SI units, cubic meters (m³) is used to measure volume. Also, meters adjusted by any of the metric prefixes, cubed, such as cubic centimeters (cm³ or cc), cubic millimeters (mm³), and cubic kilometers (km³) can be used. The capacity of a container, though, is measured in **liters** (l). One liter is roughly equal to one quart.

$$1 \text{ quart} = 0.946 \text{ liters}$$

The metric prefixes can be used to adjust the size of the unit, for example, ml.
One milliliter is the same volume as one cubic centimeter. In other words, 1000 cubic centimeters holds 1 liter.

EXAMPLE 5.5

Convert 4.8×10^5 cc into m³.

Solution cc means cm³. Use the conversion fact 100 cm = 1 m three times.

$$4.8 \times 10^5 \text{ cm}^3 \times \frac{1 \text{ m}}{100 \text{ cm}} \times \frac{1 \text{ m}}{100 \text{ cm}} \times \frac{1 \text{ m}}{100 \text{ cm}} = 0.48 \text{ m}^3$$

Here is an alternate approach.

$$4.8 \times 10^5 \, \text{cm}^3 \times \left[\frac{1 \text{ m}}{100 \text{ cm}}\right]^3 = 0.48 \text{ m}^3$$

Note that the cm³ cancels out, but the 100 in the denominator must also be cubed. Divide by 100^3 or 10^6.

In US units, any of the lengths cubed can be used to describe volume: in³, ft³, yd³, and mi³. To measure the capacity of a container, a bewildering array of other units is used.

$$1 \text{ gallon, British Imperial (gal)} = 277.4 \text{ in}^3$$
$$1 \text{ gallon, US liquid (gal)} = 231 \text{ in}^3$$
$$1 \text{ barrel} = 42 \text{ gallons of petroleum}$$
$$1 \text{ bushel (bu)} = 1.244 \text{ ft}^3$$
$$4 \text{ quarts (qt)} = 1 \text{ gallon}$$
$$2 \text{ pints (pt)} = 1 \text{ quart}$$
$$2 \text{ cups} = 1 \text{ pint}$$
$$8 \text{ fluid ounces (fl oz)} = 1 \text{ cup}$$
$$5 \text{ fifths} = 1 \text{ gallon}$$

In US units, the connection between volume and capacity is not very handy. One cubic foot holds 7.481 gallons.

A later chapter is dedicated to area and volume calculations.

EXAMPLE 5.6 How many cubic inches are in a cubic foot?

Solution

$$1 \text{ ft}^3 \times \left[\frac{12 \text{ in}}{1 \text{ ft}}\right]^3 = 1728 \text{ in}^3$$

To cube 12, the sequence on my calculator is 12 y^x 3 = . The question asks for the number of in³ in exactly 1 ft³. The 1 does not limit the answer to 1 significant digit. The 12's are also exact. The answer does not have to be rounded.

EXAMPLE 5.7 Convert 1.55 liters to in³.

Solution 1 ml is 1 cm³.

$$1.55 \text{ l} \times \frac{1000 \text{ ml}}{1 \text{ l}} \times \frac{1 \text{ cm}^3}{1 \text{ ml}} \times \left[\frac{2.54000 \text{ in}}{1 \text{ cm}}\right]^3 = 25,400 \text{ in}^3$$

Calculator sequence: 1.55 × 1000 × 2.54 y^x 3 =
Significant digits: 1.55 limits the result to an accuracy of three.

SECTION 5.4 EXERCISES

Reduce or convert.

1. How many cubic centimeters are in a cubic meter?
2. How many cubic inches are in a cubic yard?
3. How many gallons will a cubic mile hold?
4. How many cubic feet are in 7.5 cubic yards?
5. How many fluid ounces are in 1 gallon?
6. How many liters are in a gallon?

5.5 ■ QUANTITY—TIME, SYMBOL—t

Fortunately, the US and SI systems use the same unit for time, seconds (sec). Other units include minutes, hours, days, weeks, years, decades, centuries, and millenniums. To measure short periods of time, the second is subdivided using the metric prefixes: milliseconds (msec), microseconds (μsec), nanoseconds (nsec), and picoseconds (psec).

EXAMPLE 5.8 How many microseconds are in one day?

Solution

$$1 \text{ day} \times \frac{24 \text{ hr}}{1 \text{ day}} \times \frac{60 \text{ min}}{1 \text{ hr}} \times \frac{60 \text{ sec}}{1 \text{ min}} \times \frac{10^6 \text{ μs}}{1 \text{ s}} = 86\,400 \times 10^6 \text{ μs}$$

$$1 \text{ day} = 8.64 \times 10^{10} \text{ μs}$$

All factors used are exact. The result does not need to be rounded.

5.6 ■ QUANTITY—VELOCITY, SYMBOL—v

Velocity is the rate of change of position. Its units are length divided by time. In SI units, velocity is measured in meters per second (m/sec), meters per minute, kilometers per hour, centimeters per year, and so on. In US units, velocity is measured in feet per second (ft/sec), feet per minute, miles per hour (mph), nautical miles per hour (knots), miles per second, inches per week, and so on.

A velocity that appears in many equations in applied science and technology is the speed of light, c. Light and other electromagnetic waves travel in free space at 186 000 miles per second, accuracy of three digits.

EXAMPLE 5.9 Calculate the speed of light in meters per second.

Solution

$$\frac{186{,}000 \text{ mi}}{\text{sec}} \times \frac{5280 \text{ ft}}{1 \text{ mi}} \times \frac{12 \text{ in}}{1 \text{ ft}} \times \frac{2.54 \text{ cm}}{1 \text{ in}} \times \frac{1 \text{ m}}{100 \text{ cm}} =$$

$$= 2.993\,379\,8 \times 10^8 \text{ m/sec}$$

150 ■ CHAPTER 5 UNITS OF MEASURE

186,000 has an accuracy of three; 2.540 00 has an accuracy of six; and 12 and 100 are exact. Round the answer to three significant digits.

$$c = 186\,000 \text{ mi/s} = 2.99 \times 10^8 \text{ m/s}$$

c is known more accurately than this, but we started with information that only has an accuracy of three.

EXAMPLE 5.10 How far is a light-year in miles?

Solution A light-year is the distance light can travel in one year. Reduce miles per second to miles per year.

$$\frac{186\,000 \text{ mi}}{\text{sec}} \times \frac{60 \text{ sec}}{1 \text{ min}} \times \frac{60 \text{ min}}{1 \text{ hr}} \times \frac{24 \text{ hr}}{1 \text{ day}} \times \frac{364.25 \text{ days}}{1 \text{ yr}} =$$
$$= 5.853\,643\,2 \times 10^{12} \text{ miles per year}$$

186 000 has an accuracy of three and limits the result to three significant digits. Light travels 5.85×10^{12} miles in one year. One light-year is 5.85×10^{12} miles.

EXAMPLE 5.11 Is 20 knots faster than 20 mph?

Solution Convert 20 knots to mph.

$$\frac{20 \text{ nautmi}}{\text{hr}} \times \frac{6076 \text{ ft}}{1 \text{ nautmi}} \times \frac{1 \text{ mi}}{5280 \text{ ft}} = 23.015\,152 \text{ mph}$$

Assuming 20 knots has an accuracy of two, round the answer to two significant digits.

20 nautical mph = 23 mph. Yes, 20 knots is faster than 20 mph.

SECTION 5.6 EXERCISES

1. 1 mph = _____ ft/sec
2. 1 mph = _____ km/hour
3. 1 km/hour = _____ mph
4. 1 knot = _____ mph
5. 1 ft/sec = _____ km/hour
6. 1 ft/sec = _____ mph

5.7 ■ QUANTITY—ACCELERATION, SYMBOL—a

Acceleration is the rate of change of velocity. If you are cruising down a straight stretch of I-75 at a constant 70 mph, your acceleration is zero. If you speed up or accelerate, your acceleration is positive. If you apply the brakes, you decelerate and your acceleration is negative. The unit of measure of acceleration is velocity per time or length per time per time, for example, feet per second per second. This is usually written feet per second squared, ft/sec². In SI units, acceleration is expressed in meters per second per second. This is usually written meters per second squared, m/sec² or kilometers per hour squared (km/hr²), and so on.

EXAMPLE 5.12 Later, we will encounter a special acceleration, the acceleration due to gravity here on Earth, g. g = 32.2 ft/sec². Convert g to SI units.

Solution
$$\frac{32.2 \text{ ft}}{\text{sec}^2} \times \frac{12 \text{ in}}{1 \text{ ft}} \times \frac{2.540\,00 \text{ cm}}{1 \text{ in}} \times \frac{1 \text{ m}}{100 \text{ cm}} = 9.814\,56 \frac{\text{m}}{\text{sec}^2}$$

12 and 100 are exact. 32.2 is accurate to three digits, and limits the answer to three digits.

$$32.2 \text{ ft/sec}^2 = 9.81 \text{ m/sec}^2.$$

5.8 ■ QUANTITY—MASS, SYMBOL—m

Mass is the measure of an object's ability to resist a change in velocity. The more difficult it is to set an object into motion, the greater the mass. In SI units, the standard measure of mass is the **kilogram**. In US Customary units, the standard unit for mass is the **slug**. The term *slug* is not used much by the general public. You may hear someone speak of pounds of mass, but technically that is not correct in this system of units. The conversion factor between the two systems is 1 slug = 14.59 kg.

EXAMPLE 5.13 Convert 27.4 slugs to grams.

Solution
$$27.4 \text{ slugs} \times \frac{14.59 \text{ kg}}{1 \text{ slug}} \times \frac{1000 \text{ gram}}{1 \text{ kg}} = 399\,766 \text{ gram}$$

Round to three significant digits.

$$27.4 \text{ slugs} = 400\,000 \text{ grams}$$

5.9 ■ QUANTITY—FORCE, SYMBOL—F

The tension in a cable being used as a guy wire and the compression in a pad of concrete supporting a column are examples of **forces**. A person's or object's weight is also a force. Weight is the force on a body due to gravity.

In SI units, the primary unit of force is the **newton (N)**. A much smaller unit is the **dyne**.

$$10^5 \text{ dynes} = 1 \text{ newton (N)}$$

In US units, force and weight are measured in **pounds (lb)**. In mechanics, the kilopound is often used. One **kilopound (kip)** is 1000 pounds. Other units are ounce (oz), and ton. Here are the reduction factors.

$$1000 \text{ pounds} = 1 \text{ kip}$$
$$1 \text{ pound} = 16 \text{ oz}$$
$$2000 \text{ pounds} = 1 \text{ ton}$$

To convert between the two systems

$$1 \text{ newton} = 0.2248 \text{ pounds}$$
$$\text{or } 1 \text{ pound} = 4.448 \text{ newtons}$$

EXAMPLE 5.14 Convert 385 newtons to pounds.

Solution
$$385 \text{ N} \times \frac{1 \text{ lb}}{4.448 \text{ N}} = 86.555\ 75 \text{ lb}$$

385 has an accuracy of three. Round the answer to three significant digits.

$$385 \text{ N} = 86.6 \text{ lb}$$

The following two statements illustrate how units within a system work together.

If an object with a mass of one kilogram is pushed with a constant force of one newton, it will accelerate with an acceleration of one meter per second squared.

If an object with a mass of one slug is pushed with a constant force of one pound, it will accelerate with an acceleration of one foot per second squared.

5.10 ■ QUANTITY—DENSITY, SYMBOL—ρ

Density is the measure of the mass of an object per unit of volume. The units of density come from those of mass and volume. The standard SI unit for density is kilograms per cubic meter. A smaller measure is grams per cubic centimeter, gm/cm^3. The density of water is 1 gram/cm^3. Mercury is much more dense at 13.6 grams/cm^3. Lead has a density of 11.4 grams/cm^3, and gold has a density of 19.32 grams/cm^3. Ice has a density of less than one, 0.917 grams/cm^3, and maple wood has a density of about 0.70 grams/cm^3. Both ice and wood float in water as do all solids with a density of less than one. In the US system, the standard unit of density is slugs per cubic foot.

5.11 ■ QUANTITY—PRESSURE, SYMBOL—P

A force applied over an area is called a **pressure.** The units for pressure are derived from the units for force and area. In SI units, pressure is measured in newtons per square meter, which is given the name **pascals, (Pa)**. The standard US unit for pressure is **pounds per square inch (psi)**. One pressure you are familiar with is atmospheric pressure, the pressure caused by the weight of the atmosphere on our bodies. Atmospheric pressure at sea level is 14.7 psi. Each square inch of our bodies is subjected to 14.7 lb of force. This is called one **atmosphere** of pressure. For each 33 feet that you dive under salt water, another atmosphere of pressure is added to your body. At a depth of 99 feet, you are subjected to 3 atmospheres of pressure due to the weight of sea water plus another atmosphere from the air above the water, for a total of four atmospheres of pressure or 58.8 psi (4·14.7 psi).

Other than psi and atmospheres, US units include **inches of mercury (in-Hg)**, which is the pressure created by a column of mercury 1 inch high, and pounds per square foot.

$$1 \text{ atmosphere} = 14.7 \text{ psi}$$

$$1 \text{ inch mercury} = 70.73 \text{ pounds per square foot}$$

EXAMPLE 5.15 How many pascals make an atmosphere?

Solution 1 atmosphere = 14.7 psi. Convert 14.7 psi to newtons per square meter (pascals).

$$\frac{14.7 \text{ lb}}{\text{in}^2} \times \left[\frac{1 \text{ in}}{2.54 \text{ cm}}\right]^2 \times \left[\frac{100 \text{ cm}}{1 \text{ m}}\right]^2 \times \frac{4.448 \text{ N}}{1 \text{ lb}} = 101\,347.88 \text{ Pa}$$

14.7 has an accuracy of three; 4.448 has an accuracy of four. The other numbers involved are exact. Round the result to three significant digits.

$$101\,000 \text{ pascals or } 101 \text{ kpascals} = 1 \text{ atmosphere}$$

Meteorologists measure atmospheric pressure in **millibars (mb)**, where one millibar is defined as 100 pascals.

$$1 \text{ millibar} = 100 \text{ pascals}$$

Standard atmospheric pressure at sea level is taken as 1013.25 mb. Worldwide, atmospheric pressure varies from 970 to 1050 mb. In August, 1992, hurricane Andrew devastated South Florida. Its atmospheric pressure dropped to about 925 mb. The lowest ever recorded, 870 mb, was associated with Typhoon Tip Northwest of Guam in October 1979. The highest ever recorded, 1084 mb, was associated with an extremely cold air mass in Agata, Siberia, in December 1968.[1]

Another method of measuring pressure is **millimeters of mercury (mm-Hg)**, the pressure exerted by a column of mercury 1 mm high. Standard pressure is taken as 760 mm-Hg.

---

[1] Frederick K. Lutgens and Edward J. Tarbuck, *The Atmosphere,* 6th Edition, (Englewood Cliffs, NJ: Prentice Hall, 1995) 144.

5.12 ■ QUANTITY—TEMPERATURE, SYMBOL—T

In the SI system, **degrees kelvin (K)** is the standard measure of temperature. Kelvin is an absolute scale meaning that 0 K is absolute zero, the temperature at which there is no heat energy. **Degrees Celsius** (°C) is also used. Kelvin and Celsius each have 100° between the freezing point of water and the boiling point of water. So, a 5° change in Celsius is a 5° change in kelvin. Celsius is offset from kelvin by 273.15°. 0 K (absolute zero) is −273.15° C. To convert degrees Celsius to degrees kelvin, add 273.15 degrees. Water boils at 100° C and freezes at 0° C. In US units, **degrees Fahrenheit** (°F) is the standard measure of temperature. Water boils at 212°F and freezes at 32°F.

5.13 ■ QUANTITY—CHARGE, SYMBOL—Q

The basic unit of **charge** is the electron. Each electron has a charge of −1. When a substance possesses an excess of electrons, it has a negative charge. When it possesses a deficiency of electrons, it has a positive charge. A charge equivalent to 6.242×10^{18} electrons is called 1 **coulomb (C)**. The coulomb is used in the definitions of ampere and farad in the following sections.

5.14 ■ QUANTITY—CURRENT, SYMBOL—I OR i

Current is the flow of electrons. The rate of flow of electrons determines how many amperes or amps are flowing in the circuit. 6.242×10^{18} electrons (1 coulomb) flowing past a point in a circuit each second is a current of one **amp (A)**. Current is measured in amps, milliamps (mA), or microamps (μA), depending on the type and size of circuit.

5.15 ■ QUANTITY—VOLTAGE OR ELECTROMOTIVE FORCE (EMF), SYMBOL—E OR V_s

The driving force that causes electrons to flow around the circuit is called **electromotive force (EMF)**. EMF is measured in **volts (V)**. Generators, batteries, and solar cells are sources of EMF. As current flows through a resistor, a **voltage drop,** also called a **potential difference,** is created. This voltage drop is also measured in volts. The voltage drop across R_1 is often denoted V_1, and the supply voltage or source voltage for the circuit is often denoted E, E_S, or V_S. A variety of EMF devices that generate direct current (DC) voltage are shown in Figure 5-1, and devices that generate alternating current (AC) voltage are shown in Figure 5-2.

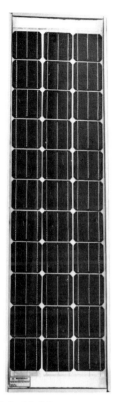

Solar Module
40 Watt, 12 Volt

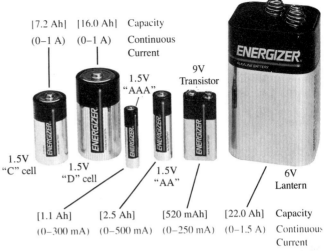

Alkaline Cells of Various Amp-Hour Ratings
Eveready Battery Co.

1.2 V   1.2 V   7.2 V   1.2 V   1.2 V
4 Ah   1.2 Ah   100 mAh   500 mAh   180 mAh

DC Power Supply Leader

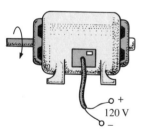

DC Generator

Eveready® BH 500 cell
1.2 V, 500 mAh

Rechargeable Ni-Cad
Nickel-Cadmium Cells
Eveready Battery Co.

Lithium-Iodine Cell for Pacemaker

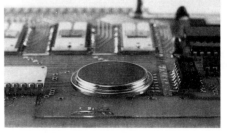

Lithium-Iodine Cell on Printed Circuit Board

FIGURE 5-1
Sources of Direct Current EMF

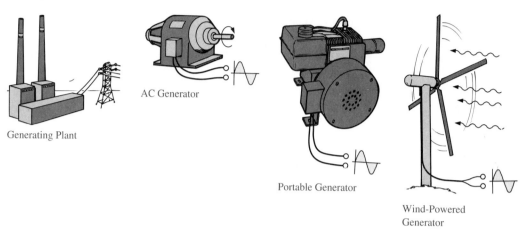

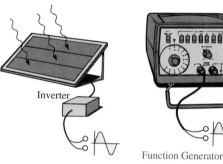

FIGURE 5-2
Sources of Alternating Current EMF

5.16 ■ QUANTITY—RESISTANCE, SYMBOL—R

As electrons flow through a substance, they collide with the electrons and atoms within the substance, and the flow is restricted. The amount of restriction or resistance to the flow of current is called the **resistance** of the material. Resistance is measured in **ohms** (Ω). Resistors are encountered in the MΩ, kΩ, and Ω ranges. A variety of resistors is shown in Figure 5-3.

To summarize the last three quantities, EMF measured in volts pushes current measured in amps through resistance measured in ohms. If one volt pushes one amp through a resistance, that resistance is assigned a value of one ohm. These three quantities are related by the powerful equation referred to as Ohm's law; $E = IR$ or $V = IR$.

FIGURE 5-3
Resistors

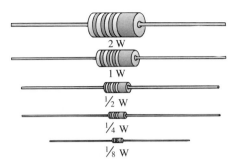

Carbon Composition Resistors of Various Power Ratings

Thick-Film Resistor Networks
Dale Electronics

Metal Film Precision Resistor Ohmite Manufacturing Company

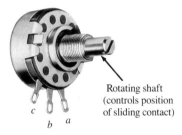

Potentiometer (Variable Resistor)
Allen Bradley

5.17 ■ QUANTITY—CONDUCTANCE, SYMBOL—G

Conductance is the reciprocal of resistance: $G = 1/R$. The units of conductance are **siemens (S)**, formerly called mhos. Resistance is a measure of the difficulty in pushing electrons through a substance, and conductance is a measure of the ease with which current will flow through a substance. mS and µS are often encountered in work with parallel circuits.

5.18 ■ QUANTITY—CAPACITANCE, SYMBOL—C

A **capacitor** in its simplest form is two parallel plates of conducting material like copper or aluminum separated by an insulator called the *dielectric*. Air, mica, mylar, and glass are examples of dielectrics. Capacitors are components found in most electronic circuits and are used in a great variety of ways. **Capacitance** is a measure of the storage capacity of a capacitor. If the voltage across a capacitor increases by 1 volt for each coulomb of charge that is stored on the capacitor, the capacitor has a capacitance of 1 **farad (F)**. A farad is a very large unit. More common are capacitors in the pF and µF range. A variety of capacitors are shown in Figure 5-4.

Monolithic Chip Capacitors
Vitramon Inc.

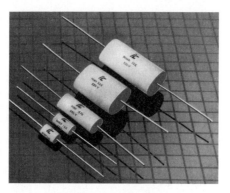

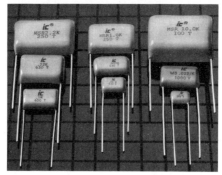

Polyester Film Capacitors
Illinois Capacitor

Solid Aluminum Electrolytic Capacitors
Phyllips Components

Ceramic Disc Capacitors

Variable Air Capacitor
Johnson Manufacturing

Variable Air Capacitor
James Millen Manufacturing

FIGURE 5-4
Capacitors

5.19 ■ QUANTITY—INDUCTANCE, SYMBOL—L

An **inductor** is basically a coil of wire. When current flows through the coil of wire, a magnetic field like that of a bar magnet is produced. An electromagnet is produced. A change in current through the coil causes a change in the magnetic field of the coil which, in turn, causes a voltage to be induced across the inductor. The measure of this ability to self-induce a voltage is called **inductance** and is measured in **henrys (H)**. A change in current of one amp per second through a one henry inductor generates a voltage across the inductor of one volt. µH, mH, and H are encountered. A variety of inductors are shown in Figure 5-5.

The schematic symbols for a battery, an AC signal generator, a resistor a capacitor, and an inductor are shown in Figure 5-6.

5.20 ■ QUANTITY—FREQUENCY, SYMBOL—f

Figure 5-7 shows a sine wave. From one point on the wave to the corresponding point on the wave is called a **cycle.** The number of cycles that occur each second is called **frequency** of the wave. Frequency is measured in **cycles per second,** which is given the special name **hertz (Hz)**. A wide range of frequencies is encountered in electronics. Here are a few examples:

1. The voltage distributed to homes by power companies has a frequency of 60 Hz.
2. FM radios range from 88 to 108 MHz.
3. VHF television channels range from 470 to 880 MHz.
4. K band microwave ranges from 20 to 40 GHz.
5. Infrared (heat) region ranges from 0.3 to 4.3 THz (Tera=10^{12}).
6. Visible light ranges from 4.3 THz to 1.0 PHz (P=Penta=10^{15}).

5.21 ■ QUANTITY—ENERGY, SYMBOL—W

Energy exists in several different forms, and different units of measure are encountered in each.

The basic unit of electrical energy is the **joule (J)**. It is discussed in the next section in relation to power.

One measure of mechanical energy is the amount of force exerted through a distance. Energy or work equals force times distance.

$$W = F \cdot d$$

In SI units, force times distance is **newton-meters.** One newton-meter is the same as one **joule (J)**. In US units, this form of energy is measured in **foot-pounds.**

Toroidal Power Inductor
1.4 µH to 5.6mH
Microtran Co.

Reels of Surface Mount
Inductor
0.1 µH to 1mH
Bell Industries

Moulded Inductor
0.1 µH to 10 µH
Dale Electronics

High Current Filter
24 µH at 60A to
500 µH at 15A
Dale Electronics
24 µH at 60A to 500 µH at 15A
Dale Electronics

Toroid Filter Inductor
40 µH to 5H
Dale Electronics

Air Core Inductors for
High Frequency
Dale Electronics

FIGURE 5-5
Inductors

FIGURE 5-6
Schematic Symbols

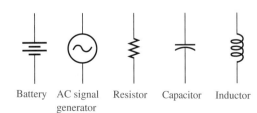

Battery   AC signal   Resistor   Capacitor   Inductor
          generator

FIGURE 5-7
Cycle

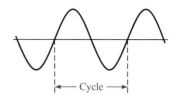

EXAMPLE 5.16   Convert 1 foot-pound to joules.

Solution   A joule is a newton-meter. Convert 1 foot-pound to newton-meters.

$$1 \text{ ft-lb} \times \frac{12 \text{ in}}{1 \text{ ft}} \times \frac{2.54 \text{ cm}}{1 \text{ in}} \times \frac{1 \text{ m}}{100 \text{ cm}} \times \frac{4.448 \text{ N}}{1 \text{ lb}} = 1.3557504 \text{ J}$$

4.448 limits the accuracy of the answer to four.

$$1 \text{ foot-pound} = 1.356 \text{ newton-meters}$$

Another form of energy is heat energy. In US units, the amount of energy needed to raise the temperature of one pound of water one degree Fahrenheit is called one **British thermal unit (Btu)**.

$$1 \text{ Btu} = 1054 \text{ joules}$$
$$3410 \text{ Btu} = 1 \text{ kilowatt-hour}$$
$$1200 \text{ Btu} = 1 \text{ ton of air-conditioning}$$

To measure huge quantities of energy such as annual energy consumption of a country, the **quad** is used. A quad is one quadrillion Btu's or 1×10^{15} Btu's or 1 000 000 000 000 000 Btu's. The United States uses about 70 quads of energy each year.

In SI units, the amount of energy needed to raise the temperature of one cubic centimeter of water one degree Celsius is called a **calorie.**

5.22 ■ QUANTITY—POWER, SYMBOL—P

Power is the rate at which energy is being expended. P=W/t. A joule of energy being expended each second is a measure of power and is given the name **watt (W)**.

$$1 \text{ watt} = 1 \text{ joule/second}$$

So, watt is a unit of power and joule is a unit of energy. If P=W/t, then W=P·t. Energy equals power times time. A joule is a watt-second. The meter on your house actually measures the energy consumed in kilowatt-hours. Kilowatt is a measure of power; kilowatt-hours is a measure of energy.

EXAMPLE 5.17 Convert 1 kilowatt-hour into joules.

Solution A joule is a watt-second. Convert kilowatt-hours to watt-seconds.

$$1 \text{ Kw-hr} \times \frac{1000 \text{ w}}{1 \text{ kw}} \times \frac{3600 \text{ s}}{1 \text{ hr}} = 3\,600\,000 \text{ watt-second}$$

$$1 \text{ kilowatt-hour} = 3.6 \times 10^6 \text{ J}$$

Horsepower (HP) is a commonly used unit of power in the US system. The corresponding unit of energy is horsepower-hours.

$$1 \text{ horsepower} = 42.44 \text{ Btu/min}$$
$$1 \text{ horsepower} = 746 \text{ watts}$$
$$550 \text{ foot-pounds per second} = 1 \text{ horsepower}$$

The terms *power* and *energy* are used interchangeably in everyday language, and some effort is required to keep them straight in a technical context. Here is a summary of the units discussed in the last two sections in the form of a list of units of energy matched up with the corresponding units of power.

Energy	Power
joule	watt
foot-pound	foot-pound per second
Btu	Btu per hour or Btu per minute
calorie	calorie per second
horsepower-hour	horsepower

SECTION 5.22 EXERCISES

Which of these units are for power and which are for energy? Put P for power and W for energy.

1. watts
2. horsepower
3. Btu
4. ft-lb/sec
5. joules
6. kilowatt-hours
7. Btu/min
8. calorie
9. newton-meter/sec
10. ft-lb
11. watt-second

Convert.

12. 1 watt = _____ ft-lb/sec
13. 1 ft-lb/sec = _____ newton-meters/sec
14. 1 Btu = _____ kilowatt-hours

SUMMARY

1. The Système International d'unités (SI) or International System and the United States Customary Units (US) are two systems of units of measure.

2. A technique called *dimensional analysis* can be used to convert from one system of units to another, or to reduce units in one system to other units within the same system.

3. In the dimensional analysis technique, the original quantity is multiplied by one or more fractions. Each fraction is equivalent to 1, so that the original quantity is not altered. The fractions are written so that the original and intermediate units cancel, leaving only the desired units.

4. Here are the quantities studied, their symbols, and their basic units in the US and SI systems of units.

Quantity	Symbol	US Unit	SI Unit
Length	s	foot	meter
Area	A	foot^2	meter^2
Volume	V	foot^3	meter^3
Time	t	second	second
Velocity	v	foot/second	meter/second
Acceleration	a	foot/second^2	meter/second^2
Mass	m	slug	kilogram
Force	F	pound	newton
Density	ρ	pound/foot^3	kilogram/meter^3
Pressure	P	pound/inch^2	pascal (newton/meter^2)
Temperature	T	°F	°C or K
Charge	Q	coulombs	coulombs
Current	I or i	amperes	amperes
Electromotive Force	E or V_s	volts	volts
Resistance	R	ohms	ohms
Conductance	G	siemens	siemens
Capacitance	C	farads	farads
Inductance	L	henrys	henrys
Frequency	f	hertz	hertz
Energy or Work	W	foot-pound	newton-meter (joule)
		Btu	calorie
		kilowatt-hour	watt-second (joule)
Power	P	horsepower	watts

SOLUTIONS TO CHAPTER 5 EXERCISES

Solutions to Section 5.2 Exercises

1. 1.609 34 km
2. 0.621 371 mi
3. 91.440 0 cm
4. 30.480 0 cm
5. 95 800 m
6. $3.937\,01 \times 10^{-5}$ in

Solutions to Section 5.3 Exercises

1. 9 ft²
2. 247.105 acres
3. 144 in²
4. 4 046.86 m²
5. 640 acres
6. 1296 in²

Solutions to Section 5.4 Exercises

1. 1×10^6 cc
2. 46 656 in³
3. $1.101\ 117\ 147 \times 10^{12}$ U.S.gal, $9.169\ 360\ 53 \times 10^{11}$ B.I.gal
4. 202.5 ft³
5. 128 fl oz
6. 3.78 l

Solutions to Section 5.6 Exercises

1. 1.467 ft/sec
2. 1.609 km/hr
3. 0.621 371 mph
4. 1.151 mph
5. 1.09728 km/hr
6. 0.6818 mph

Solutions to Section 5.22 Exercises

1. P
2. P
3. W
4. P
5. W
6. W
7. P
8. W
9. P
10. W
11. W
12. 0.737 ft-lb/sec
13. 1.356 newton-meters/sec
14. 2.928×10^{-4} kilowatt-hours

CHAPTER 5 PROBLEMS

Answer True or False

1. The SI unit of area is in².
2. The US unit of volume is liter.
3. A liter is roughly a quart.
4. The SI unit of length is the furlong.
5. The units for area are equivalent to length².
6. 1 ml is the same volume as 1 cc.
7. 1 ft³ holds 9.5 gallons.
8. The speed of light is about 3×10^8 ft/sec.
9. A light-year is a measure of time.
10. Acceleration is change in velocity with respect to time.

11. The SI unit for mass is newton.
12. 10 knots is faster than 10 mph.
13. The units for acceleration are equivalent to distance per length².
14. The US unit of force is pounds.
15. Heat energy is measured in calories.
16. A joule, a watt-second, and a newton-meter are the same amount of energy.
17. 2.54000 cm is 1 inch.
18. An angstrom is longer than a micron.
19. Acre is a measure of area.
20. Btu stands for butane thermal unit.
21. Watts is power and joules is energy.
22. The power company meterperson comes to your house to measure the power consumed.
23. The power company charges you for the kilowatt-hours consumed.
24. Current is the flow of electrons and is measured in amps.
25. Capacitance is measured in henrys.
26. Hertz is a measure of frequency.
27. Inductance is the reciprocal of conductance.
28. A change in current of 1 A/s through a 1 henry inductor induces a voltage across the inductor of 1 volt.
29. EMF means electronic measurement of frequency.
30. Volts push amps through ohms.
31. The symbol for ohms is Ω.
32. 1 amp flowing through 1 ohm resistance produces a voltage drop of 1 volt.
33. A force of 1 pound pushing on a mass of 1 kilogram causes the mass to accelerate at 1 meter per second².
34. Batteries and solar cells are sources of EMF.
35. Siemen is a measure of conductance.
36. Conductance is a measure of the ability of a component to conduct current.

Fill in the blanks. There can be more than one answer in the Units columns.

	Quantity	Symbol	Units	
			US	SI
37.	_____	_____	pound	_____
38.	Pressure	_____	_____	_____
39.	_____	s	_____	_____
40.	_____	_____	_____	kilogram

166 ■ CHAPTER 5 UNITS OF MEASURE

41. Power _____ _____ _____
42. _____ _____ mph _____
43. _____ _____ _____ m^2
44. Volume _____ _____ _____
45. _____ _____ _____ joule
46. Time _____ _____ _____
47. _____ T _____ _____
48. Density _____ _____ _____
49. _____ _____ _____ m/sec^2

Match

50. 5280 ft
51. 1 atmosphere
52. 1 pascal
53. 1 watt
54. 1 horsepower
55. 1 horsepower
56. 660 ft
57. 1 nautical mile
58. 1 rod
59. 1 angstrom
60. 1 micron
61. 1 slug
62. 1 kip
63. 1 lb
64. 640 acres
65. 1 gal
66. 1 gal
67. 1 amp
68. 1 cycle/sec
69. 1 quart
70. 1 fathom
71. 1 Btu
72. 12 in
73. 16 oz
74. lb
75. 186 000 mi per sec

a. 1 joule per second
b. 1×10^{-6} m
c. 1 coulomb/sec
d. 550 ft-lb/sec
e. 6 ft
f. $1\ mi^2$
g. 14.7 psi
h. 1 ft
i. 1×10^{-10} m
j. 1 lb
k. 6076 ft
l. 1 furlong
m. 1 mi
n. slug ft/sec^2
o. 16.5 ft
p. 14.59 kgrams
q. 746 watts
r. 16 cups
s. 1054 joules
t. 1 newton/$meter^2$
u. 0.946 l
v. 2.99×10^8 m/sec
w. 231 in^3
x. 4.448 newtons
y. 1 Hz
z. 1000 lb

Complete the chart

	Quantity	**Symbol**	**Unit**
76.	_____	f	_____
77.	_____	_____	coulombs
78.	EMF	_____	_____
79.	_____	_____	farads
80.	_____	_____	Ω
81.	Current	_____	_____
82.	_____	L	_____
83.	Energy	_____	_____
84.	_____	_____	watts

Match

	Quantity	**Units**
85.	capacitance	a. ohms
86.	resistance	b. watts
87.	electromotive force	c. coulombs
88.	current	d. joules
89.	frequency	e. farads
90.	inductance	f. hertz
91.	power	g. amps
92.	energy	h. siemens
93.	charge	i. henrys
94.	conductance	j. volts

Reduce or convert. Round answer to proper number of significant digits.

95. 1.5 yd = _____ angstrom
96. 1.00 µm = _____ angstrom
97. 115 nautical mi = _____ mi
98. 1.00 yd = _____ cm
99. 125 yd = _____ m
100. 256 kg = _____ slugs
101. 37.0 kips = _____ tons
102. 37.0 kips = _____ newtons
103. 79 ft/sec² = _____ m/sec²
104. 1.00 mi² = _____ ft²
105. 1.000×10^3 m² = _____ km²

CHAPTER 5 UNITS OF MEASURE

106. _____ oz = 1 gal US
107. 11 m³ can hold _____ liters
108. 1.5 ft³ can hold _____ gal
109. 54 slugs per ft³ = _____ kg/m³
110. 8.0 knots = _____ mi/day
111. 125 km/hr = _____ mph
112. 208 m/sec² = _____ ft/sec²
113. 3.5 atm = _____ mbars
114. 100°C = _____ K
115. 3.000 ×10³ psi = _____ Pa
116. 586 joules = _____ kwatt-hrs
117. 387 horsepower = _____ watts
118. 387 horsepower = _____ ft-lb/sec
119. 1234 Btu = _____ kwatt-hr

Unscramble these letters:

EYHNR ☐☐☐☐◯
MAGR   ◯☐☐☐
ENOWNT ☐◯☐☐☐
AADRF ☐☐◯☐☐
OSDPUN ☐☐☐☐◯☐
OJLUE ☐☐☐☐◯

Now unscramble the circled letters to
solve the puzzle.

6
Solving Problems

CHAPTER OUTLINE

- **6.1** The Eight-Step Approach to Problem Solving
- **6.2** Series Circuits
- **6.3** Gage and Absolute Pressure
- **6.4** Ohm's Law
- **6.5** Torricelli's Theorem
- **6.6** Tank Circuit
- **6.7** Coefficient of Friction
- **6.8** Two Resistors in Parallel
- **6.9** Two or More Resistors in Parallel
- **6.10** Superelevation
- **6.11** Mass and Weight
- **6.12** °Fahrenheit, °Celsius
- **6.13** Airplane Stall Speed
- **6.14** Pressure Exerted by a Column of Liquid
- **6.15** Wavelength, Period, Frequency, and Velocity of Electromagnetic Waves

OBJECTIVE

After completing this chapter, you should be able to follow the eight-step procedure to solve technical problems.

NEW TERMS TO WATCH FOR

series
gage pressure
absolute pressure
Ohm's law
Torricelli's theorem
tank circuit
resonant frequency
coefficient of friction (f)
parallel
superelevation (e)

degrees Fahrenheit (°F)
degrees Celsius (°C)
stall speed
wavelength (λ)
period (T)
cycle
frequency (f)
hertz (Hz)
refractive index (n)
index of refraction (n)

6.1 ■ THE EIGHT STEP APPROACH TO PROBLEM SOLVING

The problem solving skills developed in this chapter will be utilized numerous times during your career as a technician. Fourteen topics are presented that show in detail an orderly, logical approach to problem solving. In your technical courses, these steps can be modified slightly to suit the situation at hand, but the overall scheme will remain the same. In each example, the equation to be used is given. In your technical courses, the task of selecting the proper equation from the material being studied will become part of the process. Each equation is followed by a statement that begins with "where." Pay attention to the "where's." They define each of the variables in the equation and provide pertinent information about the units to be used. Likewise, in your work as a technician, when you pull an equation from a manual, read the "where's" to ensure that you understand the equation and any special requirements for units. If units are not mentioned, then standard units from any one system will yield a standard unit in the same system. For example, in the equation $P = I^2R$, if current is entered in amps and resistance is entered in ohms, then the resulting power is measured in watts. By working through each example, you will gain experience in the details of problem solving. It will take time and energy. Be patient. A more thorough understanding of the equations used will be gained in your technical courses.

The following steps will be used in each example in this chapter.

1. Identify and list the given information.
2. Identify and list the unknown variable. Sketch a figure if applicable. Identify the known and unknown variables in the sketch. Be sure to use the same notation in the sketch that is used in the "Given" and "Find" statements.
3. Find the equation that relates the known and unknown variables. In the following examples the equation will be given, but later in your technical courses and work some effort will be required to find a suitable equation or equations.
4. Rearrange the equation to solve for the unknown. Here you will apply your basic algebra skills.

5. Plug in the known values. Include units. If they do not agree, perform necessary conversions or reductions. Review Chapter 5 as needed for units of measure.
6. Use your calculator to find the preliminary answer.
7. Adjust the decimal point so that a metric prefix is produced, if necessary.
8. Round the answer to the proper number of significant digits. Use your knowledge of precision and accuracy. Review Chapter 2 as needed.

Do not worry or become frustrated if you do not understand the terms used in the examples such as capacitance or inductance. That knowledge comes later in your technical courses. Concentrate instead on the procedure used and the algebraic techniques used. Get pencil and paper and work these examples on paper. Develop good work habits. Make your work accurate, logical, and neat. Record steps taken to solve an equation so that you can use your work later as a review.

Within the examples, many of the steps are *followed* by comments describing what was done. Try to determine what was accomplished before reading the comment. You should grow less dependent on the comments as you proceed through the chapter.

6.2 ■ SERIES CIRCUITS

Example 6.1 involves resistors connected in **series**, end to end to form a chain, as shown in Figure 6-1. Resistor values in series are additive, that is, the total resistance is the sum of the individual resistors. The . . . in the equation indicates continuation (and so on) for any number of resistors in series.

FIGURE 6-1
Resistors Connected in Series

$$R_T = R_1 + R_2 + R_3 + \ldots$$

where R_1, R_2, and R_3 and so on are resistors connected in series, and R_T is the total connected resistance, all measured in ohms (Ω).

EXAMPLE 6.1 How much resistance, R_3, should be added in series with 330 Ω and 470 Ω to produce a total of 2 kΩ? All values are precise to the nearest 10 Ω.

Solution
1. Given: $R_1 = 330\ \Omega$, $R_2 = 470\ \Omega$, and $R_T = 2$ kΩ
2. Find: R_3 See Figure 6-2.

FIGURE 6-2
Example 6.1

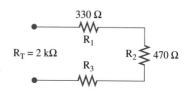

3. $R_T = R_1 + R_2 + R_3$
4. $R_T - R_1 - R_2 = R_3$
   Transpose R_1 and R_2 to isolate R_3.
   $R_3 = R_T - R_1 - R_2$
   Swap sides to get unknown on left.
5. $R_3 = 2{,}000\ \Omega - 330\ \Omega - 470\ \Omega$
   Replace k with its value of 1000.
6. $R_3 = 1200\ \Omega$
7. $R_3 = 1.200\ k\Omega$
8. Round R_3 to $1.20\ k\Omega$.
   Significant digits: The values given were precise to the nearest $10\ \Omega$.
   Round the answer to that precision.

SECTION 6.2 EXERCISES

Follow the eight-step procedure.

1. Three resistors $4.21\ M\Omega$, $3.7\ M\Omega$, and $2.2\ M\Omega$ are connected in series. Find the total resistance.
2. If a total resistance of $504\ k\Omega$ is needed, how much resistance should be added in series to $33.6\ k\Omega$ and $223\ k\Omega$ resistors?

Review the resistor color code presented in Chapter 1 as needed to work the next two problems.

3. The following resistors are connected in series: green blue red, blue red red, and violet green red. Calculate the total resistance.
4. An orange orange red, an orange black red, and an orange blue red resistor are connected in series. What is the color code of the resistor that should be connected in series with these resistors to make a total of $11{,}900\ \Omega$?

6.3 ■ GAGE AND ABSOLUTE PRESSURE

Some pressure-measuring devices compare a pressure under measurement to the surrounding atmospheric pressure. These measurements are called **gage pressures**. When a pressure under measurement is compared to a complete vacuum, the measurement is called an **absolute pressure**. Gage and absolute pressures are related by the equation

$$P_{ABS} = P_{GAGE} + P_{ATM}$$

where P_{ABS} is absolute pressure, P_{GAGE} is gage pressure, and P_{ATM} is atmospheric pressure.

EXAMPLE 6.2 For an absolute pressure of 207 kPa (kilopascals) and an atmospheric pressure of 101 kPa, calculate the corresponding gage pressure.

Solution
1. Given: $P_{ABS} = 207\ kPa$, $P_{ATM} = 101\ kPa$
2. Find: P_{GAGE}

3. $P_{ABS} = P_{GAGE} + P_{ATM}$

where P_{ABS} is absolute pressure, P_{GAGE} is gage pressure, and P_{ATM} is atmospheric pressure.

The "where" statement did not mention units. Any pressure unit can be chosen, but all three pressures must use that unit.

4. $P_{GAGE} = P_{ABS} - P_{ATM}$

Transpose P_{ATM} to isolate P_{GAGE}.

5. $P_{GAGE} = 207 \text{ kPa} - 101 \text{ kPa}$
6. $P_{GAGE} = 106 \text{ kPa}$
7. $P_{GAGE} = 106 \text{ kPa}$

No adjustment of the decimal is needed. Both pressures are in kPa, so the answer is kPa.

8. $P_{GAGE} = 106 \text{ kPa}$

This is a subtraction problem. The precisions of the given pressures determine the precision of the result. Both are precise to the nearest thousand, as is the result.

SECTION 6.3 EXERCISES

Follow the eight-step procedure.

1. If a gage pressure reads 309 psi and atmospheric pressure measures 14.5 psi, find the absolute pressure.

2. If absolute pressure is 294.6 psi and gage pressure measures 280 psi, what is the atmospheric pressure?

6.4 ■ OHM'S LAW

In Example 6.3, a voltage source or battery, V_S, is connected to the circuit as shown in Figure 6-3. A complete path from the negative terminal of the battery to the positive terminal is formed. The battery or applied voltage measured in **volts (V)** pushes **current** measured in **amps (A)** through **resistance** measured in **ohms (Ω)**. The amount of current can be calculated using **Ohm's law**.

$$V_S = IR$$

where V_S is the source voltage measured in volts (V), I is the current in amps (A), and R is the resistance in ohms (Ω).

FIGURE 6-3
Example 6.3

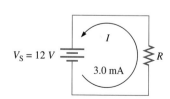

EXAMPLE 6.3 How much resistance will allow 3.0 mA to flow in a circuit if 12 V is applied?

Solution
1. Given: $I = 3.0$ mA and $V_S = 12$ V
2. Find: R See Figure 6-3.
3. $V_S = IR$ (Ohm's law)
4. $V_S/I = R$
   Tranpose I.
   $R = V_S/I$
5. $R = 12\ V/(3.0 \times 10^{-3}\ A)$
   Replace m with its value, 10^{-3}.
6. $R = 4 \times 10^3\ \Omega$
   This calculation can be done without reaching for the calculator. 10^{-3} in the denominator becomes 10^3 in the numerator.
7. $R = 4\ k\Omega$
   No adjustment of the decimal point is necessary. 10^3 is replaced with its symbol, k.
8. Round R to 4.0 kΩ.
   Significant digits: This is a division problem. The accuracy of the quotient is limited by the accuracy of V_S and I. V_S and I each have an accuracy of two. Round the answer to two significant digits.

SECTION 6.4 EXERCISES

Follow the eight-step procedure. Sketch a figure.

1. How much voltage must be applied to cause 12.6 mA to flow through an 8.2 kΩ resistor?
2. How much current will flow in a circuit that has 50 V applied to 4.7 MΩ?
3. How much resistance will allow 56 mA to flow through it when 110 volts is applied across the resistor?
4. How much current will flow when 13 V is applied across a resistor whose color code is yellow violet yellow?

6.5 ■ TORRICELLI'S THEOREM

In the study of fluid mechanics, **Torricelli's theorem** predicts the velocity at which a fluid in a reservoir will flow out of a nozzle. The equation is written for an "ideal fluid," but we don't worry about that definition.

EXAMPLE 6.4 Calculate the velocity (v) of flow of fluid from a nozzle in a tank if the nozzle is 10 meters (accuracy of 2) below the surface of the liquid (h).

Solution
1. Given: $h = 10$ m
   Start thinking SI units.
2. Find: v See Figure 6-4.

FIGURE 6-4
Example 6.4

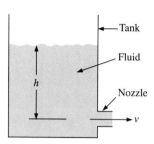

3. $v = (2gh)^{1/2}$ (Torricelli's Theorem)
   where v is the fluid velocity in meters per second, g is the acceleration due to gravity (9.81 m/sec²), and h is the height of fluid above the center of the nozzle.
4. $v = (2gh)^{1/2}$
   No rearrangement is necessary.
5. $v = (2 \cdot 9.81 \text{ m/sec}^2 \cdot 10 \text{ m})^{1/2}$
   Notice how the units work out.
   m/sec² · m = m²/sec². (m²/sec²)^{1/2} = m/sec. v is measured in m/sec.
6. $v = 14.007141$ m/sec
7. $v = 14.007141$ m/sec
   No adjustment of the decimal point is necessary.
8. $v = 14$ m/sec
   Significant digits: g has an accuracy of three, h of two. 2 is exact and does not limit the result. The radicand has an accuracy of two, so the square root has an accuracy of two. Round the result to two significant digits.

SECTION 6.5 EXERCISES

Follow the eight-step procedure.

1. Calculate the velocity of flow of a fluid from a nozzle 50 ft, accurate to the nearest foot, below the surface of the liquid.
2. Calculate the height of fluid in a tank that will cause the water to flow from a nozzle at 30 ft/sec (accuracy of two).

6.6 ■ TANK CIRCUIT

When a capacitor C is connected in parallel with an inductor L, the combo is called a **tank circuit**. See Figure 6-5.

FIGURE 6-5
Tank Circuit

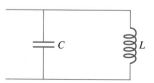

176 ■ CHAPTER 6 SOLVING PROBLEMS

The circuit has a natural frequency of oscillation, f, that is also called **resonant frequency**. The resonant frequency is calculated by the following equation.

$$f_r = 1/[2 \pi (LC)^{1/2}]$$

or

$$f_r = \frac{1}{2\pi\sqrt{LC}}$$

where f_r is the resonant frequency measured in hertz (Hz), L is inductance measured in henrys (H), and C is capacitance measured in farads (F). Note that only the product LC is raised to the 1/2 power, not the whole denominator.

EXAMPLE 6.5 Design a tank circuit that will oscillate at a frequency of 4.5 kH. Use an inductor of 15.2 µH and the appropriate capacitor.

FIGURE 6-6
Example 6.5

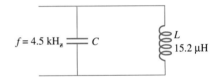

Solution
1. Given: $L = 15.2 \ \mu H$, $f_r = 4.5$ kHz
2. Find: C See Figure 6-6.
3. $f_r = 1/[2 \pi (LC)^{1/2}]$
4. $f_r^2 = 1/(4 \pi^2 LC)$
   C is removed from the square root radical by squaring both sides of the equation. Note that the 2 and the π are also squared in the process. This equation was solved in the last chapter using a different procedure.
   $Cf_r^2 = 1/(4 \pi^2 L)$
   Multiply both sides of the equation by C.
   $C = 1/(4 \pi^2 L f_r^2)$
   Transpose to isolate C.
5. $C = 1/[4 \pi^2 \ 15.2 \times 10^{-6} \ H \ (4.5 \times 10^3 \ Hz)^2]$
   µ is replaced with 10^{-6} and k is replaced with 10^3. π is not replaced with a numerical value because the π key on the calculator will be used.
6. $C = 8.2294659 \times 10^{-5}$ farad
   A calculator is in order. Here is a sequence that does not require the use of parentheses or memory. 4 × πx^2 × 15.2 EE 6 +/− × 4.5 EE 3 x^2 = 1/x
   Unitwise, what is 1 divided by a henry times a hertz squared? The answer is not at all obvious. We have to trust the "where" statement that says C will be in farads when L is in henrys and f is in hertz.
7. $C = 82.294659 \times 10^{-6}$ farad
   Shift the decimal one place to the right in engineering notation.

$C = 82.294659 \, \mu F$
Replace 10^{-6} with μ.
8. Round C to 82 μF.
Significant digits: This is a multiply/divide-type problem. Since your calculator has π stored to 8 or 10 digits, the result is limited by the accuracy of f and L. L has an accuracy of three and f and f^2 have an accuracy of two. Limit the result to two significant digits.

SECTION 6.6 EXERCISES

Follow the eight-step procedure.

1. Calculate the frequency of oscillation of a tank circuit that has an inductor of 36 mH and a capacitor of 0.1 μF.
2. Design a circuit (calculate L) that will oscillate at 500 kHz. Use a capacitor of 0.01 μF.
3. Design a circuit (calculate C) that will oscillate at 1.2 MHz. Use an inductor of 13.0 μH.

6.7 ■ COEFFICIENT OF FRICTION

Suppose a 100-pound object is sitting at rest on a horizontal, flat, smooth surface. If it takes 10 pounds of force to start the object sliding, then the frictional force is 10 pounds and the ratio between frictional force and weight is 10/100 or 0.1. This ratio is called the **coefficient of friction**. So, the coefficient of friction is a dimensionless number that ranges from 0 to 1. In the following example, the coefficient of friction between the tires of an automobile and the highway is calculated.

EXAMPLE 6.6   A car traveling 55 mph comes to a stop in 200 ft (nearest foot). Calculate the coefficient of friction (f) between the tires and the roadway.

Solution
1. Given: d (braking distance) = 200 ft, v (velocity) = 55 mph
2. Find: f
3. $d = v^2/(2fg)$
   where d = braking distance in feet, v = velocity of car in ft/sec, f = coefficient of friction between the tires and the roadway, and g = acceleration due to gravity (32.2 ft/sec²).
4. $d \cdot f = v^2 / (2g)$
   Transpose f to get the unknown out of the denominator.
   $f = v^2/(2dg)$
   Transpose d to isolate f.

Let's do a unit check to determine the units of f.

$$(ft/sec)^2/(ft \cdot ft/sec^2) = (ft^2/sec^2)/(ft^2/sec^2) = 1$$

All units cancel. f is dimensionless.

5. The units miles per hour are not consistent with feet per second squared (g) and feet (d). Convert mph to ft/sec.

$$\frac{55 \text{mi}}{\text{hr}} \cdot \frac{5280 \text{ ft}}{1 \text{ mi}} \cdot \frac{1 \text{ hr}}{3600 \text{ sec}} = 80.666667 \text{ ft per sec}$$

55 mph = 81 ft/sec.

Significant digits: 55 has an accuracy of two; the result has an accuracy of two also. Round to 81 ft/sec.

$f = v^2/(2dg)$

$f = (81 \text{ ft/sec})^2/(2 \cdot 200 \text{ ft} \cdot 32.2 \text{ ft/sec}^2)$

6. $f = 0.5093944$

Here is a calculator sequence that does not use parentheses or memory:
81 x^2 / 2 / 200 / 32.2 =

7. $f = 0.5093944$

No adjustment of decimal necessary.

8. $f = 0.51$

81 has an accuracy of two. 2 is exact. 200 and 32.2 each have an accuracy of three. Round the result to two places.

SECTION 6.7 EXERCISES

Follow the eight-step procedure.

1. For a coefficient of friction of 0.4, calculate the distance required to stop when traveling 30 m/sec.

2. For a coefficient of friction of 0.39, how fast is a car traveling to stop in 100 m?

6.8 ■ TWO RESISTORS IN PARALLEL

When two or more resistors or other components are connected together at each end, they are connected in **parallel**. Figure 6-7 shows two resistors connected in parallel. Resistors connected in parallel behave differently than resistors connected in series, and their total or equivalent cannot be found by simply adding them. In a parallel circuit, the total or equivalent resistance of two resistors (only two) can be found by dividing their product by their sum.

FIGURE 6-7
Resistors in Parallel

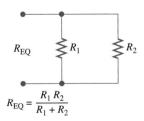

$$R_{EQ} = \frac{R_1 R_2}{R_1 + R_2}$$

6.8 TWO RESISTORS IN PARALLEL

$$R_{EQ} = R_1 R_2/(R_1 + R_2)$$

where R_1 and R_2 are connected in parallel and R_{EQ} is their equivalent resistance, all measured in Ω.

EXAMPLE 6.7 How much resistance should be placed in parallel with 12.0 kΩ to produce an equivalent resistance of 4.0 kΩ?

Solution
1. Given: $R_1 = 12.0$ kΩ, and $R_{EQ} = 4.0$ kΩ
2. Find: R_2 See Figure 6-8.
3. $R_{EQ} = R_1 R_2/(R_1 + R_2)$
4. $R_{EQ}(R_1 + R_2) = R_1 R_2$

   R_2 is in both the numerator and denominator. Eliminate the fraction by multiplying each side by $(R_1 + R_2)$.

   $R_{EQ}R_1 + R_{EQ}R_2 = R_1 R_2$

   Clear the parentheses.

   $R_{EQ}R_1 = R_1 R_2 - R_{EQ}R_2$

   Transpose $R_{EQ}R_2$ to get all terms containing R_2 on the same side and all other terms on the other.

   $R_{EQ}R_1 = R_2(R_1 - R_{EQ})$

   R_2 is a common factor on the right. Factor it out.

   $R_2 = R_{EQ}R_1/(R_1 - R_{EQ})$

   Divide both sides by $(R_1 - R_{EQ})$ to isolate R_2. Swap sides.

5. $R_2 = 12.0 \times 10^3 \, \Omega \cdot 4.0 \times 10^3 \, \Omega/(12.0 \times 10^3 \, \Omega - 4.0 \times 10^3 \, \Omega)$

   Replace k with 10^3.

6. $R_2 = 48 \times 10^6 \, \Omega^2/(8.0 \times 10^3 \, \Omega)$
   $R_2 = 6 \times 10^3 \, \Omega$

   There is no need for a calculator here.

7. $R_2 = 6$ kΩ
8. Round R_2 to 6.0 kΩ.

   Significant digits: The subtraction step yields 8.0 kΩ, precise to the nearest 100 Ω, so 8.0 kΩ has an accuracy of two. 48×10^6 also has an accuracy of two. The final result is limited to an accuracy of two.

FIGURE 6-8
Example 6.7

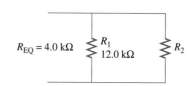

SECTION 6.8 EXERCISES

Follow the eight-step procedure.

1. 1.5 MΩ is placed in parallel with 1.5 MΩ. Calculate the equivalent resistance.
2. How much resistance must be connected in parallel with 100 kΩ to produce an equivalent resistance of 33 kΩ?
3. Two 10 MΩ resistors are connected in parallel. Calculate the equivalent resistance.
4. Two 120 Ω resistors are connected in parallel. Calculate the equivalent resistance.
5. From the results of Problems 3 and 4, develop a rule for the equivalent resistance of two equal resistors connected in parallel.

6.9 ■ TWO OR MORE RESISTORS IN PARALLEL

The product over the sum equation used in the last example only holds for two resistors. The following equation holds for any number of resistors. The ... in the equation indicates continuation (and so on) for any number of resistors in parallel.

$$R_{EQ} = \frac{1}{\frac{1}{R_1} + \frac{1}{R_2} + \frac{1}{R_3} + \ldots}$$

where R_{EQ} is the equivalent resistance, and R_1, R_2, and R_3 and so on are resistors connected in parallel, all measured in Ω.

EXAMPLE 6.8 A 56 MΩ resistor is connected in parallel with a 68 MΩ resistor and a 82 MΩ resistor. How much resistance should be connected in parallel with these three resistors to produce an equivalent resistance of 17 MΩ?

Solution

1. Given: $R_1 = 56$ MΩ, $R_2 = 68$ MΩ, $R_3 = 82$ MΩ, and $R_{EQ} = 17$ MΩ
2. Find: R_4 See Figure 6-9.
3. $R_{EQ} = \dfrac{1}{\dfrac{1}{R_1} + \dfrac{1}{R_2} + \dfrac{1}{R_3} + \dfrac{1}{R_4}}$
4. $\dfrac{1}{R_{EQ}} = \dfrac{1}{R_1} + \dfrac{1}{R_2} + \dfrac{1}{R_3} + \dfrac{1}{R_4}$

   Invert both sides.

   $\dfrac{1}{R_4} = \dfrac{1}{R_{EQ}} - \dfrac{1}{R_1} - \dfrac{1}{R_2} - \dfrac{1}{R_3}$

   Transpose $1/R_1$, $1/R_2$, $1/R_3$

   $R_4 = \dfrac{1}{\dfrac{1}{R_{EQ}} - \dfrac{1}{R_1} - \dfrac{1}{R_2} - \dfrac{1}{R_3}}$

   Invert each side.

5. $R_4 = \dfrac{1}{\dfrac{1}{17} - \dfrac{1}{56} - \dfrac{1}{68} - \dfrac{1}{82}}$

Shortcut: Since all resistors are given in MΩ, omit the M in the calculation and insert an M in the result.

6. and 7. $R_4 = 71.096539$ MΩ

8. Round the result to 71 MΩ.

FIGURE 6-9
Example 6.8

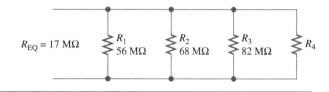

SECTION 6.9 EXERCISES

Use the eight-step procedure.

1. It is suspected that a 120 kΩ resistor in a circuit has increased in value to 980 kΩ. How much resistance should be connected in parallel to return the circuit to normal operation?
2. If six 100 Ω precision resistors (+/− 1%) are connected in parallel, what is their equivalent resistance?
3. Use your knowledge of algebra to rearrange the equation in this section for two resistors in parallel into the "product over the sum" format used in Section 6.4.

6.10 ■ SUPERELEVATION

To decrease the tendency to slide when going around a curve, the roadway is banked (outside edge raised). The amount of bank is called **superelevation** (e), and is measured in the amount of rise per foot of width of roadway. The following example relates the superelevation, the coefficient of friction between the tires of an automobile and the roadway, the velocity of the automobile, and the radius of curvature of a curve in a roadway.

EXAMPLE 6.9 In this problem, the curve is to be limited to a superelevation (e) of 0.06 feet per foot width and a maximum coefficient of friction (f) of 0.14. Calculate the minimum safe radius of curvature (R) for a vehicle traveling 50 mph.

Solution
1. Given: $e = 0.06$ ft per ft, $f = 0.14$, and $v = 50$ mph
2. Find: R
3. $e + f = v^2 / (gR)$
   where e is superelevation in ft per ft, f is coefficient of side friction, g is acceleration due to gravity (32.2 ft per sec²), R is the radius of curve in ft, and v is the velocity of the vehicle in ft/sec.

4. $R(e + f) = v^2 / g$
   Remove the unknown R from the denominator by multiplying both sides of the equation by R.

   $R = v^2 / [g(e+f)]$

   To isolate R, transpose $(e + f)$.

5. mph is not compatible with the other units. Convert to ft/sec.

   $$\frac{50 \text{ mi}}{\text{hr}} \cdot \frac{5280 \text{ ft}}{1 \text{ mi}} \cdot \frac{1 \text{ hr}}{3600 \text{ sec}} = 73.3333333 \text{ ft/sec}$$

   50 mph = 73 ft/sec
   $R = v^2 / [g(e + f)]$
   $R = (73 \text{ ft/sec})^2 / [32.2 \text{ ft/sec}^2 (0.06+0.14)]$

6. $R = 827.48447$ ft
   Unit check!
   (ft per sec)^2 / ft per sec^2 = ft^2 per sec^2 / ft per sec^2 = ft
   e and f are both dimensionless.

7. $R = 827.48447$ ft.
   No adjustment of decimal is needed.

8. $R = 830$ ft.
   Significant digits: (0.06 + 0.14) is precise to the hundredths place, so 0.20 has two significant digits, as does 73. Limit the result to two significant digits.

SECTION 6.10 EXERCISES

Follow the eight-step procedure.

1. Find the maximum safe velocity for a 1910-ft radius curve in a highway whose superelevation is 0.08. Assume a coefficient of friction of 0.10.

2. To safely round a 231-ft radius curve at 30 mph with a side friction of 0.16, what superelevation is required?

6.11 ■ MASS AND WEIGHT

A fundamental equation in physics says that force equals mass times acceleration. A more specific form of the equation says that the weight of an object equals its mass times the acceleration due to gravity. Remember that in US units, mass is measured in slugs and weight is measured in pounds. In SI units, mass is measured in kilograms and weight is measured in newtons. Also, the acceleration due to gravity (g) is 32.2 ft/sec^2 in US units and 9.81 m/sec^2 in SI units.

EXAMPLE 6.10 If an object weighs 175 lbs on Earth, calculate its mass.

Solution
1. Given: $F = 175$ lbs
2. Find: m

3. $F = mg$
     where F is weight in lbs, m is mass in slugs, and g is acceleration due to gravity
  4. $m = F/g$
     Multiply both sides by $1/g$ to isolate m.
  5. $m = 175 \text{ lb} / 32.2 \text{ ft/sec}^2$ or $175/32.2 \text{ lb·sec}^2/\text{ft}$.
     Pounds is a US unit, so 32.2 ft/sec^2 is used for g.
     lb sec^2/ft is a slug.
  6. $m = 5.434782609$ slugs
  7. No adjustment of decimal.
  8. $m = 5.43$ slugs
     Significant digits: 32.2 and 175 limit the answer to an accuracy of three. If this object travels to the moon, its weight will decrease by a factor of 6. Its mass will remain the same.

SECTION 6.11 EXERCISES

Follow the eight-step procedure.

1. Calculate the weight of an object whose mass is 56 kg.
2. Calculate the mass of an object whose weight is 111 newtons.

6.12 ■ °FAHRENHEIT, °CELSIUS

The standard unit for temperature in US units is the **degree Fahrenheit** (°F). On the Fahrenheit scale water freezes at 32°, and water boils at 212°. The standard unit for temperature in SI units is the **degree Celsius** (°C), and *Kelvin* (*K*). On the Celsius scale, water freezes at 0° and water boils at 100°. The equation for converting Fahrenheit to Celsius and vice versa is developed by equating the following quantity in each temperature scale: the temperature minus the freezing point of water divided by the difference between the boiling point and freezing point of water.

$$(°F - 32°)/(212° - 32°) = (°C - 0)/(100° - 0)$$
$$(°F - 32)/180 = °C/100$$

The denominators can be reduced by multiplying each side by 20.

$$(°F - 32)/9 = °C/5$$

EXAMPLE 6.11 Convert 78°F to °C.

Solution
1. Given: 78°F
2. Find: °C
3. $(°F-32)/9 = °C/5$
4. $5(°F-32)/9 = °C$
   Multiply both sides by 5 to isolate °C.
   $°C = 5(°F-32)/9$

5. °C = 5(78° − 32°)/9
6. °C = 25.55555556°
7. No adjustment of decimal.
8. °C = 26°
   Significant digits: 5 and 9 are exact. 78 has an accuracy of two. Round the result to two digits.

SECTION 6.12 EXERCISES

1. Convert −85°F to °C.
2. Convert 211°C to °F.
3. At what temperature are °F and °C equal?

6.13 ■ AIRPLANE STALL SPEED

Air flowing over the wings of an airplane creates lift. The velocity or the plane at which insufficient lift is created to offset the weight of the plane is called the **stall speed** of the plane. If the stall speed of a plane is known for a given gross weight, it can be calculated for any other gross weight by the following equation.

$$V_{S2} / V_{S1} = (W_2 / W_1)^{1/2}$$

where V_{S1} is the stall speed at gross weight W_1 and V_{S2} is the stall speed at gross weight W_2.

EXAMPLE 6.12 If the stall speed of an airplane is 105 knots at 11 000 lb (to nearest 100 lb), what is its stall speed at 14 400 lb?

Solution
1. Given: V_{S1} = 105 knots, W_1 = 11 000 lb, and W_2 = 14 400 lb
2. Find: V_{S2}
3. $V_{S2}/V_{S1} = (W_2/W_1)^{1/2}$
4. $V_{S2} = V_{S1}(W_2/W_1)^{1/2}$
5. V_{S2} = 105 kt (14 400 lb / 11 000 lb)$^{1/2}$
6. V_{S2} = 120.1 knots
7. No decimal point adjustment is needed.
8. V_{S2} = 120 knots

SECTION 6.13 EXERCISES

1. A plane whose gross weight is 12 200 lb stalls at 123 knots. Calculate its gross weight at a stall speed of 105 knots.
2. A plane whose gross weight is 10 200 lb stalls at 105 knots. What is its stall speed at 12,200 lb?

6.14 ■ PRESSURE EXERTED BY A COLUMN OF LIQUID

The density of a liquid is its mass per unit volume.

$$\rho \text{ (kg/m}^3\text{)}$$

If the height of a column of liquid is multiplied by the density of the liquid, the product is mass per area.

$$h \cdot \rho \text{ (m} \cdot \text{kg/m}^3 \text{ or kg/m}^2\text{)}$$

Mass can be converted to weight by multiplying by the acceleration due to gravity. The result is force or weight per area, which is pressure.

$$h \cdot \rho \cdot g \text{ (kg/m}^2 \cdot \text{m/sec}^2 \text{ or newton/m}^2 \text{ or pascal)}$$

In other words, the pressure exerted by a column of a liquid can be calculated by multiplying the height of the column of liquid by the density of the liquid by the acceleration due to gravity.

$$P = h \cdot \rho \cdot g$$

where P is the pressure caused by a column of liquid, ρ is the density of the liquid (mass per volume) and g is the acceleration due to gravity.

EXAMPLE 6.13 Calculate the pressure in pascal exerted by a column of mercury 1.0 m high.

Solution
1. Given: $h = 1.0$ m, density of mercury $= \rho = 13.60$ g/cm^3
2. Find: P
3. $P = h \cdot \rho \cdot g$
4. $P = 1\text{m} \cdot 13.60\text{g/cm}^3 \cdot 9.81\text{m/sec}^2$

The cm and m do not match. Convert cm^3 to m^3.

$$P = 1 \text{ m} \cdot \frac{13.60 \text{ g}}{\text{cm}^3} \cdot \frac{100^3 \text{ cm}^3}{1 \text{ m}^3} \cdot \frac{9.81 \text{ m}}{\text{sec}^2}$$

5. $P = 133\ 416\ 000$ g m^2/(m^3 sec^2)

From the equation $F = ma$, we find that force in newtons is mass in kilograms · acceleration in meters/sec^2. In other words, a newton is a kg m/sec^2. Our preliminary result, 133 416 000 g m^2/(m^3 sec^2) can be rewritten 133 416 kg m^2/(m^3 sec^2) or 133 416 kg m/sec^2 · m/m^3 or 133 416 kg m/sec^2 · 1/m^2 or 133 416 newton/m^2 or 133 416 pascal.

6. $P = 133\ 416$ Pa
7. $P = 133.416$ kPa
8. $P = 130$ kPa

Significant digits: The height of 1.0 m limits the result to two significant digits.

SECTION 6.14 EXERCISES

1. Calculate the pressure in millibars equal to 760 mm Hg.
2. Calculate standard pressure in inches of mercury.

6.15 ■ WAVELENGTH, PERIOD, FREQUENCY, AND VELOCITY OF ELECTROMAGNETIC WAVES

The following discussion about electromagnetic waves provides information needed to work the last four examples of this chapter. Electromagnetic waves include gamma rays, X rays, ultraviolet, visible light, infrared, radar, microwave, TV, and radio communication waves. These signals are classified according to their frequency or wavelength. These terms are defined and discussed here.

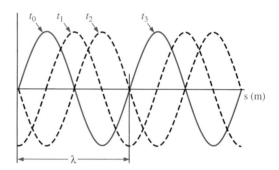

FIGURE 6-10
Electromagnetic Wave

Figure 6-10 shows a wave at time t_0 traveling horizontally to the right. At times t_1, t_2, and t_3, the wave has advanced to the right to the positions shown. Note that the horizontal axis is labeled "s" for distance. The t_3 wave is the same as the original t_0 wave. The distance a wave has traveled going from position t_0 to position t_3 is called the **wavelength.** The symbol for the wavelength is the Greek letter lambda (λ). The time it takes for the wave to move from position t_0 to position t_3 is called the **period (T).** From any point on the waveform in position t_0 to the corresponding point on the waveform in position t_3 is called a **cycle.** The number of cycles that occur each second is called the **frequency (f)** of the waveform. Frequency is measured in **cycles per second,** which is given the special name **hertz.** If frequency is measured in cycles per second and period is measured in seconds per cycle, they must be the reciprocals of each other.

$$T = 1/f \quad \text{or} \quad f = 1/T$$

where T is the period of the waveform measured in seconds, and f is the frequency of the waveform measured in hertz.

Light travels through a vacuum with a velocity of 3.0×10^8 meters each second or 300 Mm/sec. This constant is given the symbol **c.** The distance equation is

$$d = rt$$

where d is distance that can be traveled at rate r in time t.

6.15 WAVELENGTH, PERIOD, FREQUENCY, AND VELOCITY OF ELECTROMAGNETIC WAVES

For light, this equation becomes

$$\lambda = cT$$

where λ is the wavelength of a light wave, c is the velocity of light in a vacuum, and t is the period of the waveform.

The units of measure were not mentioned in the last "where" statement. The units of c must be consistent with the units of λ and T. For example, if λ is in meters and T is in seconds, c must be expressed in meters per second.

Since $T = 1/f$, $\lambda = cT$ can be rewritten as

$$\lambda = c/f$$

where λ is the wavelength of a light wave, c is the velocity of light in a vacuum, and f is the frequency of the waveform measured in hertz.

As light travels through any medium other than a vacuum (free space), it slows down. Its velocity in any other medium is denoted by v.

$$v = c/n$$

where n is the **refractive index** or **index of refraction** of the medium that light is passing through.

Since the velocity of light through any medium other than a vacuum is less than c, n for any other medium must always be greater than one. Here are some indices of refraction of a few materials.

Material	Refractive Index	Velocity of Light Traveling Through the Material
	(n)	(m/s)
Water	1.33	226×10^6
Glass	1.46 – 1.96	$205 – 151 \times 10^6$
Diamond	2.42	123.97×10^6

EXAMPLE 6.14 A green light wave has a frequency of 550 THz. Calculate its period.

Solution
1. Given: $f = 550$ THz
2. Find: T
3. $T = 1/f$
4. No rearrangement is necessary.
5. $T = 1 (550 \times 10^{12}$ cycles per sec)
6. $T = 0.181818 \times 10^{-14}$ sec per cycle
7. $T = 1.81818 \times 10^{-15}$ sec or 1.81818 fsec (femto seconds)
8. Round T to 1.82 fsec.

EXAMPLE 6.15 A green light wave has a frequency of 550 THz. Calculate its wavelength in free space.

Solution
1. Given: $f = 550$ THz and $c = 300$ Mm/sec
2. Find: λ
3. $\lambda = c/f$
4. No rearrangement is necessary.
5. $\lambda = 300 \times 10^6$ m per sec/$(550 \times 10^{12}$ cycles per sec)
6. $\lambda = 0.545 \times 10^{-6}$ m per cycle
   units: m/sec / cycles/sec = m/sec · sec/cycle = m/cycle
7. $\lambda = 545$ nm
8. Round λ to 545 nm.

EXAMPLE 6.16 Calculate the velocity of green light, $f = 550$ THz as it passes through glass whose index of refraction is 1.65.

Solution
1. Given: $f = 550$ THz and $c = 300 \times 10^6$ m/sec
2. Find: v
3. $v = c/n$
4. No rearrangement is necessary.
5. $v = 300 \times 10^6$ m/sec / 1.65
6. $v = 181\ 818\ 181$ m/sec
7. $v = 181.\ 818\ 181 \times 10^6$ m/sec or 181.818 181 Mm/sec
8. Round v to 180 Mm/sec.
   Significant digits: c has an accuracy of two as does the result.

EXAMPLE 6.17 Calculate the wavelength of the same green light as it travels through glass whose index of refraction is 1.65. (The frequency of the light remains the same: the wavelength and velocity change.)

Solution
1. Given: $f = 550$ THz and $v = 180$ Mm/sec (from last example)
2. Find: λ
3. and 4. $\lambda = v/f$
5. $\lambda = 180 \times 10^6$ m/sec / 550×10^{12} cycles/sec
   Note that the units for frequency were entered as cycles/sec instead of hertz to show how the units work together. (m/sec)/(cycles/sec) = m/sec · sec/cycle = m/cycle. The result will tell us how many meters the wave travels during each cycle.
6. $\lambda = 3.272\ 727 \times 10^{-7}$ m
7. $\lambda = 327.272\ 7 \times 10^{-9}$ m
8. Round λ to 330 nm.

SECTION 6.15 EXERCISES

If a beam of red light has a wavelength of 700 nm in a vacuum,

1. calculate its frequency.
2. calculate its period.
3. calculate its velocity through water whose index of refraction is 1.33.
4. calculate its wavelength through water.

■ SUMMARY

An organized approach to problem solving is provided by these eight steps.

1. Identify and list the given information.
2. Identify and list the unknown variable. Sketch a figure if applicable. Identify the known and unknown variables in the sketch. Be sure to use the same notation in the sketch that is used in the "Given" and "Find" statements.
3. Find the equation that relates the known and unknown variables.
4. Rearrange the equation to solve for the unknown.
5. Plug in the known values. Include units. If they do not agree, perform necessary conversions or reductions.
6. Use your calculator to find the preliminary answer.
7. Adjust the decimal point so that a metric prefix is produced, if necessary.
8. Round the answer to the proper number of significant digits.

■ SOLUTIONS TO CHAPTER 6 SECTION EXERCISES

Solutions to Section 6.2 Exercises

1. 10.1 MΩ   2. 247 kΩ   3. 19.3 kΩ   4. red black red

Solutions to Section 6.3 Exercises

1. 324 psi   2. 15 psi

Solutions to Section 6.4 Exercises

1. 100 V   2. 11 µA   3. 2.0 Ω   4. 28 µA

Solutions to Section 6.5 Exercises

1. 57 ft/sec   2. 14 ft

Solutions to Section 6.6 Exercises

1. 3000 Hz   2. 10 µH   3. 0.0014 µF

Solutions to Section 6.7 Exercises

1. 100 m   2. 28 m/sec

Solutions to Section 6.8 Exercises

1. 750 kΩ
2. 49 kΩ
3. 5.0 MΩ
4. 60 Ω
5. When two resistors of equal resistance are connected in parallel, the equivalent resistance is half the resistance of one of the resistors.

Solutions to Section 6.9 Exercises

1. 140 kΩ
2. 17 Ω
3. $\dfrac{R_1 R_2}{R_1 + R_2}$

Solutions to Section 6.10 Exercises

1. 110 ft/sec
2. 0.10 ft per ft

Solutions to Section 6.11 Exercises

1. 550 newtons
2. 11.3 kg

Solutions to Section 6.12 Exercises

1. −65.0°C
2. 412°F
3. −40°

Solutions to Section 6.13 Exercises

1. 8890 lb
2. 115 knots

Solutions to Section 6.14 Exercises

1. 1010 mb
2. 29.9 in-Hg

Solutions to Section 6.15 Exercises

1. 430 THz
2. 2.33 fsec
3. 230 Mm/sec
4. 530 nm

■ CHAPTER 6 PROBLEMS

All of the following problems can be solved using the equations presented in this chapter. Use the eight-step approach.

1. How much resistance should be added in series with 56 Ω, 39 Ω, and 22 Ω to produce a total of 275 Ω?

2. Three resistors 4.7 MΩ, 9.1 MΩ, and 1.0 MΩ are connected in series. Find the total resistance.

3. If a total resistance of 94 kΩ is needed, how much resistance should be added in series to 27 kΩ and 43 kΩ resistors?

4. The following resistors are connected in series: red yellow brown, brown blue black, green blue brown, grey red black. Calculate the total resistance.

5. White brown red, orange orange orange, and yellow orange red resistors are connected in series. What is the color code of the resistor that should be connected in series with these resistors to make a total of 50 kΩ ?

6. If a gauge pressure reads 483 psi and atmospheric pressure measures 14.7 psi, find the absolute pressure.

7. If absolute pressure is 284 kPa and gauge pressure measures 183 kPa, what is the atmospheric pressure?

8. How much resistance will allow 17.0 μA to flow in a circuit if 25.0 V are applied?

9. How much voltage must be applied to cause 56.6 mA to flow through a 7.1 MΩ resistor?

10. How much current will flow in a circuit that has 45 mV applied to 100 MΩ?

11. How much resistance will allow 5.6 A to flow when 11.0 volts is applied across the resistor?

12. How much current will flow when 25.5 V is applied across a resistor whose color code is orange orange red?

13. Calculate the velocity of flow of a liquid from a nozzle 36 ft below the surface of the liquid.

14. Calculate the height of liquid in a tank that will cause the liquid to flow from a nozzle at 30 mps.

15. Design a tank circuit that will oscillate at a frequency of 98.0 kH. Use an inductor of 15.2 mH and the appropriate capacitor.

16. Calculate the frequency of oscillation of a tank circuit that has an inductor of 2.7 mH and a capacitor of 0.022 μF.

17. Design a circuit (calculate L) that will oscillate at 750 kHz. Use a capacitor of 0.047 μF.

18. Design a circuit (calculate C) that will oscillate at 0.75 MHz. Use an inductor of 6.8 μH.

19. For a coefficient of friction of 0.40, calculate the distance required to stop when traveling 30 mph. (Watch your units!)

20. For a coefficient of friction of 0.39, how fast is a car traveling to stop in 70 m?

21. How much resistance should be placed in parallel with 56.0 kΩ to produce an equivalent resistance of 18.0 kΩ?

22. A brown black red resistor is placed in parallel with a yellow violet red resistor. Calculate the equivalent resistance.

23. How much resistance must be connected in parallel with a brown grey green resistor to produce an equivalent resistance of 1.2 MΩ?

24. Two 75 MΩ resistors are connected in parallel. Calculate the equivalent resistance.

25. Two 19.1 MΩ resistors are connected in parallel. Calculate the equivalent resistance.
26. A 39 kΩ resistor is connected in parallel with a 62 kΩ resistor and a 43 kΩ resistor. How much resistance should be connected in parallel with these three resistors to produce an equivalent resistance of 7 kΩ?
27. It is suspected that a 470 Ω resistor in a circuit has increased in value to 1.2 kΩ. How much resistance should be connected in parallel to return the circuit to normal operation?
28. If eight 41.2 Ω precision resistors (+/− 1%) are connected in parallel, what is their equivalent resistance?
29. Find the maximum safe velocity for a 500-ft radius curve in a highway whose superelevation is 0.1. Assume a coefficient of friction of 0.10.
30. To safely round a 400-ft radius curve at 50 mph with a side friction of 0.16, what superelevation is required?
31. Calculate the weight of an object whose mass is 835 kg.
32. Calculate the mass of an object whose weight is 1566 newtons.
33. Convert 212 °F to °C.
34. Convert 212 °C to °F.
35. A plane whose gross weight is 1.34×10^4 lb stalls at 145 knots. Calculate its gross weight at a stall speed of 115 knots.
36. A plane whose gross weight is 1.02×10^4 lb stalls at 105 knots. What is its stall speed with an additional 185 lb passenger?
37. Calculate the pressure in millibars equal to 315 mm-Hg.
38. Calculate the pressure in inches of mercury equal to 315 mm-Hg.
39. A wave in the infrared region has a wavelength of 770 nm (nanometers). Calculate its frequency.
40. Calculate the period of the wave in Problem 39.
41. Calculate the velocity of the infrared wave in Problem 39 as it passes through glass whose index of refraction is 1.59.
42. Calculate the wavelength of the same infrared wave (Problem 39) as it travels through glass whose index of refraction is 1.59. (The frequency of the light remains the same; the wavelength and velocity change.)

If a beam of blue light has a wavelength of 475 nm in a vacuum,

43. calculate its frequency.
44. calculate its period.
45. calculate its velocity through water whose index of refraction is 1.33.
46. calculate its wavelength as it travels through water.

7
Geometry

CHAPTER OUTLINE

- **7.1** Angles
- **7.2** Triangles
- **7.3** Pythagorean Theorem
- **7.4** Area of Triangles
- **7.5** Quadrilaterals
- **7.6** Area of Quadrilaterals
- **7.7** Circles
- **7.8** Degrees, Minutes, and Seconds
- **7.9** Radian Measure
- **7.10** Area of a Sector
- **7.11** Arc Length
- **7.12** Angular Velocity
- **7.13** Angular Displacement
- **7.14** Linear Velocity
- **7.15** Volume

OBJECTIVES

After completing this chapter, you should be able to

1. classify angles according to measure of angle (acute, obtuse, right).
2. classify angles according to length of side (scalene, isosceles, equilateral).
3. apply the Pythagorean theorem.
4. calculate the area of triangles.
5. calculate the area of quadrilaterals.
6. calculate the area of circles.
7. calculate the area of complex shapes.
8. calculate the circumference of a circle.
9. convert degrees to degrees, minutes, seconds, and vice versa.
10. convert from degrees to radians and vice versa.
11. calculate arc length, s.
12. calculate angular velocity, ω (omega).
13. calculate angular displacement, θ.
14. calculate linear velocity, v.
15. calculate volume.

NEW TERMS TO WATCH FOR

vertex	trapezoid
theta (θ)	radius
alpha (α)	diameter
beta (β)	central angle
phi (ϕ)	sector
acute	arc
obtuse	secant
right	chord
triangle	circumference
right triangle	degrees, minutes, and seconds
acute triangle	radian
obtuse triangle	arc length
scalene	angular velocity (ω)
isosceles	angular displacement
equilateral	linear velocity (v)
complementary angles	simple gear train
hypotenuse	compound gear train
Pythagorean theorem	tangent
quadrilateral	transversal
parallelogram	alternate interior angles
rhombus	corresponding angles
rectangle	vertical angles
square	

7.1 ■ ANGLES

An angle is formed by two straight lines called **sides** that intersect at a point called the **vertex**. The angle can be identified by several methods. In Figure 7-1A, the angle is identified by a single letter placed at the vertex. The symbol $\angle$ is sometimes used for angle, $\angle$A. Angles are also identified by a Greek letter written on the interior of the angle. See Figure 7-1B. Common choices are **theta (θ), alpha (α), beta (β), and phi (ϕ)**. If the angle is part of a more complex figure, it can be singled out by a sequence of three letters with the middle letter being the vertex of the angle in question. See Figure 7-1C.

Angles are classified according to their measure. Angles that measure less than 90° are called **acute**. See Figure 7-2. Angles that measure more than 90° are called **obtuse**. Angles that measure 90° are called **right** angles. Right angles are often marked with a small square.

7.1 ANGLES

FIGURE 7-1
Identifying Angles

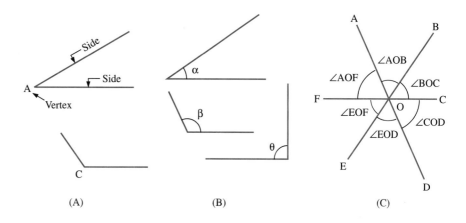

FIGURE 7-2
Classifying Angles

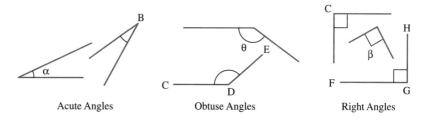

SECTION 7.1 EXERCISES

1. Which angles in Figure 7-3 are acute?
2. Which angles in Figure 7-3 are obtuse?
3. Which angles in Figure 7-3 are right?

FIGURE 7-3
Section 7.1 Exercises, Problems 1–3

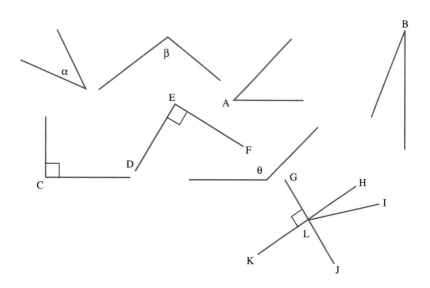

7.2 ■ TRIANGLES

Any three-sided figure is called a **triangle**. A point at which any two sides of the triangle intersect is called a **vertex**. Often, each vertex is labeled with an uppercase letter. Then the triangle is referred to by the three letters, such as triangle ABC or PQR. See Figure 7-4.

Triangles are classified according to their three interior angles. If one of the angles of the triangle is a 90° angle, the triangle is a **right triangle**. If all three angles are less than 90°, the triangle is an **acute triangle**. If one of the angles is greater than 90°, the triangle is an **obtuse triangle**. See Figure 7-4.

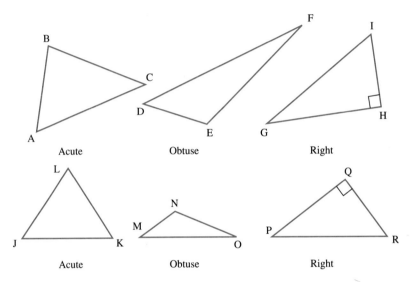

FIGURE 7-4
Classifying Triangles by Angles

Triangles are also classified according to the lengths of their sides. If all three sides are different lengths, the triangle is **scalene**. If two sides are the same in length, the triangle is **isosceles**. If all three sides are the same length, the triangle is **equilateral**. See Figure 7-5.

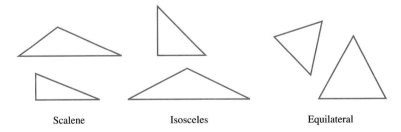

FIGURE 7-5
Classifying Triangles by Length of Sides

The sum of the angles in any triangle is 180°. See Figure 7-6.

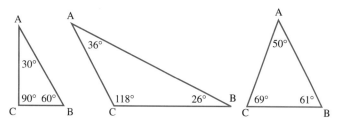

FIGURE 7-6
∠ A + ∠ B + ∠ C = 180°

In a right triangle, the right angle is 90°; that leaves 90° for the other two angles to share. Two angles whose measures add to 90° are called **complementary angles.**

SECTION 7.2 EXERCISES

1. Label each triangle in Figure 7-7 as right, acute, or obtuse and as scalene, isosceles, or equilateral.
2. Can a right triangle also contain an obtuse angle? Explain.
3. Draw a scalene right triangle.
4. Draw an isosceles right triangle.
5. Draw an equilateral right triangle.
6. Draw an acute scalene triangle.
7. Draw an acute isosceles triangle.
8. Draw an obtuse scalene triangle.
9. Draw an obtuse isosceles triangle.

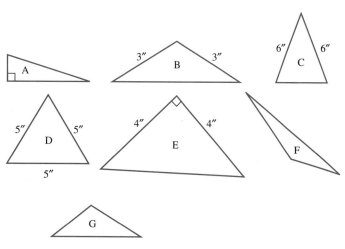

FIGURE 7-7
Section 7.2 Exercises

7.3 ■ PYTHAGOREAN THEOREM

Figure 7-8 shows right triangle ABC. The right angle C is marked with a small square. The side opposite the right angle is called the **hypotenuse** and is labeled c. The other two sides are labeled a and b. Note that the side opposite $\angle A$ is labeled a; and the side opposite $\angle B$ is labeled b. The **Pythagorean theorem** states that in a right triangle, the hypotenuse squared is equal to the sum of the squares of the other two sides.

$$c^2 = a^2 + b^2$$

where c is the hypotenuse of a right triangle, and a and b are the other two sides.

If you do not already know this theorem, learn it right now. Given any two sides of a right triangle, the Pythagorean theorem can be used to find the remaining side.

FIGURE 7-8
Pythagorean Theorem

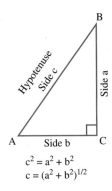

$c^2 = a^2 + b^2$
$c = (a^2 + b^2)^{1/2}$

EXAMPLE 7.1   If the two sides of a right triangle are 5.0 m and 8.0 m, find the hypotenuse.

Solution
1. **Given:** right triangle ABC, a=5.0 m, b=8.0 m
2. **Find:** c See Figure 7-9.

FIGURE 7-9
Example 7.1

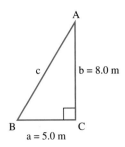

3. $c^2 = a^2 + b^2$
4. $c = (a^2 + b^2)^{1/2}$

Take the square root of both sides.

5. c = ((5.0 m)² + (8.0 m)²)^(1/2)

Unit check! (m² + m²)^(1/2) = (m²)^(1/2) = m
a and b were given in meters, the result c is in meters.

6. c = 9.4339811 m

Calculator sequence used: 5 INV x² + 8 INV x² = √

7. No adjustment of decimal is needed.

8. c = 9.4 m

5.0 and 8.0 have an accuracy of two. 5.0² and 8.0² have an accuracy to two. 25 and 64 are precise to the unit's place, so is their sum 89. The square root of 89 will likewise contain two significant digits. Round the final result to two significant digits.

In the next example, the right triangle is called DEF. The Pythagorean theorem is rewritten to use sides d, e, and f. Follow the pattern. It is always the hypotenuse squared that equals the sum of the squares of the other two sides, and the hypotenuse is always opposite the right angle.

EXAMPLE 7.2 If the hypotenuse of a right triangle is 2'3" and one side is 14", find the other side.

Solution **1. Given:** right triangle DEF, e = 2'3", f = 14"
2. Find: d See Figure 7-10.

FIGURE 7-10
Example 7.2

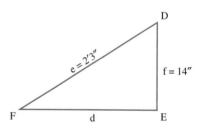

3. e² = d² + f²
4. e² − f² = d²
Transpose
d² = e² − f²
d = (e² − f²)^(1/2)
Take the square root of both sides.
5. 2'3" = 24" + 3" = 27"
Make units compatible.
d = (27² − 14²)^(1/2)
6. d = 23.086793"
Calculator sequence used: 27 x² − 14 x² = √

7. Not necessary to shift the decimal.
8. Round d to 23".
e and f have an accuracy of two, as do e² and f², 730 and 200. Their difference, 730 − 200, is precise to the ten's place, two significant digits. The square root will also have an accuracy of two. Round the answer to 23".

The Pythagorean theorem is used in construction to check the accuracy of right angles. One side of the angle is marked off at 3 ft and the other at 4 ft. If the angle is 90°, then the line connecting the two marks is the hypotenuse of a right triangle. Its length should measure $(3^2 + 4^2)^{1/2}$ or $(9 + 16)^{1/2} = 25^{1/2} = 5$ ft. This triangle is referred to as a 3, 4, 5 right triangle. Multiples of 3, 4, 5 also form right triangles, such as 6, 8, 10 and 30, 40, 50.

Another handy right triangle has sides of 5 and 12. The hypotenuse is $(5^2 + 12^2)^{1/2}$ or $(25 + 144)^{1/2}$ or $169^{1/2}$ or 13. This is a 5, 12, 13 right triangle. Multiples of 5, 12, 13 also form right triangles, such as 20, 48, 52 and 50, 120, 130.

SECTION 7.3 EXERCISES

1–4. Calculate the length of the third side of each triangle in Figure 7-11. Use the eight-step approach.

5. If a triangle has sides of 7, 8, 10, does it form a right triangle?

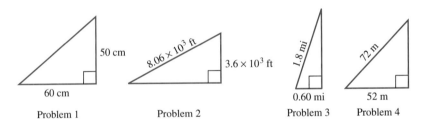

Problem 1   Problem 2   Problem 3   Problem 4

FIGURE 7-11
Section 7.3 Exercises

7.4 ■ AREA OF TRIANGLES

The unit of measure for area is any unit of length squared. In the US system of units, area can be measured in square inches (in²), square feet (ft²), square miles (mi²), and so on. Land area is often measured in acres (640 acres = 1 square mile). In the SI system of units, area can be measured in square meters (m²), square centimeters (cm²), and so on.

The area of a triangle is

$$A = bh/2$$

where A is the area of the triangle, b is the length of the base, and h is the perpendicular distance from the base to the vertex opposite the base.

The "where" statement made no mention of units. The b and h must be consistent in units and A will be measured in that unit squared. If the triangle is a right triangle as shown in Figure 7-12, then one side is chosen as the base and the other side (not the hypotenuse) becomes the height.

FIGURE 7-12
Example 7.3

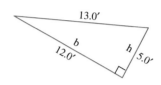

EXAMPLE 7.3   Calculate the area of the triangle in Figure 7-12.

Solution   **Given:**   b=12.0', h=5.0'
Find: A

$$A = bh/2$$
$$A = 12.0' \cdot 5.0'/2$$
$$A = 30 \text{ ft}^2$$

What if the triangle is not a right triangle? The height will no longer be one of the sides of the triangle.

EXAMPLE 7.4   Calculate the area of the triangle DEF in Figure 7-13.

Solution   When the base line is extended to the right, a line that passes through the opposite vertex perpendicular to the base line can be drawn. That line is the height of the triangle. The area is calculated by the same equation.

Given: b = 76.5 m, h = 25.3 m
Find: A

$$A = bh/2$$
$$A = (76.5 \text{ m})(25.3 \text{ m})/2$$
$$A = 967.725 \text{ m}^2$$

Round A to 968 m².

FIGURE 7-13
Example 7.4

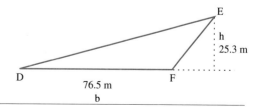

EXAMPLE 7.5 Calculate the length of line k in triangle GHI in Figure 7-14. The area of the triangle is 786 cm².

FIGURE 7-14
Example 7.5

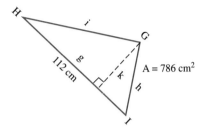

Solution k is perpendicular to side b and can be found using the area equation.

Given: Triangle GHI, g=112 cm A=786 m²
Find: k (Note that h is the side opposite angle H. k is the height of the triangle.)

$$A = gk/2$$
$$2A = gk$$
$$2A/g = k$$
$$k = 2A/g$$
$$k = 2 \cdot 786 \text{ cm}^2/112 \text{ cm}$$
$$k = 14.0357 \text{ cm}$$
Round k to 14.0 cm.

SECTION 7.4 EXERCISES

1–3. Calculate the area of each triangle in Figure 7-15.

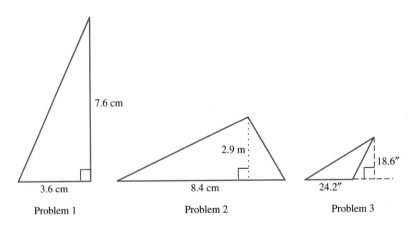

FIGURE 7-15
Section 7.4 Exercises

7.5 ■ QUADRILATERALS

Any plane (flat), four-sided figure is called a **quadrilateral**. A **parallelogram** is a special quadrilateral in which opposite sides are equal and parallel. A **rhombus** is a special parallelogram in which all four sides are equal. A **rectangle** is also a special parallelogram in which all interior angles are right angles. A **square** is a special rectangle in which all four sides are equal. A **trapezoid** is a special quadrilateral in which only two sides are parallel.

Figure 7-16 is an organizational chart that shows the relationship between these terms, with the most general term, quadrilateral at the top, and the most specific term, square, at the bottom. A square belongs to all of the categories that are above it in its section of the flow chart. A square is a rhombus because all of its sides are equal and opposite sides are parallel, but a rhombus is not a square unless all interior angles are right angles. A square is also a rectangle, but not all rectangles are squares. All the properties for one of the quadrilaterals will also be true for any quadrilateral below it on the organizational chart.

FIGURE 7-16
Quadrilaterals

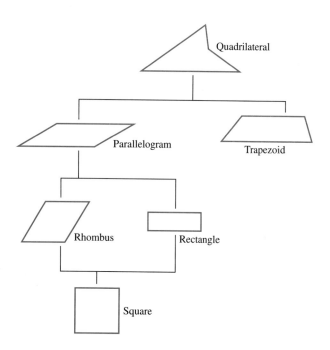

SECTION 7.5 EXERCISES

Use the definitions and organizational chart to answer True or False.

1. A trapezoid is a special parallelogram.
2. A rectangle is a special parallelogram.
3. All four sides of a parallelogram are equal.
4. A rhombus is a rectangle.
5. A square is a parallelogram.

7.6 ■ AREA OF QUADRILATERALS

RECTANGLE: The area of a rectangle is equal to its length times its width.

$$A = l \cdot w \quad \text{See Figure 7-17.}$$

FIGURE 7-17
Area of a Rectangle

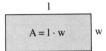

EXAMPLE 7.6   How many acres are contained in a rectangular plot of land that measures 896 ft by 645 ft?

Solution   Given: l=896 ft, w=645 ft
Find: A in acres.

$$A = l \cdot w$$

$$A = 896 \text{ ft} \times 645 \text{ ft} \times \frac{(1 \text{ mi})^2}{(5280 \text{ ft})^2} \times \frac{640 \text{ acre}}{1 \text{ mi}^2}$$

Review units of measure as needed.

$$A = 13.2672 \text{ acres}$$

Round A to 13.3 acres.

The reduction units are exact. The measurements given have an accuracy of three. Round the result to three significant digits.

SQUARE: The equation for the area of a rectangle also applies to a square since the square is below the rectangle on the organizational chart. In a square, the length and width are equal, so the area of a square can be written as length squared.

$$A = l \cdot w$$
$$A = l^2 \quad \text{See Figure 7-18.}$$

FIGURE 7-18
Area of a Square

EXAMPLE 7.7   If the area contained between the base paths of a baseball diamond is 8100 ft², find the distance between the bases.

Solution   Given: A=8100 ft²
Find: l

$$A = l^2$$
$$l = A^{1/2}$$
$$l = (8100 \text{ ft}^2)^{1/2}$$
$$l = 90 \text{ ft}$$

PARALLELOGRAM: The parallelogram in Figure 7-19 has a length called b (base) and a width w. Since the sides are not perpendicular, the equation $A = b \cdot w$ does not hold. The parallelogram can be modified to show how its area can be calculated. Right triangle T has been created by drawing line h (height) perpendicular to the base b. Triangle T can be removed and repositioned at the right end of the parallelogram, shown as T'. This creates a rectangle of length b and width h. The area of the new rectangle is the same as the original parallelogram.

FIGURE 7-19
Area of a Parallelogram

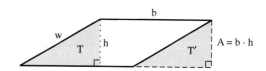

$$A = b \cdot h$$

TRAPEZOID: The derivation of the equation for the area of a trapezoid requires the use of the equations for the area of triangles and rectangles. Figure 7-20A shows a trapezoid with parallel bases a and b. The trapezoid has been divided into a rectangle of length a and width h, a triangle on the left of height h and base c, and a triangle on the right of height h and base d. Note that the overall base b has been divided into $c + a + d$. The area of the trapezoid is the sum of the area of the rectangle and the two triangles.

$$A = a \cdot h + 1/2 \cdot c \cdot h + 1/2 \cdot d \cdot h$$

This equation is tedious and can be simplified as follows. h is a common factor; factor it out.

$$A = h(a + 1/2\,c + 1/2\,d)$$

1/2 is common to the last two terms; factor it out.

$$A = h[a + 1/2(c + d)]$$

Since $b = c + a + d$, then $b - a = c + d$. Substitute $b - a$ for $c + d$ in the last equation.

$$A = h[a + 1/2(b - a)]$$

Clear the parentheses.

$$A = h[a + b/2 - a/2]$$

$a - a/2 = a/2$

$$A = h[a/2 + b/2]$$

1/2 is common to each term; factor it out.

$$A = h(a + b)/2$$

And there you have it: the area of a trapezoid is 1/2 the height times the sum of the bases.

In the last equation, $(a+b)/2$ is the average of the lengths of the two bases. So, the area of a trapezoid is the height times the average length of the two bases. As shown in Figure 7-20B, the average length of the two bases occurs at one-half the height. Right triangle A can be trimmed off and repositioned as B. Likewise, C can be trimmed off and repositioned at D. The resulting shape is a rectangle of width x and original height h, where x is the average length of the two bases.

FIGURE 7-20
Area of a Trapezoid

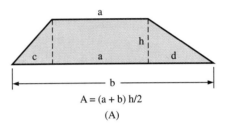

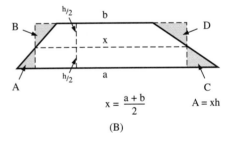

SECTION 7.6 EXERCISES

1. For the pattern in Figure 7-21, calculate the area of shape A, B, and A_T, the total area. Each grid represents 1 inch.

FIGURE 7-21
Section 7.6 Exercises, Problem 1

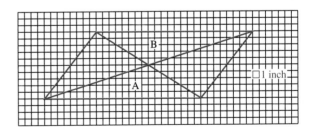

2. For the pattern in Figure 7-22, calculate the area of shapes A, B, and A_T, the total area. Each grid represents 10 feet.

3. For the pattern in Figure 7-23, calculate the area of shape A, B, and C. Each grid represents 12 cm.

FIGURE 7-22
Section 7.6 Exercises, Problem 2

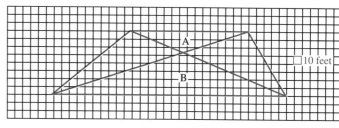

FIGURE 7-23
Section 7.6 Exercises, Problem 3

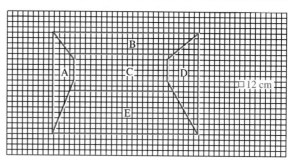

7.7 ■ CIRCLES

Refer to the circles in Figure 7-24 as you review the following definitions. The **radius** is the distance from the center to a point on the circle, and the **diameter** is the distance across the circle through the center.

$$d = 2 \cdot r$$

where r is the radius of the circle and d is the diameter.

FIGURE 7-24
Circles

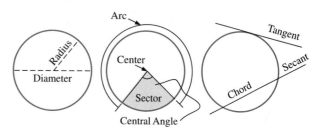

An angle formed by two radii is called a **central angle**. The area bounded by the two radii and the circle is called a **sector** of the circle. The portion of the circle between the two radii is called an **arc**.

A straight line that intersects a circle in two points is called a **secant**. The line segment that lies within the circle is called a **chord**.

The area of a circle is given by the equation

$$A = \pi r^2$$

where A is the area of a circle of radius r.

This equation can be rewritten in terms of the diameter of the circle by substituting d/2 for r.

$$A = \pi(d/2)^2$$
$$A = \pi d^2/4$$

where A is the area of a circle of diameter d.

EXAMPLE 7.8   Calculate the diameter of a circle whose area is 55.7 m².

Solution **Given:** A = 55.7 m²
Find: d

$$A = d^2/4$$
$$4A = d^2$$
$$(4A)^{1/2} = d$$
$$d = (4A)^{1/2}$$
$$d = 2A^{1/2}$$
$$d = 2(55.7 \text{ m}^2)^{1/2}$$
$$d = 14.92648 \text{ m}$$
Round d to 14.9 m.

Often a complex shape can be broken down into triangles and quadrilaterals that we have already studied. The area of the complex shape can be calculated as the sum or difference of areas of simpler shapes.

EXAMPLE 7.9   Calculate the area of the lot shown in Figure 7-25A.

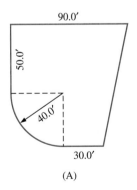 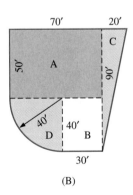

FIGURE 7-25
Example 7.9

(A)   (B)

Solution   The lot can be divided into four simple shapes as shown in Figure 7-25B.

Area A   A = lw
A = 70.0 ft · 50.0 ft
A = 3500 ft²

Area B A = lw
A = 40.0 ft · 30.0 ft
A = 1200 ft²

Area C A = 1/2 bh
A = 1/2 20.0 ft · 90.0 ft
A = 900 ft²

Area D A = π r²/4 (1/4 of a circle)
A = π (40.0 ft)²/4
A = 1260 ft²

Total Area = Area A + Area B + Area C + Area D
Total Area = 3500 ft² + 1200 ft² + 900 ft² + 1260 ft²
Total Area = 6,860 ft²

EXAMPLE 7.10 The backyard shown in Figure 7-26 is to be sodded. Calculate the amount of sod needed in square feet.

FIGURE 7-26
Example 7.10

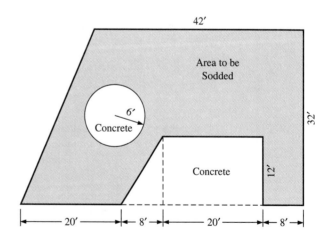

Solution The overall area to be sodded is a trapezoid. From the area of the trapezoid, the areas of the circle, the rectangle, and the triangle must be subtracted.

Area of trapezoid: A = (a + b) h/2
= (42' + 56') 32'/2
= 1568 ft²

Area of circle: A = π r²
= π (6')²
= 113 ft²

$$\text{Area of rectangle: } A = l \cdot w$$
$$= 20' \cdot 12'$$
$$= 240 \text{ ft}^2$$
$$\text{Area of triangle: } A = b \cdot h/2$$
$$= 8' \cdot 12'/2$$
$$= 48 \text{ ft}^2$$

Area to be sodded = Area of trapezoid − Area of circle − Area of rectangle − Area of triangle.

$$\text{Area to be sodded} = 1568 \text{ ft}^2 - 113 \text{ ft}^2 - 240 \text{ ft}^2 - 48 \text{ ft}^2$$
$$= 1167 \text{ ft}^2$$

The perimeter of a circle is called the **circumference.** The circumference c is calculated by

$$c = 2 \cdot \pi \cdot r$$

where c is the circumference of a circle of radius r.

Or, in terms of the diameter,

$$c = \pi d$$

where c is the circumference of a circle of diameter d.

EXAMPLE 7.11 Calculate the circumference of a circle whose area is 999 mm².

Solution **Given:** A = 999 mm²
Find: c

The equation for circumference calls for diameter or radius, neither of which is given in the problem. First calculate d and then solve the problem.

$$A = d^2/4$$
$$4A = d^2$$
$$(4A)^{1/2} = d$$
$$d = (4A)^{1/2}$$
$$d = 2A^{1/2}$$
$$d = 2(999 \text{ mm}^2)^{1/2}$$
$$d = 63.2139225 \text{ mm}$$

Round d to 63.2 mm.

SECTION 7.7 EXERCISES Calculate the revolutions that the tires on your car have made.

7.8 ■ DEGREES, MINUTES, AND SECONDS

So far, we have measured all our angles in degrees and tenths of degrees. More precise measurements are sometimes made in **degrees** (°), **minutes** ('), and **seconds** (").

$$360° = 1 \text{ revolution}$$
$$1° = 60'$$
$$1' = 60''$$

These conversion factors can be used with our dimensional analysis procedure to reduce degrees to degrees, minutes, and seconds.

EXAMPLE 7.12 Convert 49.86° into degrees, minutes, and seconds.

Solution Convert 0.86° into minutes and then convert any fractional part of a minute into seconds.

$$.86° \times \frac{60'}{1°} = 51.6' \qquad .6' \times \frac{60''}{1'} = 36''$$

$$49.86° = 49°51'36''$$

EXAMPLE 7.13 Convert 123°29'43" into degrees.

Solution

$$1° = 60' = 60' \times \frac{60''}{1'} = 3600''$$

$$123°29'43'' = 123° + 29' \times \frac{1°}{60'} + 43'' \times \frac{1°}{3600''}$$

$$= 123° + .48333° + .011944°$$

$$= 123.495274°$$

What about significant digits? I thought you might ask. An angle measured to the nearest second, say 43" as in our last example, ranges between 42.5" and 43.5".

$$42.5'' = 42.5/3600° = 0.0118055556$$
$$43'' = 1/3600° = 0.0119444444$$
$$43.5'' = 43.5/3600° = 0.0120833333$$

These bounds differ in the ten-thousandth place. Round the result to the nearest ten-thousandth place.

$$123°29'43'' = 123.4953°$$

EXAMPLE 7.14 Calculate the complementary angle to $38°27'18''$.

Solution $90° = 89°59'60''$

$$89°59'60'' - 38°27'18'' = 51°32'42''$$

7.9 ■ RADIAN MEASURE

Figure 7-27 shows a circle of radius r and a central angle θ. If the arc intercepted by the central angle θ is equal to the radius, then the central angle has a measure of one **radian**. Since the circumference of the circle is $2\pi r$, it takes 2π radians to complete a full revolution, $360°$. In other words, 2π radians $= 360°$. Divide both sides by 2 to get

$$\pi \text{ radians} = 180°$$

FIGURE 7-27
Radian Measure

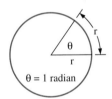

EXAMPLE 7.15 Express $74°$ in radians.

Solution

$$74° \cdot \frac{\pi \text{ radians}}{180°} = 1.29 \text{ radians}$$

Significant digits: $180°$ is exact, and π on your calculator has an accuracy of eight or ten. The accuracy of the result is determined by 74, two digits.

$$74° = 1.3 \text{ radians}$$

EXAMPLE 7.16 Express $30°$ in radians in terms of π.

Solution

$$30° \cdot \frac{\pi \text{ radians}}{180°} = \pi/6 \text{ radians}$$

Since radians is a ratio or arc length to radius, and since arc length and radius are in the same units, radians is a dimensionless unit. When an angle is expressed without units, it is understood to be in radians.

SECTIONS 7.8 AND 7.9 EXERCISES

Express in degrees, minutes, and seconds.

1. 34.367° **2.** 276.56°

Express in degrees.

3. 239°24′3″

4. 56°38′17″

5. Express each angle θ in radians in terms of π.

$\theta°$	θ(radians)
30	
60	
90	
120	
150	
180	
210	
240	
270	
300	
330	
360	

To use radians as the measure of angles being studied, put your calculator into the radian mode. On a Casio, press the mode key and then 5. A small RAD will appear in the display, indicating that the calculator is in the radian mode.

In the next three equations discussed, the angle used must be expressed in radians.

7.10 ■ AREA OF A SECTOR

The area of a sector of a circle is given by the equation

$$A = r^2 \cdot \theta/2$$

where r is the radius of the circle, θ is a central angle expressed in radians, and A is the area of the sector bounded by angle θ.

See Figure 7-28.

FIGURE 7-28
Area of a Sector

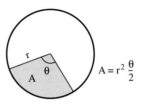

EXAMPLE 7.17   A 2′8″ door swings through an angle of 120°. Calculate the area of floor that it covers.

Solution   Given: r = 2′8″, θ = 120°
Find: A See Figure 7-29.

$$A = r^2 \cdot \theta/2$$

$$A = [2\tfrac{8'}{12}]^2 \times \frac{120°}{2} \times \frac{\pi \text{ radians}}{180°}$$

$$A = 7.4467 \text{ ft}^2$$

FIGURE 7-29
Example 7.17

The units are ft² · radians. The radians are dimensionless, which leaves ft². The calculator sequence used is

$$2 + 8/12 = x^2 \times 120/2 \times \pi/180$$

Significant digits: The accuracy of the result is limited by the accuracy of the width of the door: 2 8/12′ = 2.666666′. Suppose the door is measured to the nearest 1/64 of an inch. 1/64 is approximately 0.02″. The width of the door has an accuracy of three, 2.67′. Round the answer to three significant digits, 2.37 ft².

7.11 ■ ARC LENGTH

An **arc** is a portion of a circle along the circumference of the circle. **Arc length** is the length of an arc along its curved path. The measure of a central angle in radians is equal to the ratio of the arc length intercepted by the angle and the radius of the circle. See Figure 7-30.

$$\theta = s/r$$

where s is the arc length intercepted, r is the radius of the circle,
θ is the central angle in radians.

FIGURE 7-30
Arc Length, s

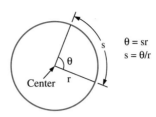

$\theta = sr$
$s = \theta/r$

To calculate arc length, solve this equation for s.

$$s = r \cdot \theta$$

EXAMPLE 7.18 A circular curve in a highway sweeps a central angle of 38.6°. If the distance from the center of the circle to the center line of the highway is 486′, calculate the arc length of the highway at the centerline.

Solution: **Given:** $\theta = 38.6°$, $r = 486'$
Find: s

$$s = r\,\theta$$
$$s = 486' \times 38.6° \times \frac{\pi \text{ radians}}{180°}$$
$$s = 327.4167'$$

Round s to 327′.

Significant digits: The radius and angle each have an accuracy of three. 180° is exact.

7.12 ■ ANGULAR VELOCITY

The rate at which an object is rotating is called its **angular velocity, ω (omega)**. Angular velocity can be measured in such units as degrees per second, radians per second, and revolutions per minute (rpm). One revolution is 360° or 2π radians.

EXAMPLE 7.19 A synchronous generator is rotating at 3600 rpm. Calculate its angular velocity in radians per second.

Solution

$$\frac{3600 \text{ revolutions}}{\text{min}} \times \frac{1 \text{ min}}{60 \text{ sec}} \times \frac{2\pi \text{ radians}}{1 \text{ rev}} = 377.0 \text{ rad/sec}$$

7.13 ■ ANGULAR DISPLACEMENT

The angle through which a body turns during a certain time period is called its **angular displacement**. Since angular displacement is an angle, its unit of measure is degrees, radians, or revolutions. The angular displacement is the product of the angular velocity and the time.

$$\theta = \omega t$$

where ω is the angular velocity, t is the elapsed time, and θ is the angular displacement.

EXAMPLE 7.20 How long will it take a wheel spinning at 34 radians per second to turn through 4 revolutions?

Solution **Given:** ω = 34 radians per second, θ = 4 revolutions
Find: t

$$\theta = \omega t$$
$$t = \theta/\omega$$
$$t = 4 \text{ rev} \times \frac{2\pi \text{ rad}}{\text{rev}} \times \frac{1}{34 \text{ rad/sec}}$$

Units: Invert the rad per sec in the denominator of the last fraction and multiply. The units become sec / rad. All units cancel except seconds.

$$t = 0.739198 \text{ sec}$$

Round t to 0.74 sec.

Significant digits: Assuming the problem means exactly 4 revolutions, the result is limited to two significant digits by ω.

7.14 ■ LINEAR VELOCITY

If an object is rotating at a certain angular velocity, the velocity at which a point on the object is moving is called its **linear velocity, v.** The greater the distance from the center of the object, the faster the point moves.

$$v = \omega r$$

where v is the linear velocity of a point on a rotating body, ω is the angular velocity of the rotating body in radians per second, and r is the radius from the center of the rotating body to the point.

EXAMPLE 7.21 A merry-go-round is spinning at an angular velocity of 1 rev every 4 seconds. The diameter of the merry-go-round is 15 ft. How fast is a child who is standing on the outer edge moving? How fast is a child who is standing 3 ft from the center moving?

Solution **Given:** $\omega = 1$ rev every 4 s
$r = 7.5$ ft and $r = 3$ ft
Find: v

$$v = \omega r$$
$$\omega = 2\pi \text{ radians}/4 \text{ sec} = 0.5\pi \text{ radians/sec}$$

$v = 0.5\pi$ rad/sec · 7.5 ft $v = 0.5\pi$ rad/sec · 3 ft

$v = 12$ ft/sec $v = 4.7$ ft/sec

Radians are dimensionless so the unit of the result is ft/sec.

SECTIONS 7.10 THROUGH 7.14 EXERCISES

1. For a circle of radius 18″, calculate the area of a sector whose central angle is 1.5 radians.
2. For a circle of radius 63.5 cm, calculate the area of a sector whose central angle is 98.6°.
3. A circular curve in a road has a central angle of 0.895 radian. If the length of the arc is 258 ft, calculate the radius of the curve.
4. An arc on a circle measures 785 cm. If the diameter of the circle is 655 cm, calculate the central angle of the arc.
5. If a gear is rotating at 33 rpm, what is its angular velocity in radians per second?
6. If a synchronous motor is rotating at 188.5 rad per second, what is its angular velocity in revolutions per minute?
7. How long will it take a gear rotating at 2.5 rad per sec to rotate through 720°?
8. What is the angular velocity in radians per second of the hour hand on a clock?
9. How fast is the tip of a second hand on a large clock moving if the second hand is 12″ long?
10. To remain fixed over a position on Earth (geosynchronous orbit), satellites are placed into the Clarke Orbit, which is directly over the equator. A satellite in geosynchronous orbit travels 6874 miles per hour to complete an orbit in one day. Calculate the height of the satellite above sea level. The diameter of the Earth at the equator is 7926 miles.

7.15 ■ VOLUME

The units for volume are any of the lengths cubed, such as cubic centimeters, cc or cm³, and cubic feet, ft³. In each solid in which the top and bottom bases are identical (congruent) and the sides of the solid are perpendicular to the bases, the volume is the area of the base times the height.

$$V = A \cdot h$$

where A is the area of the base, h is the height, and V is the volume of the solid.

EXAMPLE 7.22 Calculate the volume contained in the interior of the roof shown in Figure 7-31.

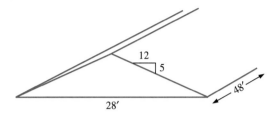

FIGURE 7-31
Example 7.22

Solution The roof has a slope of 5/12. For each foot of run, the roof rises 5". For a run of 14', the roof rises 14·5" or 70" or 5.833'.
   For a triangle,

$$A = 1/2\, b \cdot h$$
$$A = 1/2 \cdot 28 \text{ ft} \cdot 5.833 \text{ ft}$$
$$A = 81.662 \text{ ft}^2$$

Round A to 81.7 ft² (one extra significant digit).
The volume is this area times the length of the roof.

$$V = A \cdot h$$
$$V = 81.7 \text{ ft}^2 \cdot 48 \text{ ft}$$
$$V = 3921.6 \text{ ft}^3$$

Round V to 3900 ft³.

EXAMPLE 7.23 How many loads of shell will it take to cover the driveway and parking area shown in Figure 7-32 to an average depth of 5"? Each load is 7.5 yd³.

FIGURE 7-32
Example 7.23

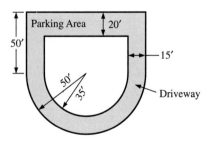

Solution First find the area of the driveway and parking area, then multiply by the depth of shell to get the volume. Finally, convert to yd³ to calculate the number of loads.
   Area 1: circular part.
   Calculate the area of the large semicircle and subtract the area of the small semicircle.

Area of circular drive = Area large semicircle − Area small semicircle = $\pi r_1^2/2 - \pi r_2^2/2$

$$= \pi (r_1^2 - r_2^2)/2$$
$$= \pi (50^2 - 35^2)/2$$
$$= 2002.76$$

Round area to 2000 ft².

Area of parking = l · w

$$= (100 - 15 - 15) \cdot 20$$
$$= 70 \cdot 20 = 1400 \text{ ft}^2$$

Area of straight drive = (l · w) 2

$$= 50 \cdot 15 \cdot 2$$
$$= 1500 \text{ ft}^2$$

Total Area = 2000 ft² + 1400 ft² + 1500 ft² = 4900 ft²

Volume = Area of base·height

$$= 4900 \text{ ft}^2 \cdot 5/12 \text{ ft} = 2042 \text{ ft}^3$$

Convert to cubic yards.

$$= 2042 \text{ ft}^3 * [\frac{1 \text{ yd}}{3 \text{ ft}}]^3$$

Volume = 75.629 yd³

Round volume to 76 yd³.

Loads = Volume in yd³/7.5 yd³ per load
= 76 yd³/7.5 yd³ per load
= 10.1333 loads

Order 10 loads.

EXAMPLE 7.24 An 85-ft-long trench is to be dug. The trench is to be 5.0 ft deep and 4.0 ft wide at the bottom. The sides must slope at 70° to keep the compacted soil from caving in. See Figure 7-33. The two ends of the trench will be shored and need not be sloped. Calculate the volume of soil in yd³ to be excavated.

Solution Calculate the cross-sectional area of the trench and multiply by the length of the trench to get the volume.

For a trapezoid,

$$A = h (a + b)/2$$

where a and b are the bases and h is the height.

FIGURE 7-33
Example 7.24

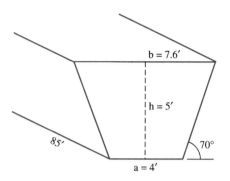

To use this equation to find the cross-sectional area of the trench, we need to know both bases and the height. Using the trigonometry that will be studied in a later chapter, the base b at the top of the trench can be calculated to be 7.6 ft.

$$A = h(a + b)/2$$
$$A = 5(4 + 7.6)/2$$
$$A = 29 \text{ ft}^2$$

$$V = A \cdot h$$
$$V = 29 \text{ ft}^2 \cdot 85 \text{ ft}$$
$$V = 2465 \text{ ft}^3 \text{ or } 91.296 \text{ yd}^3$$

The accuracy of the result is limited by the accuracy of a and h, two significant digits.

Round V to 91 yd³.

EXAMPLE 7.25 A concrete footing is to be poured for the building in Figure 7-34. The footing is 16″ wide at the bottom, 8″ deep, and the sides slope as shown in Figure 7-34B. How many yd³ of concrete are required?

Solution The dimensions shown in Figure 7-34A on the outside of the perimeter denote the outside edge of the building. The centerline of the footing to be poured will lie inside that boundary by one-half the thickness of the wall. Since the wall is to be concrete block, 8″ thick, the centerline of the footing will fall 4″ inside the outer perimeter. The centerline of the footing is shown with a dotted line.

First let's calculate the distance around the foundation at the centerline of the footing, and then we will calculate the cross-sectional area of the footing. The volume of concrete needed is the product of the two. Figure 7-34A shows the length of the centerline along each edge, inside the dotted line. Note that in some cases the overall dimension was reduced at each end by 4″ and in other cases it was not reduced at all, only shifted. The total centerline length is

$$cl = 23' 4'' + 49' 4'' + 23' 4'' + 10' + 12' + 29' 4'' + 12' + 10'$$
$$cl = 168' 16'' = 169' 4'' = 169.33'$$

FIGURE 7-34
Example 7.25

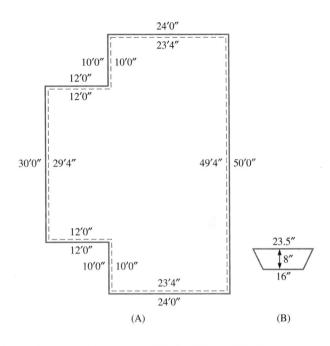

The cross-section of the footing to be poured is a trapezoid. See Figure 7-34B.

$$A = h \cdot (a + b)/2$$
$$A = 8" \cdot (16" + 23.5")/2$$
$$A = 158 \text{ in}^2$$

The required volume is the product of the centerline distance and the cross-sectional area.

$$V = A \cdot cl$$
$$V = 158 \text{ in}^2 \cdot 169.33 \text{ ft} \quad \text{Dimension mismatch.}$$
$$V = 158 \text{ in}^2 \times \frac{(1 \text{ ft})^2}{(12 \text{ in})^2} \times 169.33 \text{ ft}$$
$$V = 185.7926 \text{ ft}^3$$
Round V to 200 ft³.

Significant digits: The dimensions for the trench are only precise to the nearest inch, an accuracy of 1 for the height. The volume should be rounded to one significant digit.

$$V = 200 \text{ ft}^3$$

How many yards (cubic yards) of concrete should be ordered?

$$V = 200 \text{ ft}^3 \times \frac{(1 \text{ yd})^3}{(3 \text{ ft})^3} = 7.4 \text{ yd}^3$$

SECTION 7.15 EXERCISES

1. Calculate the length of each side of a cube whose volume is 1259 m³.
2. Calculate the volume contained in the interior of the roof shown in Figure 7-35.

FIGURE 7-35
Section 7.15 Exercises, Problem 2

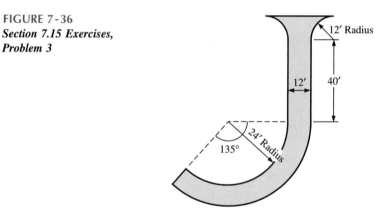

3. How many loads of shell will it take to cover the driveway shown in Figure 7-36 to an average depth of 3"? Each load is 7.5 yd³.

FIGURE 7-36
Section 7.15 Exercises, Problem 3

4. A 32-ft trench is to be dug. The trench is to be 3.0 ft deep, 5.0 ft wide at the bottom, and 6 ft wide at the top. The two ends of the trench will be shored and need not be sloped. Calculate the volume of soil to be excavated in yd³.

FIGURE 7-37
Section 7.15 Exercises, Problem 5

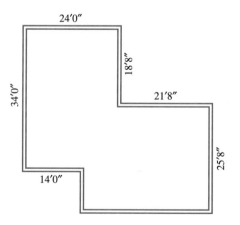

5. A concrete footing is to be poured for the building in Figure 7-37. The footing is 16″ wide at the bottom, 8″ deep, and 24″ wide at the top. Place the center line of the footing 4″ inside the outer perimeter of the building. How many yd³ of concrete are required?

SUMMARY

1. Angles that measure less than 90° are called acute; angles that measure more than 90° are called obtuse; and angles that measure 90° are called right angles.

2. If one of the angles of the triangle is a 90° angle, the triangle is a right triangle. If all three angles are less than 90°, the triangle is acute. If one of the angles is greater than 90°, the triangle is obtuse.

3. If all three sides are different lengths, the triangle is scalene; if two sides are the same length, the triangle is isosceles; if all three sides are the same length, the triangle is equilateral.

4. The sum of the angles in any triangle is 180°.

5. Two angles whose measures add to 90° are called complementary angles.

6. In a right triangle, the side opposite the right angle is called the hypotenuse.

7. If the hypotenuse of a right triangle is labeled c and the other two sides are labeled a and b, then the Pythagorean theorem states that

$$c^2 = a^2 + b^2$$

8. The area of a triangle is

$$A = bh/2 \text{ where b is the base and h is the height.}$$

9. Any plane (flat), four-sided figure is called a quadrilateral.

10. A parallelogram is a special quadrilateral in which opposite sides are equal and parallel. A rhombus is a special parallelogram in which all four sides are equal. A rectangle is also a special parallelogram in which all interior angles are right angles. A square is a special rectangle in which all four sides are equal. A trapezoid is a special quadrilateral in which two sides are parallel.

11.

Geometric Shape	Figure	Area
Rectangle	Figure 7-17	$A = l \cdot w$
Square	Figure 7-18	$A = l^2$
Parallelogram	Figure 7-19	$A = b \cdot h$
Trapezoid	Figure 7-20A	$A = h(a + b)/2$
Circle	Figure 7-24	$A = \pi r^2$ $A = \pi d^2/4$

12. The perimeter or circumference of a circle is calculated by

$$c = 2 \cdot \pi \cdot r$$

13. Angles can be measured in degrees (°), minutes ('), and seconds (").

$$360° = 1 \text{ revolution}$$
$$1° = 60'$$
$$1' = 60''$$

14. If the arc intercepted by the central angle is equal to the radius, then the central angle has a measure of one radian. See Figure 7-29.
15. $360° = 2\pi$ radians
16. The area of a sector of a circle is given by the equation

$$A = r^2 \cdot \theta/2$$

where r is the radius of the circle, θ is a central angle expressed in radians, and A is the area of the sector bounded by angle θ.

17. An arc is a portion of a circle along the circumference of the circle. Arc length is the length of an arc along its curved path.

$$\theta = s/r$$

where s is the arc length intercepted, r is the radius of the circle, and θ is the central angle in radians.

18. The rate at which an object is rotating is called its angular velocity, ω (omega). Angular velocity can be measured in such units as degrees per second, radians per second, and revolutions per minute (rpm).
19. The angle through which a body turns during a certain time period is called its angular displacement and is measured in degrees, radians, or revolutions. The angular displacement is the product of the angular velocity and the time.

$$\theta = \omega t$$

where ω is the angular velocity, t is the elapsed time, and θ is the angular displacement.

20. If an object is rotating at a certain angular velocity, the velocity at which a point on the object is moving is called its linear velocity, v. The greater the distance from the center of the object, the faster the point moves.

$$v = \omega r$$

where v is the linear velocity of a point on a rotating body, ω is the angular velocity of the rotating body in radians per second, and r is the radius from the center of the rotating body to the point.

21. In each solid in which the top and bottom bases are identical (congruent) and the sides of the solid are perpendicular to the bases, the volume is the area of the base times the height.

$$V = A \cdot h$$

where A is the area of the base, h is the height, and V is the volume of the solid.

SOLUTIONS TO CHAPTER 7 SECTION EXERCISES

Solutions to Section 7.1 Exercises

1. α, A, B, ∠ILJ, ∠HLI
2. β, φ, ∠GLI, ∠KLI
3. C, E, ∠GLH, ∠GLK, ∠KLJ, ∠JLH

Solutions to Section 7.2 Exercises

1. A. right scalene, B. obtuse isosceles, C. acute isosceles, D. acute equilateral, E. right isosceles, F. obtuse scalene, G. obtuse scalene
2. No.
3–9. See Figure 7-38.

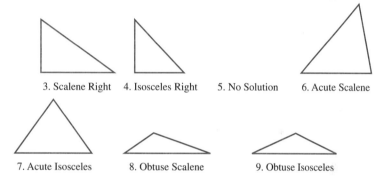

FIGURE 7-38
Solution to Section 7.2 Exercises, Problems 3 through 9

Solutions to Section 7.3 Exercises

1. 78 cm
2. 7.2×10^3 ft
3. 1.7 mi
4. 50 m
5. No. $7^2 + 8^2 \neq 10^2$

Solutions to Section 7.4 Exercises

1. 13.7 cm²
2. 12.2 m²
3. 225 m²

Solutions to Section 7.5 Exercises

1. F
2. T
3. F
4. F
5. T

Solutions to Section 7.6 Exercises

1. $A_A = 60$ in²
$A_B = 60$ in²
$A_T = 240$ in²

2. $A_A = 3300$ ft²
$A_B = 14\,000$ ft²
$A_T = 30\,500$ ft²

3. $A_A = 6620$ cm²
$A_B = 13\,800$ cm²
$A_T = 76\,600$ cm²
$A_D = 9900$ cm²
$A_E = 27\,600$ cm²
$A_C = 18\,700$ cm²

Solution to Section 7.7 Exercise

From the odometer, determine the miles that your tires have traveled. Determine the radius of your wheel either by measurement or by calculation (wheel size plus tire size). Calculate the circumference of your tire. Convert the circumference into miles. Divide the miles traveled by the miles per revolution to get the revolutions of the tires.

Solutions to Sections 7.8 and 7.9 Exercises

1. 34° 22′ 1″
2. 276° 33′ 36″
3. 239.4008°
4. 56.6381°.
5.

θ	θ (radians)
30	$1/6\ \pi$
60	$1/3\ \pi$
90	$1/2\ \pi$
120	$2/3\ \pi$
150	$5/6\ \pi$
180	π
210	$7/6\ \pi$
240	$4/3\ \pi$
270	$3/2\ \pi$
300	$5/3\ \pi$
330	$11/6\ \pi$
360	$2\ \pi$

Solutions to Sections 7.10 through 7.14 Exercises

1. 230 in^2
2. 3470 in^2
3. 288 ft
4. 2.40 rad
5. 3.5 rad/sec
6. 1800 rpm
7. 5.0 sec
8. 145 μ rad per sec
9. 0.4 in per sec
10. 22 290 mi

Solutions to Section 7.15 Exercises

1. 10.80 cm
2. 7300 ft^3
3. Order 2 loads.
4. 20 yd^3
5. 7.3 yd^3

■ CHAPTER 7 PROBLEMS

1. Which angles in Figure 7-39 are acute?
2. Which angles in Figure 7-39 are obtuse?

FIGURE 7-39
Chapter 7 Problems, Problems 1 through 3

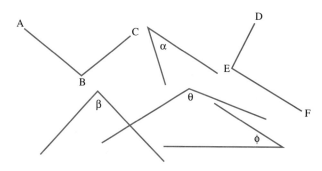

3. Which angles in Figure 7-39 are right?

4. Label each triangle in Figure 7-40 as right, acute, or obtuse and as scalene, isosceles, or equilateral.

FIGURE 7-40
Chapter 7 Problems, Problem 4

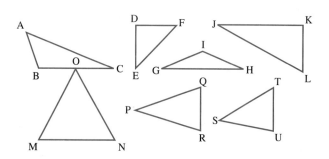

5. Can a right triangle be equilateral? Explain.
6. Draw a scalene right triangle.
7. Draw an isosceles right triangle.
8. Draw an acute scalene triangle.
9. Draw an acute isosceles triangle.
10. Draw an obtuse scalene triangle.
11. Draw an obtuse isosceles triangle.
12. If the two sides of a right triangle are 32.8 m and 56.1 m, find the hypotenuse.
13. If the hypotenuse of a right triangle is 5′4″ and one side is 3′6″, find the other side.
14. Calculate the length of the third side of each triangle in Figure 7-41.
15. If a triangle has sides of 10, 24, and 30, does it form a right triangle?
16. If a triangle has sides of 12.5, 30.0, and 32.5, does it form a right triangle?
17. Calculate the area of triangle ABC in Figure 7-42.
18. Calculate the length of line h in triangle ADE in Figure 7-42. The area of the triangle is 39.0 cm².

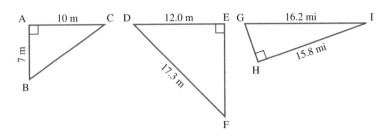

FIGURE 7-41
Chapter 7 Problems, Problem 14

FIGURE 7-42
Chapter 7 Problems, Problems 17 through 20

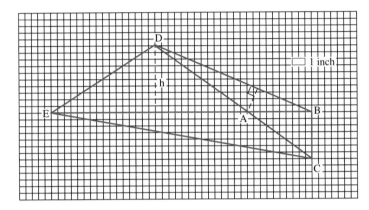

19. Calculate the area of triangle ABD in Figure 7-42.
20. Calculate the area of figure BCED in Figure 7-42.
21. How many acres are contained in a rectangular plot of land that measures 555 ft by 666 ft?
22. For the pattern in Figure 7-43, calculate the area of shape A, B, and A_T, the total area. Each grid represents 1 inch.

FIGURE 7-43
Chapter 7 Problems, Problem 22

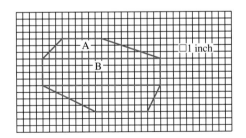

23. For the pattern in Figure 7-44, calculate the area of the shaded area. Each grid represents 10 ft.
24. For the pattern in Figure 7-45, calculate the enclosed area. Each grid represents 12 cm.

FIGURE 7-44
Chapter 7 Problems, Problem 23

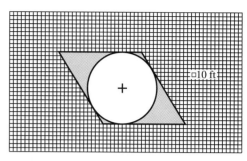

FIGURE 7-45
Chapter 7 Problems, Problem 24

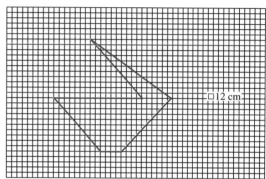

25. Calculate the diameter of a circle whose area is 674.2 m².
26. Calculate the circumference of a circle whose area is 429 mm².
27. Calculate the area of the lot shown in Figure 7-46 in acres. Each grid represents 20 ft.

FIGURE 7-46
Chapter 7 Problems, Problem 27

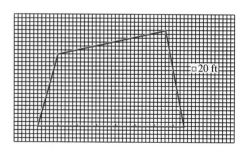

28. In the backyard shown in Figure 7-47, a fence is to be built to separate the garden from the lawn. Calculate the length of fence needed.
29. For the backyard shown in Figure 7-47, calculate the amount of sod needed in square feet.
30. For the backyard shown in Figure 7-47, calculate the cubic yards of concrete needed to pour the patio and pad to a depth of 4″.
31. Convert 74.83° into degrees, minutes, and seconds.

FIGURE 7-47
Chapter 7 Problems, Problems 28 through 30

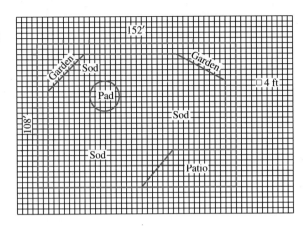

32. Convert 32.847° into degrees, minutes, and seconds.
33. Convert 64°39'39" into degrees.
34. Convert 95°9'48" into degrees.
35. Calculate the complementary angle to 27°57'46".
36. Calculate the complementary angle to 45°22'53".
37. Express 65° in radians.
38. Express 183° in radians.
39. Express 150° in radians in terms of π.
40. Express 300° in radians in terms of π.
41. A 3'0" door swings through an angle of 87°. Calculate the area of floor that it covers.
42. A 3'0" door covers an area of 12 ft² as it swings. Calculate the angle of swing.
43. A circular curve in a highway sweeps a central angle of 68.4°. If the distance from the center of the circle to the centerline of the highway is 335', calculate the arc length of the highway at the centerline.
44. In a circular curve in a highway, the centerline sweeps an arc length of 207'. If the distance from the center of the circle to the center line of the highway is 238', calculate the central angle of the arc.
45. A synchronous generator is rotating at 1800 rpm. Calculate its angular velocity in radians per second.
46. A synchronous generator is rotating at 3600 rpm. Calculate its angular velocity in degrees per second.
47. How long will it take a wheel spinning at 29 radians per second to turn through 10 revolutions?
48. If an auto is traveling at 75 mph, calculate the angular velocity of the wheels. Assume the auto has 15" wheels.
49. A merry-go-round is spinning at an angular velocity of 0.4 revolutions every second. The diameter of the merry-go-round is 12 ft. How fast is a child who is standing on

the outer edge moving? How fast is a child who is standing 3 ft from the center moving?

50. How fast is the tip of a second hand on a clock moving if the second hand is 1.5" long?

51. For a circle of radius 5.5", calculate the area of a sector whose central angle is 2.5 radians.

52. For a circle of radius 3.5 cm, calculate the area of a sector whose central angle is 37.6°.

53. An arc on a circle measures 38.9 cm. If the diameter of the circle is 55.0 cm, calculate the central angle of the arc.

54. If a gear is rotating at 33 rpm, what is its angular velocity in radians per second?

55. If a synchronous motor is rotating at 188.5 rad per second, what is its angular velocity in revolutions per second?

56. If a gear whose radius is 2.6" is rotating at 446 rpm, calculate its angular velocity in radians per second.

57. How long will it take a gear rotating at 33 rad per sec to rotate through 720°?

58. If a gear is rotating at 5.6 radians per second, how long will it take to complete 10 revolutions?

59. Calculate the volume contained in the interior of the roof shown in Figure 7-48.

FIGURE 7-48
Chapter 7 Problems, Problem 59

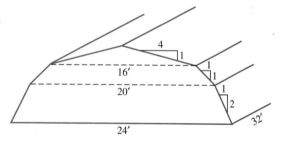

60. How many loads of shell will it take to cover the driveway in Figure 7-49 to an average depth of 5"? Each load is 10 yd³.

FIGURE 7-49
Chapter 7 Problems, Problem 60

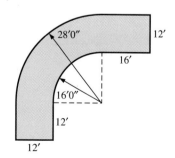

FIGURE 7-50
Chapter 7 Problems, Problems 62 and 63

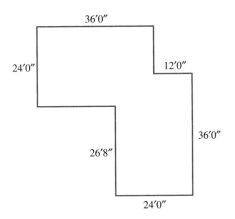

61. A 130-ft-long drainage ditch is to be dug. The ditch is to be 6.5 ft deep and 4.5 ft wide at the bottom and 6.0 ft wide at the top. The two ends of the trench will be shored and need not be sloped. Calculate the volume of soil in yd³ to be excavated.

62. A concrete footing is to be poured for the building in Figure 7-50. The footing is 16″ wide at the bottom, 20″ wide at the top, and 8″ deep. The centerline of the footing lies 4″ inside the building perimeter shown in the figure. How many cubic yards of concrete are required?

63. For the building shown in Figure 7-50, calculate the cubic yards of cement required to pour a slab 4″ thick.

8

Linear Equations and Graphing

CHAPTER OUTLINE

- 8.1 The Cartesian Coordinate System
- 8.2 Distance between Two Points
- 8.3 Slope of a Line
- 8.4 Parallel Lines
- 8.5 Perpendicular Lines
- 8.6 Grade or Gradient
- 8.7 Midpoint of a Line Segment
- 8.8 Linear Equations
- 8.9 Point-Slope Form of Equation
- 8.10 Slope-Intercept Form of Equation
- 8.11 General Form of Equation
- 8.12 Graphing Experimental Data
- 8.13 Graphing Trig Functions
- 8.14 Trig Identities

OBJECTIVES

After completing this chapter, you should be able to

1. describe the Cartesian coordinate system.
2. graph points on the coordinate plane.
3. calculate the distance between two points.
4. calculate the slope of a line.
5. determine whether two lines are parallel.
6. determine whether two lines are perpendicular.
7. calculate the midpoint between two points.
8. define linear equation.
9. write the equation of a line in point-slope form.
10. write the equation of a line in slope-intercept form.
11. write the equation of a line in general form.
12. graph experimental data.
13. graph trig functions.

NEW TERMS TO WATCH FOR

Cartesian coordinate plane	gradient
rectangular coordinate plane	linear equation
x-axis	degree
y-axis	reference angle
origin	midpoint
quadrant	x and y intercepts
ordered pair	point-slope form
abscissa	slope-intercept form
ordinate	general form
slope	independent variable
rise	dependent variable
run	asymptote
pitch	asymptotically
parallel	leads
negative reciprocal	lags
perpendicular	trig identity
grade	

Graphs are a means of communication. They are used in technical fields to show the relationship between two or more variables. Complex relationships are often easier to understand in graph form than through equations or tables of data. A technician must be able to represent data in graph form and to interpret graphs that have been published in data books or journals or graphs that have been drawn by colleagues. In this chapter, we will study the fundamentals of linear equations and graphing and then concentrate on graphing experimental data and interpreting graphs.

8.1 ■ THE CARTESIAN COORDINATE SYSTEM

A number line was shown and discussed in Chapter 3. Figure 8-1 shows a horizontal number line and a vertical number line crossing at their zero points. The two intersecting lines determine a plane called the **Cartesian coordinate plane** or **rectangular coordinate plane**. The horizontal number line is called the ***x*-axis** and the vertical number line is called the ***y*-axis**. The point at which the two axes cross is called the **origin**.

The x and y axes divide the graph into four regions called **quadrants**. The upper right quadrant is called Quadrant I; the upper left is Quadrant II; the lower left is Quadrant III, and the lower right is Quadrant IV. A point in the plane is specified by an

8.1 THE CARTESIAN COORDINATE SYSTEM

FIGURE 8-1
Cartesian Coordinate Plane

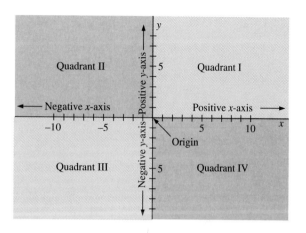

ordered pair of numbers (the order of the numbers is critical) written in parentheses, with the x coordinate first and then the y coordinate, (x, y). A point is located on the x–y coordinate plane by moving from the origin either left or right as specified by the x-coordinate. If x is positive, move to the right; if x is negative, move to the left. Then move up or down according to the y-coordinate. If y is positive, move up; if y is negative, move down. The y-coordinate of a point on the x-axis is 0. The x-coordinate of a point on the y-axis is 0. The x coordinate of a point in the plane is called the **abscissa**, and the y-coordinate is called the **ordinate**.

EXAMPLE 8.1 Locate these points on the x–y coordinate axis: A (4, 6), B (−5, 3), C (−5, −4), D (3, −5), E (3, 0), F (0, 4), G (−6, 0), and H (0, −2).

Solution See Figure 8-2.

FIGURE 8-2
Example 8.1

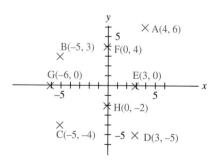

Note that points whose coordinates are both positive lie in Quadrant I. Negative x, positive y always falls in Quadrant II. Points whose coordinates are both negative lie in Quadrant III. Positive x, negative y always falls in Quadrant IV.

EXAMPLE 8.2   Write the coordinates of points A through H in Figure 8-3.

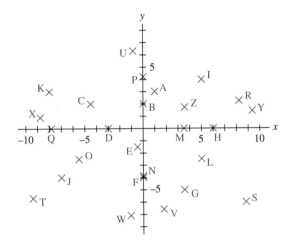

FIGURE 8-3
Example 8.2 and Section 8.1
Exercises, Problem 2

Solution   A(1, 3), B(0, 2), C (−4.5, 2), D (−3, 0), E (−0.5, −1.5), F (0, −4), G (3.5, −5), and H (6, 0)

SECTION 8.1 EXERCISES

1. Graph these points: A (5, 7), B (−4, 6), C (−6, −7), D (8, −6), E (3.5, 2.7), F (−5.2, 1.9), G (−3.1, −3.8), H (4.3, −3.6), I (0, 7.4), J (0, −4.3), K (6.4, 0), and L (−4.6, 0)

2. Write the coordinates of points I through Z in Figure 8-3.

8.2 ■ DISTANCE BETWEEN TWO POINTS

To calculate the distance between two points, create a right triangle and make use of the Pythagorean theorem. To calculate the distance between points A (x_1, y_1) and B(x_2, y_2) in Figure 8-4, create a right triangle with one side parallel to the x-axis, one side parallel to the y-axis, and with A and B at either end of the hypotenuse. The length of the vertical side is $y_2 - y_1$. The length of the horizontal side is $x_2 - x_1$. The length of the hypotenuse, d, is found using the Pythagorean theorem.

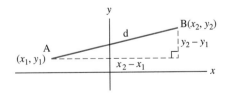

FIGURE 8-4
Distance between Two Points

$$d = [(y_2 - y_1)^2 + (x_2 - x_1)^2]^{1/2}$$

where d is the distance between two points (x_1, y_1) and (x_2, y_2).

EXAMPLE 8.3

Calculate the distance between points A and B in Figure 8-5.

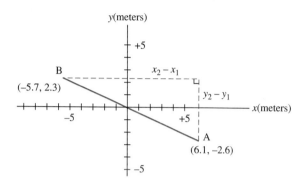

FIGURE 8-5
Example 8.3

Solution 1   **Given:**   A(6.1, −2.6), B(−5.7, 2.3)
             Find: d

$$d = [(y_2 - y_1)^2 + (x_2 - x_1)^2]^{1/2}$$
$$d = [(-2.6 - 2.3)^2 + (6.1 - -5.7)^2]^{1/2}$$
$$d = 12.77693 \text{ m}$$

Round d to 12.8 m.

These numbers were determined from the graph to the nearest tenth of a meter. Round the result to the nearest tenth of a meter.

Point A was assigned the coordinates (x_2, y_2). The problem will be reworked using B as (x_2, y_2) to show that the result is the same either way.

Solution 2
$$d = [(y_2 - y_1)^2 + (x_2 - x_1)^2]^{1/2}$$
$$d = [(2.3 - -2.6)^2 + (-5.7 - 6.1)^2]^{1/2}$$
$$d = 12.8 \text{ m}$$

SECTION 8.2 EXERCISES

In Figure 8-2, find the distance between the following points. Check to see that your answer is reasonable. Assume that the points in the figure are graphed to the nearest tenth of a meter and round your results to the nearest tenth of a meter.

1. AH
2. BD
3. CA
4. CF
5. GE
6. Is FD longer than BD?
7. Find the shortest path that connects all the dots.

8.3 ■ SLOPE OF A LINE

The **slope** of a line is determined by how much it rises or falls (**rise**) over a known horizontal distance (**run**). See Figure 8-6. Slope equals rise divided by the run. If a slope rises to the right, it is said to be positive. Line AB in Figure 8-6 has a positive slope. If a line falls to the right, its slope is negative.

FIGURE 8-6
Slope of a Line

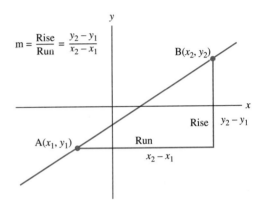

Line AB in Figure 8-7 has a negative slope. The rise of a line connecting two points $A(x_2, y_2)$ and $B(x_1, y_1)$ is the difference in y values, $y_2 - y_1$. The run of the line is $x_2 - x_1$. The slope, m, of a line connecting the two points $A(x_2, y_2)$ and $B(x_1, y_1)$ is the quotient of $y_2 - y_1$ and $x_2 - x_1$.

$$m = (y_2 - y_1)/(x_2 - x_1)$$

where m is the slope of the line connecting the points (x_1, y_1) and (x_2, y_2).

FIGURE 8-7
Negative Slope

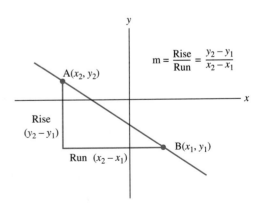

8.3 SLOPE OF A LINE ■ 239

EXAMPLE 8.4   Find the slope of the line connecting points A and B in Figure 8-8.

FIGURE 8-8
Example 8.4

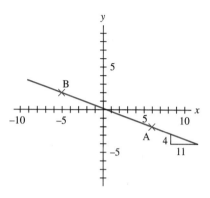

Solution 1   Given: A $(6, -2)$ B $(-5, 2)$
Find: m

$$m = (y_2 - y_1) / (x_2 - x_1)$$
$$m = (2 - -2) / (-5 - 6)$$

Point A was assigned the coordinates (x_1, y_1). In Solution 2, point B will be taken as (x_1, y_1).

$$m = 4/-11$$
$$m = -0.36$$

The $-$ sign indicates that the line is falling to the right. The result was rounded to two places arbitrarily.

Solution 2

$$m = (y_2 - y_1)/(x_2 - x_1)$$
$$m = (-2 - 2)/(6 - -5)$$
$$m = -4/11$$
$$m = -0.36$$

The steeper the line, the greater the magnitude of the slope. A line with a slope of 4 rises more sharply than a line with a slope of 3. Likewise, a line with a slope of -2 falls more rapidly than a line with a slope of -1. A line with a slope of 1 rises at 45°. See Figure 8-9. Note that the slope of a horizontal line is 0, and the slope of a vertical line is undefined.

A roof with a slope or **pitch** of 4/12 rises 4″ for each foot of horizontal run.

CHAPTER 8 LINEAR EQUATIONS AND GRAPHING

FIGURE 8-9
Slope

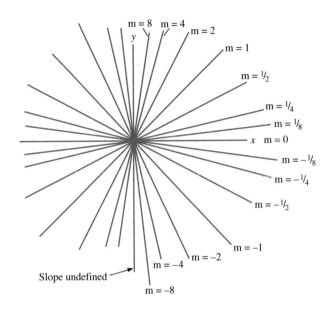

EXAMPLE 8.5 A roof with a pitch (slope) of 4/12 has a horizontal run of 12 ft from edge to center. How high does it rise?

Solution **Given:** m = 4/12, run = 12 ft
Find: rise

$$m = \text{rise} / \text{run}$$
$$\text{rise} = \text{run} \cdot m$$
$$\text{rise} = 12 \text{ ft} \cdot 4/12$$
$$\text{rise} = 4 \text{ ft}$$

SECTION 8.3 EXERCISES

1. In Figure 8-2, which of the lines connecting point B to each of the other points in the figure have a positive slope?
2. Find the slope of each line connecting point B to each of the other points in Figure 8-2.
3. Through point (2, 1), draw lines with these slopes:

   A, m = −1
   B, m = 2
   C, m = −3.5
   D, m = 0
   E, m = 1/3
   F, m = −1/4

8.4 ■ PARALLEL LINES

Lines with the same slope are **parallel**, and lines that are parallel to each other have the same slope.

Figure 8-10 shows five parallel lines, each with the same slope. Since they fall to the right, their slope is negative. For each line, the rise is twice the run. In each case, the slope is −2. The slope is indicated with a small triangle showing a rise of 2 and a run of 1. The symbol for "is parallel to" is ∥. In Figure 8-10, AB ∥ CD ∥ EF ∥ GH ∥ IJ.

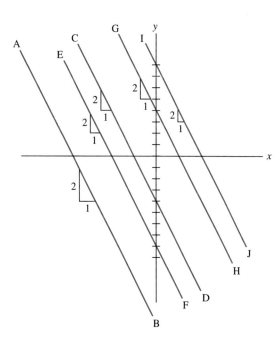

FIGURE 8-10
Parallel Lines

EXAMPLE 8.6 In Figure 8-2, is the line connecting points F and G parallel to the line connecting E and H?

Solution

$$m = (y_2 - y_1)/(x_2 - x_1)$$
$$\text{FG } m = (4 - 0)/(0 - -6) = 4/6 = 2/3$$
$$\text{EH } m = (0 - -2)/(3 - 0) = 2/3$$

Yes, FG is parallel to EH because their slopes are equal.

SECTION 8.4 EXERCISES

Plot these points. A(−4.5, 3.8), B(2.7, −4.2), C(6.1, 3.1), and D(−7.2, −2.3)

1. Graph a line through A parallel to CD.
2. Graph a line through C parallel to AB.

3. Graph a line through B parallel to CD.
4. Graph a line through D parallel to AB.

8.5 ■ PERPENDICULAR LINES

If two lines lie in the same plane and the slope of one is the **negative reciprocal** of the other, the lines are **perpendicular**; if two lines are perpendicular, their slopes are negative reciprocals. To form the negative reciprocal, change the sign and invert the magnitude. The negative reciprocal of 2/3 is −3/2. The negative reciprocal of −8 is 1/8. Figure 8-11 shows line AB and line CD. The slope of AB is −3/2, and the slope of CD is 2/3. Since their slopes are negative reciprocals, they are perpendicular. The symbol for "is perpendicular to" is ⊥. In Figure 8-11, AB ⊥ CD. EF has a slope of 2/3. So, EF ∥ CD, and EF ⊥ AB.

FIGURE 8-11
Perpendicular Lines

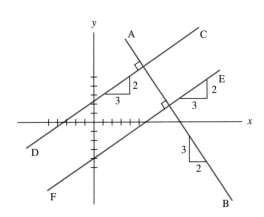

EXAMPLE 8.7 In Figure 8-2, is the line connecting points F and E perpendicular to the line connecting E and H?

Solution
$$m = (y_2 - y_1)/(x_2 - x_1)$$
$$\text{FE } m = (4 - 0)/(0 - 3) = -4/3$$
$$\text{EH } m = (0 - -2)/(3 - 0) = 2/3$$

No, −4/3 and 2/3 are not negative reciprocals.

SECTION 8.5 EXERCISES

Plot these points. A(−4.5, 3.8), B(2.7, −4.2), C(6.1, 3.1), and D(−7.2, −2.3)

1. Graph the line through A perpendicular to CD.
2. Graph the line through C perpendicular to AB.

3. Graph the line through B perpendicular to CD.

4. Graph the line through D perpendicular to AB.

8.6 ■ GRADE OR GRADIENT

In surveying, the **grade** or **gradient** is the rise (positive or negative) of a surface per 100 units of run (horizontal distance). For example, if a field rises 23 ft in 100 ft, it has a grade of 23% and a slope of 0.23. The grade or gradient is the slope expressed as a percent.

$$\text{grade} = \text{slope} \cdot 100$$

EXAMPLE 8.8 An approach to a bridge rises 10 ft along a 200-ft horizontal run. What is its grade and slope?

Solution **Given:** rise = 10 ft, run = 200 ft
Find: slope and grade

$$m = \text{rise} / \text{run}$$
$$m = 10 / 200$$
$$m = 0.05$$
$$\text{grade} = m \cdot 100\%$$
$$\text{grade} = 0.05 \cdot 100\%$$
$$\text{grade} = 5\%$$

EXAMPLE 8.9 A stretch of highway is to have a grade of 5%. How many feet should it rise for a run of one mile?

Solution **Given:** grade = 5%, run = 5280 ft
Find: rise

$$\text{grade} = m \cdot 100$$
$$\text{grade} = \text{rise/run} \cdot 100$$
$$\text{rise/run} = \text{grade}/100$$
$$\text{rise} = \text{run} \cdot \text{grade}/100$$
$$\text{rise} = 5280 \cdot 5/100$$
$$\text{rise} = 264 \text{ ft}$$

SECTION 8.6 EXERCISES

1. If the grade of a road averages 3% over a distance of 3500 feet, calculate the change in elevation.

2. Calculate the gradient of a road that rises 370 feet over a distance of three miles.

8.7 ■ MIDPOINT OF A LINE SEGMENT

How would you find the midpoint of a line segment connecting two points? Find the average of the x-coordinates of the two endpoints, and find the average of the y-coordinates of the two endpoints. The two averages are the coordinates of the midpoint of the line connecting the two points. See Figure 8-12. The midpoint of the line connecting $A(x_2, y_2)$ and $B(x_1, y_1)$ is

$$\text{midpoint} = [(x_2 + x_1)/2, (y_2 + y_1)/2]$$

FIGURE 8-12
Midpoint of a Line Segment

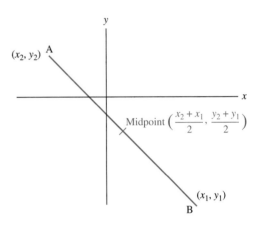

EXAMPLE 8.10 Find the midpoint of the line connecting points A and B in Figure 8-13.

FIGURE 8-13
Example 8.10

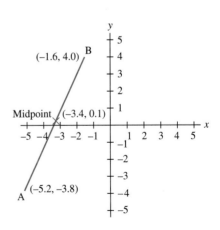

Solution **Given:** $A(-5.2, -3.8)$, $B(-1.6, 4.0)$. The coordinates are estimations from the graph.

Find: the midpoint

$$\text{midpoint} = [(x_2 + x_1)/2, (y_2 + y_1)/2]$$
$$\text{midpoint} = [(-5.2 + -1.6)/2, (-3.8 + 4.0)/2]$$
$$\text{midpoint} = [(-6.8)/2 , (0.2/2)]$$
$$\text{midpoint} = (-3.4, 0.1)$$

The answers were read from the graph to the nearest tenth. -6.8 is precise to the nearest tenth, 2 significant digits. -3.4 has an accuracy of two.

SECTION 8.7 EXERCISES

In Figure 8-2, find the midpoint of the lines connecting the following points. Does your answer appear to lie on the line about halfway between the endpoints?

1. AH
2. BD
3. CA
4. CF
5. GE

8.8 ■ LINEAR EQUATIONS

If the graph of an equation is a straight line, then the equation is called a **linear equation**. A more formal definition involves the concept of **degree**. The degree of a term is the sum of the exponents of the variables or unknowns in the term. $33x^3y^2$ is of degree $3 + 2$ or 5. $10xy$ is of degree $1 + 1$ or 2 or second degree. The highest degree of all the terms in an equation is also the degree of the equation. $25x^2 + 49y^2 = 1$ is a second-degree equation. $4x + 5y = 10$ is a first-degree equation. Equations of the first degree are called linear equations.

EXAMPLE 8.11

Which of these equations are linear?

1. $y = x^2 - 6x + 12$
2. $y = 3x + 7$
3. $xy = -1$
4. $x = 2$
5. $y = -9$
6. $y - 4 = 3(x + 5)$
7. $12x + 5y = -4$

Solution
1. No. x^2 term (parabola).
2. Yes.
3. No. xy term (hyperbola).
4. Yes (vertical line).
5. Yes (horizontal line).
6. Yes.
7. Yes.

Let's create a graph that contains all the x, y ordered pairs such that the difference between x and y is 4. $x - y = 4$. What are some values of x and y that qualify for the graph? $7 - 3 = 4, 4 - 0 = 4, 3 - -1 = 4, -7 - -11 = 4$. Organize the data in a data chart.

x	y
7	3
4	0
3	-1
-7	-11

Plot the points $(7, 3)$, $(4, 0)$, $(3, -1)$, and $(-7, -11)$. See Figure 8-14.

FIGURE 8-14
$x - y = 4$

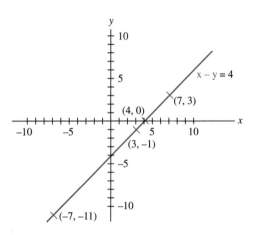

Draw a straight line through the points. Label the graph.

In each case, all four points fell on the same straight line. In fact, there are an infinite number of points that qualify for this graph, and all of them fall on the graph. All the points that satisfy the equation must fall on the graph, and, conversely, all the points that fall on the graph must satisfy the equation.

What would happen if we made the difference between y and x equal to 4, $y - x = 4$? Will the graph be the same? Is subtraction commutative? The values of x and y are swapped.

x	y
3	7
0	4
-1	3
-11	-7

Graph the points $(3, 7)$, $(0, 4)$, $(-1, 3)$, and $(-11, -7)$, and draw a straight line through the points. See Figure 8-15.

FIGURE 8-15
$y - x = 4$

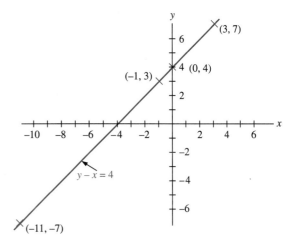

In each case, four points were determined and used to graph the linear equation. In reality, only two points are required. Two points determine a straight line. A third point is usually used as a check. A straight line through two of the points must also cross through the third.

When choosing points to graph, the x and y intercepts, the points where the graph crosses the x and y axes, are good points to choose because they are easy to calculate. Here are some general rules to follow.

Step 1. To find the x-intercept, set y equal to zero and solve for x. For example, in the last equation

$$y - x = 4$$
$$0 - x = 4$$
$$-x = 4$$
$$x = -4 \text{ (multiply each side by } -1\text{)}$$
$$(-4, 0) \; x\text{-intercept}$$

Step 2. To find the y-intercept, set x equal to zero and solve for y. For example, in the last equation

$$y - x = 4$$
$$y - 0 = 4$$
$$y = 4$$
$$(0, 4) \; y\text{-intercept}$$

Step 3. For the third point, choose any non-zero value for x or y and solve for the other. A graph can be drawn more accurately if the points are spread out. Don't be afraid to choose negative coordinates. Take advantage of the space you have to work with; spread your work out. If your three points are too close together, choose a few more. So far, for the last equation, we have points for x values of -4 and 0. Let's calculate a third point at $x = 4$.

$$y - x = 4$$
$$y - 4 = 4$$
$$y = 8$$
(4, 8) third point

Step 4. Draw a straight line through the points. Be sure to label the *x* and *y* axes and label the graph.

EXAMPLE 8.12   Graph $2x + 6y = 21$.

Solution   To find the *x*-intercept, set $y = 0$.

$$2x + 6y = 21,\ 2x + 6 \cdot 0 = 21,\ 2x = 21,\ x = 21/2 = 10\ 1/2$$

The *x* intercept is (21/2, 0).

To find the *y*-intercept, set $x = 0$.

$$2 \cdot 0 + 6y = 21,\ 6y = 21,\ y = 21/6 = 7/2 = 3\ 1/2$$

The *y* intercept is (0, 7/2).

Find a third point. Let $x = -3$.

$$2 \cdot -3 + 6y = 21,\ 6y = 27,\ y = 27/6 = 9/2$$

The third point is $(-3, 9/2)$.

The graph is drawn in Figure 8-16.

FIGURE 8-16
Example 8.12

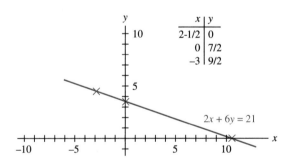

SECTION 8.8 EXERCISES

Graph the following equations by finding the *x* and *y* intercepts and a third point. Draw all graphs on the same axes.

1. $x - y = 6$
2. $x + y = -6$
3. $x - y = -6$
4. $x + y = 6$
5. $x + 2y = 6$
6. $x - 2y = -6$
7. $x - 2y = 6$
8. $x + 2y = -6$

8.9 ■ POINT-SLOPE FORM OF EQUATION

If a line passes through a known point and if the slope of the line is also known, the equation of the line can be written using the **point-slope form** of equation.

$$y - y_1 = m(x - x_1)$$

where (x_1, y_1) is a point the line passes through and m is the slope of the line.

EXAMPLE 8.13 Write the equation of the line that passes through point (1, 4) with a slope of 2/3. Graph the equation.

Solution **Given:** $(x_1, y_1) = (1, 4)$, m = 2/3
Find: equation

$$y - y_1 = m(x - x_1)$$
$$y - 4 = 2/3(x - 1)$$

See Figure 8-17.

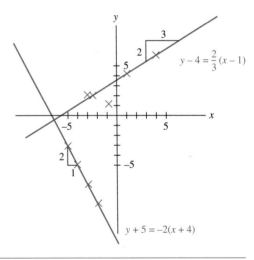

FIGURE 8-17
Examples 8.13 and 8.14

EXAMPLE 8.14 Write the equation of the line that passes through the point (−4, −5) with a slope of −2. Graph the equation.

Solution **Given:** $(x_1, y_1) = (-4, -5)$, m = −2
Find: equation

$$y - y_1 = m(x - x_1)$$
$$y - -5 = -2(x - -4) \text{ or } y + 5 = -2(x + 4)$$

See Figure 8-17.

SECTION 8.9 EXERCISES

Graph the line that passes through the given point with the given slope and write the equation of the line.

1. Point $(-3, 2)$, $m = 2.5$
2. Point $(4, -2)$, $m = -2$
3. Point $(-1, -4)$, $m = -3.5$

8.10 ■ SLOPE-INTERCEPT FORM OF EQUATION

Suppose a line passes through a known point on the y-axis. The x-coordinate of that point is 0. The y-coordinate is given the special name **b**. The point (0, b) is the y-intercept of the line. Substitute (0, b) into the point-slope equation for (x_1, y_1).

$$y - y_1 = m(x - x_1)$$
$$y - b = m(x - 0)$$

Solve for y.

$$y = mx + b$$

where m is the slope and b is the y-intercept.

This is called the **slope-intercept form** of equation.

EXAMPLE 8.15

Write the equation of a line that cuts the y-axis at $(0, -7)$ and has a slope of 3/4. Graph the equation.

Solution Given: m=3/4, b=−7

Find: equation and graph. See Figure 8-18.

$$y = mx + b$$
$$y = 3/4\ x - 7$$

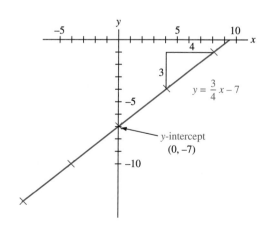

FIGURE 8-18
Example 8.15

8.10 SLOPE-INTERCEPT FORM OF EQUATION ▪ 251

To graph the equation, locate the intercept $(0, -7)$ and put a mark there. Go up 3 and over 4 to the right and put a mark. Repeat. Draw a straight line through the marks.

EXAMPLE 8.16 Write the equation of the line that passes through the point $(-3, 5)$ and is parallel to the line $y + 3 = 2(x + 3)$. Graph the two lines.

Solution **Given:** point $(-3, 5)$
Find: equation of line. See Figure 8-19.

FIGURE 8-19
Example 8.16

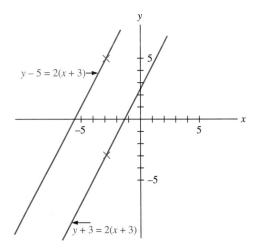

$y + 3 = 2(x + 3)$ has a slope of 2. Parallel lines have the same slope, so make the slope of the new line 2. We know a point that the line passes through and its slope. Use the point-slope form of equation.

$$y - y_1 = m(x - x_1)$$
$$y - 5 = 2(x - -3)$$
$$y - 5 = 2(x + 3)$$

EXAMPLE 8.17 Write the equation of a line that intercepts the y-axis at -1 and is perpendicular to the line $y = 2x - 5$. Graph the two lines.

Solution **Given:** $y = 2x - 5$, and $b = -1$
Find: equation of line through $(0, -1)$ perpendicular to $y = 2x - 5$. See Figure 8-20. The slope of the given line is 2. The slope of a line perpendicular to it is $-1/2$. Use the slope-intercept form of equation.

$$y = mx + b$$
$$y = -1/2\,x - 1$$

FIGURE 8-20
Example 8.17

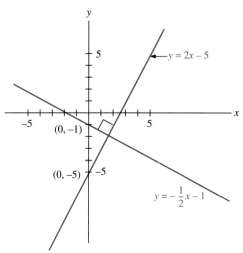

SECTION 8.10 EXERCISES

1. Write the equation for each line in Figure 8-21.

FIGURE 8-21
Section 8.10 Exercises, Problem 1

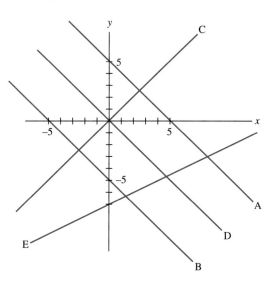

2. Write the equation of the line that passes through the point $(3, -4)$ and is parallel to the line $y - 5 = -1/2(x + 4)$. Graph the two lines.

3. Write the equation of the line that intercepts the y-axis at 4 and is perpendicular to the line $y = -2x + 6$. Graph the two lines.

8.11 ■ GENERAL FORM OF EQUATION

Besides the point-slope form of equation and the slope-intercept form of equation, a linear equation can also be expressed in **general form**.

$$ax + by + c = 0 \text{ is a linear equation in general form.}$$

EXAMPLE 8.18 Express $y = -3x - 12$ in general form.

Solution
$$y = -3x - 12 \quad \text{(slope-intercept form)}$$
$$3x + y + 12 = 0 \quad \text{(general form)}$$

EXAMPLE 8.19 Express $y - 4 = -2(x - 4)$ in general form.

Solution
$$y - 4 = -2(x - 4) \quad \text{(point-slope form)}$$
$$y - 4 = -2x + 8$$
$$2x + y - 12 = 0 \quad \text{(general form)}$$

An efficient way to graph an equation written in general form is to rearrange the equation into slope-intercept form or point-slope form and proceed as above.

EXAMPLE 8.20 Graph $7x + 3y - 6 = 0$.

Solution Solve the equation for y.
$$7x + 3y - 6 = 0$$
$$3y = -7x + 6$$
$$y = -7/3\,x + 2$$

The equation is now in slope-intercept form, $y = mx + b$, where the slope is $-7/3$ and the y-intercept is 2. Locate the intercept. See Figure 8-22. From the intercept $(0, 2)$, go down 7 and over 3 to the right. From the intercept, go up 7 and 3 to the left. These three points should lie on the graph.

FIGURE 8-22
Example 8.20

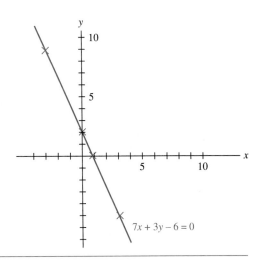

254 ■ CHAPTER 8 LINEAR EQUATIONS AND GRAPHING

**SECTION 8.11
EXERCISES**

Graph by locating points.

A. $5x - 7y + 4 = 0$ **B.** $3y + 4x - 8 = 0$
C. $-2x + 5y = -7$ **D.** $9x - 4y - 3 = 0$

Graph by converting to the slope-intercept form.

E. $2x - 3y + 7 = 0$ **F.** $5x + 7y - 9 = 0$ **G.** $-3x - 8y - 4 = 0$

8.12 ■ GRAPHING EXPERIMENTAL DATA

In many lab exercises in your technical courses, one variable will be varied and its effect on another variable will be measured. The first is called the **independent variable,** and the second is called the **dependent variable.** The collected data forms a set of points that can be graphed to show the relationship between the two variables. The independent variable is often graphed on the horizontal axis, and the dependent variable is graphed on the vertical axis. If the independent variable is called x and is graphed horizontally, and the dependent variable is called y and is graphed vertically, then y depends on x. **y is a function of x.**

In the following example, the applied voltage is varied and the effect on the current is measured. The applied voltage is the independent variable and the current is the dependent variable. The current is a function of the applied voltage.

EXAMPLE 8.21 For the circuit shown in Figure 8-23, the applied voltage V_S was varied in one volt steps. The resulting current, I, was measured with an ammeter.

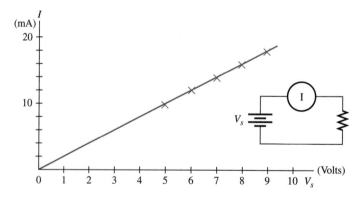

FIGURE 8-23
Example 8.21

V_s	5	6	7	8	9	volts
I	9.9	12.2	14.2	15.8	18.0	mA

a. Graph the results.
b. Write the equation of the resulting line.

c. Use the equation to estimate the current that will flow when the applied voltage is 3 V.

d. Use the equation to estimate the current that will flow when the applied voltage is 27 V.

Solution   See Figure 8-23.

a. Make V_S the horizontal axis. Label it with both the quantity and unit. I will be vertical. Label it. Choose scales for the horizontal and vertical axes that will spread the data out. A common error is to use only a small portion of the space available. Graph the experimental data. The data is linear. **Instead of connecting the dots, draw the "best fit" straight line through the dots that most closely represents all the data.** That is, if a single straight line cannot pass directly through all points, draw the line so that the points are balanced on both sides of the line.

b. Calculate the slope of the line.

$$m = (y_2 - y_1)/(x_2 - x_1)$$
$$m = (18 - 9.9) \times 10^{-3} / (9 - 5)$$
$$m = 2.025 \times 10^{-3}$$

Round m to 2.0×10^{-3}.

The graph passes through the origin. Use the slope-intercept form of equation.

$$y = mx + b$$
$$I = mV_S + b$$
$$I = 2.0 \times 10^{-3} V_S + 0 = 2.0 V_S \text{ mA}$$

c.
$$I = 2.0 V_S \text{ mA}$$
$$I = 2.0 \cdot 3 \text{ mA} = 6.0 \text{ mA}$$

This value falls on the graph in Figure 8-23 and can be confirmed.

d.
$$I = 2.0 V_S \text{ mA}$$
$$I = (2.0 \cdot 27) \text{ mA} = 54 \text{ mA}$$

In the last example, the graph of the current in the circuit as a function of the applied voltage is linear. This means that if the applied voltage doubles, the current that flows also doubles. If the applied voltage is reduced by a factor of 3 (divided by 3), the current is also reduced by a factor of 3.

EXAMPLE 8.22

a. Use the following data to graph the output power in mW versus the current through the device in (mA) for a light-emitting diode (LED) and for an injection laser diode (ILD).

b. From your graph, contrast the response of these two devices to the drive current.

c. How much current is required to produce an output power of 1.4 mW in each device?
d. Calculate the slope of the portion of the laser curve past the threshold (region where laser is conducting).
e. Write the equation of the portion of the laser curve past the threshold.
f. Use your equation to predict the output power of the laser with a drive current of 300 mA.

Drive Current (mA)	0	50	100	150	200	210	220	225	250
LED	0	0.5	1.0	1.2	1.5	1.6	1.6	1.6	1.6
ILD	0	0.1	0.2	0.3	0.4	0.5	0.6	0.9	3.2
				Output Power (mW)					

Solution **a.** See Figure 8-24.

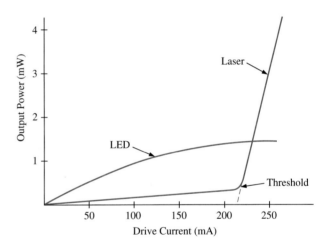

FIGURE 8-24
Example 8.22

b. Laser: The drive current has to increase beyond 200 mA before the output power begins to increase significantly. LED: As the current increases from zero to 150 mA, the output power also increases nearly linearly. At 150 mA, the LED is approaching its maximum output power. Past 150 mA drive current, the increase in LED output power is constant for a large increase of drive current.

c. LED: 200 mA; Laser: 230 mA

d. m=rise/run=(3.2−0.9mW)/(250−225mA)=0.092. The horizontal and vertical units are different and cause the slope to be less than one even though the line is steep.

e. The curve passes through the point (250, 3.2). Use the point-slope form of equation.

$$y - y_1 = m(x - x_1)$$
$$y - 3.2 = 0.092(x - 250)$$

f.
$$y - 3.2 = 0.092(x - 250)$$
$$y - 3.2 = 0.092(300 - 250)$$
$$y = 0.092(50) + 3.2$$
$$y = 7.8 \text{ mW}$$

EXAMPLE 8.23 The voltage amplification of a circuit was measured at a variety of frequencies (hertz).

a. Graph the voltage amplification as a function of frequencies. (Voltage amplification is the ratio of output voltage to input voltage. It has no units.)

b. From your graph, estimate the voltage amplification at 16 250 Hz.

c. From your graph, estimate the voltage amplification at 14 500 Hz.

d. From the graph, predict the frequencies at which the circuit will have a voltage amplification of 9.2.

Frequency in kHz	10	11	12	13	14	15	16	17	18
Voltage gain	10.0	9.9	9.7	8.2	7.1	3.6	5.7	9.8	10.0

Solution **a.** See Figure 8-25.
b. 7
c. 5
d. 12 200 and 16 750 Hz

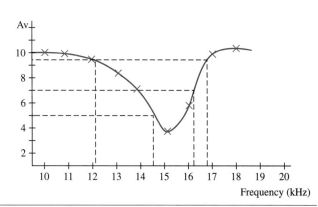

FIGURE 8-25
Example 8.23

SECTION 8.12 EXERCISES

1. MFE910 is a switching transistor. The drain to source voltage, V_{DS}, was varied from 0–60 V in 5 V increments. At each setting, the capacitance of C_{iss} (input capacitance), C_{rss} (reverse transfer capacitance), and C_{oss} (output capacitance) was measured. Here is a data table listing capacitances in picofarads (pF).

V_{DS} (volts)	0	5	10	15	20	25	30	35	40	45	50	55	60
C_{iss}		60	37	34	33	32	31	30	30	30	30	30	30	30
C_{oss}		80	44	30	20	19	18	17	16	15	15	15	15	15
C_{rss}		30	10	4	3	2	2	2	2	2	2	2	2	2

Capacitance in pF

a. Graph the three sets of data on the same coordinate axes. The voltage is the independent variable; assign it to the horizontal axis. Choose a suitable scale that will spread 0–60 volts across the paper with increments that are easy to deal with. Study the data to determine the range of the output data, and choose a scale for the vertical axis. Graph the three sets of data. Draw a smooth curve through the points lightly in pencil. When you are satisfied with the curves, draw them in ink. Label the axes and curves.

From your graphs, answer these questions.
b. At what voltage are C_{iss} and C_{oss} equal?
c. What is the capacitance of C_{oss} at a V_{DS} of 12 volts?
d. At what voltage does $C_{iss} = 43$ pF?

2. A 2N3904 general-purpose transistor was biased with different values of base current from 100–500 μamp in 100 μA steps. At each bias, the voltage from collector to emitter V_{CE} was varied from 0 to 40 volts in 5 V steps. The collector current I_C was measured at each voltage setting. This data table lists I_C in mA.

I_B (μA) V_{CE} (V)	100	200	300	400	500
0	0	0	0	0	0
5	20	35	50	62	71
10	21	37	53	66	74
15	22	40	56	68	78
20	23	42	58	72	82
25	24	44	61	76	85
30	25	46	63	78	88
35	26	48	66	81	92
40	27	50	68	83	95

I_C (mA)

Graph the output characteristic curves for this transistor. Assign V_{CE} to the horizontal axis and I_C to the vertical axis. Draw one curve for each value of I_B.

3. The maximum operating frequency of a 4569 CMOS Programmable, Divide by N, Dual 4-Bit Binary/BCD Down Counter depends on the ambient temperature and the DC supply voltage. To study this relationship, the temperature was varied from −40° to 100°C in 40° increments. At each temperature, the maximum frequency was measured with supply voltages V_{DD} of 15, 10, and 5 volts. The frequencies are listed in this data table in Mhertz.

T_A (°C)	−40	−20	0	20	40	60	80
V_{DD}							
15 V	17.0	15.7	14.5	13.2	12.3	11.5	10.8
10 V	13.2	12.0	10.6	9.5	8.6	7.9	7.2
5 V	4.6	4.2	3.6	3.3	3.0	2.9	2.8

f in MHz

    a. Graph frequency (vertical axis) versus ambient temperature (horizontal axis). You should have one curve for each supply voltage.
        From your graph, find the answers to these questions.
    b. With an operating voltage of 10 V and an ambient temperature of 50°, what is the maximum operating frequency?
    c. At what temperature must the 4569 be maintained to expect an operating frequency of 4 MHz if a 5 V power supply is used?

4. One of the tests performed on a freshly mixed batch of concrete is the slump test. The wet mixture is packed into a cone-shaped form. The form is removed and the distance that the concrete sags or "slumps" is measured. That distance is called the slump. Two batches of cement were mixed, one with 5" slump and one with 2 1/4" slump. Three cylinders were poured from each batch. The cylinders were cured for 7, 14, and 28 days before breaking. The pressure required to crush each cylinder is listed in the data table in psi.

Time in days	7	14	28
5" slump	3236	3625	4316
2 1/4" slump	3183	3590	4209

Breaking strength in psi

    a. Graph the breaking strength or compression strength (vertical axis) for each slump as a function of curing time (horizontal axis).
    b. From your graph, estimate the compression strength of 2 1/4" slump concrete after 21 days of curing.
    c. What is the difference in compression strength in 2 1/4 and 5" slump concrete?
    d. How long does 5" slump concrete have to cure to have a compression strength of 3,750 psi?

5. The voltage gain (G_v) in decibels (dB) for an active notch filter was measured over a range of frequencies. The gain is shown in the following data table.

f(kHz)	3.00	3.05	3.10	3.15	3.20	3.23	3.25	3.30	3.35	3.40	3.45	3.50
G_v (dB)	10	8	7	3	−8	−28	−7	4	7	9	10	11

    a. Graph G_v versus frequency in kHz.
    b. At what frequency is the output a minimum?
    c. At what frequencies will the voltage gain be 0?
    d. What is the gain at 3.40 kHz?

8.13 ■ GRAPHING TRIG FUNCTIONS

Trigonometric functions are studied in detail in Chapter 10, and are applied in many of the following chapters. They are introduced here by observing their graphs. Three trig functions, **sine (sin), cosine (cos),** and **tangent (tan)** are included on scientific calculators. With the use of a scientific calculator, trig functions are easy to graph. To make a graph of the trig function sine, make a table of amplitudes for different angles from 0 to 360°. Enter an angle into your calculator and press the sin key. The corresponding amplitude will appear in the display. Plot degrees on the x-axis and amplitude on the y-axis. Include enough values to determine the range of the amplitude so that a proper scale can be chosen for the y-axis. Label the horizontal axis in degrees, 0 to 360 for one complete cycle of the waveform. Label the vertical axis in units of amplitude: volts if the waveform is voltage; amps if the waveform is current; or, in this case, just generic units to represent amplitude. Once the initial points have been graphed, find the amplitude at additional angles to fill in the gaps. Add the additional information to the data table.

EXAMPLE 8.24   Graph $y = \sin \theta$.

Solution   See Figure 8-26.

$\theta°$	y
0	0
30	.5
60	.87
90	1
120	.87
150	.5
180	0
210	−.5
240	−.86
270	−1
300	−.86
330	−.5
360	0

The amplitude ranges from +1 to −1. Choose a vertical scale that will spread the graph out vertically. The curve changes gradually around the positive and negative peaks. Locate a few more points in those regions.

$\theta°$	y
75	.97
255	−.97

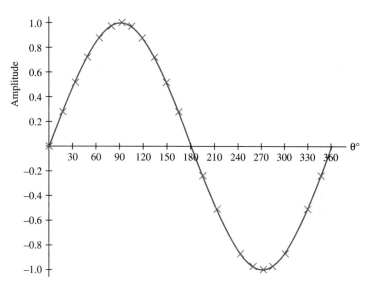

FIGURE 8-26
Example 8.24

Due to the symmetry of the curve, .97 can also be plotted at 105° and −.97 can be plotted at 285°. Add any other points needed to fully define the curve.

θ°	y
15	.26
45	.7

These magnitudes can also be used at 135, 175, 195, 225, 315, and 345°.

SECTION 8.13 EXERCISES, Part 1

Graph these equations.

1. $y = \cos \theta$
2. $y = \tan \theta$

The graph of tangent θ is different from sine θ and cosine θ in several respects. Tan θ completes a cycle in 180°, while sin θ and cos θ require 360°. Sin θ and cos θ range between +1 and −1. Tan θ increases without bound at 90° and 270°.

The vertical lines through $\theta = 90°$ and $\theta = 270°$ that the graph approaches are called **asymptotes**. Tan θ approaches these lines **asymptotically**.

The amplitude of the trig functions can be modified by multiplying the function by a constant. $y = a \cdot \sin \theta$ will have peak amplitudes of +a and −a.

EXAMPLE 8.25

Graph $y = 10 \sin \theta$ and $y = 5 \sin \theta$ and $y = \sin \theta$ on the same axes.

Solution

θ	sin θ	5 sin θ	10 sin θ
0	0	0	0
15	.26	1.3	2.6
30	.5	2.5	5
45	.7	3.5	7
60	.87	4.4	8.7
75	.97	4.8	9.7
90	1	5	10

Because of the symmetry of the sine wave, these values will be repeated in the remainder of the curve. See Figure 8-27.

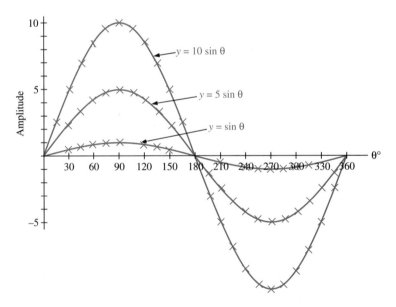

FIGURE 8-27
Example 8.25

A trig function can be shifted to the left or right by modifying the angle. $y = \sin(\theta + A°)$ shifts the function to the left by A°. $y = \sin(\theta + A°)$ is said to **lead** $y = \sin \theta$ by A°. In general, the waveform on the left leads the waveform on the right. $y = \sin(\theta - A°)$ shifts the function to the right by A°. $y = \sin(\theta - A°)$ **lags** $y = \sin \theta$ by A°. In general, a waveform on the right lags the waveform on the left.

8.14 TRIG IDENTITIES

EXAMPLE 8.26 Graph $y = \sin \theta$, $y = \sin (\theta + 30°)$, and $y = \sin (\theta - 30°)$ on the same graph.

Solution

θ	$\sin \theta$	$\theta + 30°$	$\sin(\theta + 30°)$	$\theta - 30°$	$\sin(\theta - 30°)$
0	0	30	.5	−30	−.5
30	.5	60	.87	0	0
60	.87	90	1	30	.5
90	1	120	.87	60	.87
120	.87	150	.5	90	1

For each angle θ, graph $\sin \theta$, $\sin (\theta + 30°)$, and $\sin (\theta - 30°)$. Use symmetry to complete the graphs through a complete cycle. See Figure 8-28.

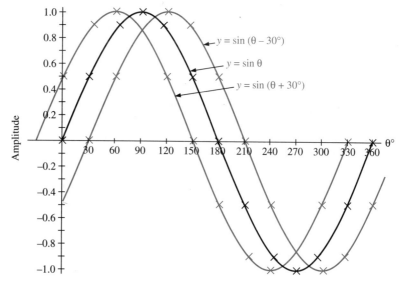

FIGURE 8-28
Example 8.26

SECTION 8.13 EXERCISES, Part 2

3. Graph on the same axes: $y = \cos \theta$, $y = 2 \cos \theta$, and $y = 1/2 \cos \theta$.
4. Graph on the same axes: $y = \cos \theta$, $y = \cos (\theta - 15°)$, and $y = \cos (\theta - 30°)$.
5. Graph: $y = 3 \sin (\theta + 30°)$.

8.14 ■ TRIG IDENTITIES

Whereas equations hold true only for specific values of the variables, **trig identities** are true for any value of the variable. Trig identities are numerous, and only a few are studied here as graphing exercises.

EXAMPLE 8.27 Graph $y = \sin^2 \theta + \cos^2 \theta$.

Solution This equation is read "y equals sine squared θ +cosine squared θ." Choose a value of θ and use your calculator to find the corresponding value of y. For 30°, the sequence used is 3 0 sin x^2 + 3 0 cos x^2 = . Several values were calculated, graphed, and plotted in Figure 8-29. As can be seen in the graph, for any θ, $\sin^2 \theta + \cos^2 \theta$ always equals 1.

$$\sin^2 \theta + \cos^2 \theta \equiv 1$$

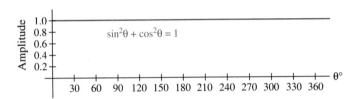

FIGURE 8-29
Example 8.27

There are three more trig functions that are not included on scientific calculators: **secant (sec), cosecant (csc),** and **cotangent (cot).** Here are three examples of trig identities.

$$\cot \theta \equiv 1/\tan \theta$$
$$\sec \theta \equiv 1/\cos \theta$$
$$\csc \theta \equiv 1/\sin \theta$$

These statements are true for any value of θ.

SECTION 8.14 EXERCISES

1. Graph $y = \sin \theta / \cos \theta$. From the result write, a trig identity.
2. Graph $y = \cot \theta$.
3. Graph $y = \sec \theta$.
4. Graph $y = \csc \theta$.

■ **SUMMARY**

1. The x and y axes divide the Cartesian coordinate plane into four quadrants. The upper right quadrant is called Quadrant I, the upper left is Quadrant II, the lower left is Quadrant III, and the lower right is Quadrant IV.

2. A point in the plane is specified by an ordered pair of numbers (the order of the numbers is critical) written in parentheses, with the x-coordinate first and then the y-coordinate, (x, y).

3. The distance d between two points (x_1, y_1) and (x_2, y_2) is calculated by the equation

$$d = [(y_2 - y_1)^2 + (x_2 - x_1)^2]^{1/2}$$

4. The slope m of a line is the vertical rise divided by the horizontal run. A line that rises to the right has a positive slope, and a line that falls to the right has a negative slope.

5. The slope of a line that passes through the points (x_2, y_2) and (x_1, y_1) is calculated by

$$m = (y_2 - y_1)/(x_2 - x_1)$$

6. The steeper the line, the greater the magnitude of the slope. A line with a slope of 2 is steeper than a line with a slope of 1/2.
7. Lines with the same slope are parallel, and lines that are parallel to each other have the same slope.
8. If two lines lie in the same plane and the slope of one is the negative reciprocal of the other, the lines are perpendicular; if two lines are perpendicular, their slopes are negative reciprocals. To form the negative reciprocal, change the sign and invert the magnitude.
9. The midpoint of the line connecting $A(x_2, y_2)$ and $B(x_1, y_1)$ is

$$\text{midpoint} = [(x_2 + x_1)/2 , (y_2 + y_1)/2]$$

10. The degree of a term is the sum of the exponents of the variables or unknowns in the term.
11. The highest degree of all the terms in an equation is also the degree of the equation.
12. All the points that satisfy the equation must fall on the graph, and, conversely, all the points that fall on the graph must satisfy the equation.
13. The x and y intercepts, the points where the graph crosses the x and y axes, are good points to graph because they are easy to calculate. To find the x-intercept, set y equal to zero and solve for x. To find the y-intercept, set x equal to zero and solve for y.
14. The equation of a line that passes through point (x_1, y_1) with slope m is given by the point-slope form of equation.

$$y - y_1 = m(x - x_1)$$

15. The equation of a line that intercepts the y-axis at b with slope m is given by the slope-intercept form of equation

$$y = mx + b$$

16. A linear equation can be expressed in general form.

$$ax + by + c = 0$$

17. In many lab exercises, one variable will be varied, and its effect on another variable will be measured. The first is called the independent variable, and the second is called the dependent variable.
18. The graph of $y = \sin \theta$ and $y = \cos \theta$ ranges from $+1$ to -1 and completes a cycle in $360°$.
19. The graph of $y = \tan \theta$ increases without bound at $90°$ and $270°$ and completes a cycle in $180°$.
20. A trig function can be shifted to the left or right by modifying the angle. $y = \sin(\theta + A°)$ shifts the function to the left by $A°$. $y = \sin(\theta + A°)$ is said to lead $y = \sin \theta$ by $A°$.

21. $y = \sin(\theta - A°)$ shifts the function to the right by $A°$. $y = \sin(\theta - A°)$ lags $y = \sin \theta$ by $A°$.

22. A trig identity is true for any value of θ. Here are two trig identities.

$$\sin^2 \theta + \cos^2 \theta = 1$$

$$\sin \theta / \cos \theta = \tan \theta$$

SOLUTIONS TO CHAPTER 8 SECTION EXERCISES

Solutions to Section 8.1 Exercises

1. See Figure 8-30.

FIGURE 8-30
Section 8.1 Exercises, Problem 1

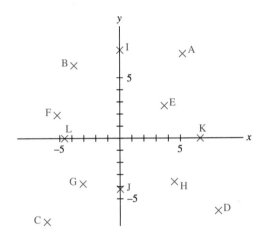

2. I(5, 4), J(−7, −4), K(−8, 3), L(5, −2.5), M(3.5, 0), N(0, −3.8), O(−5.5, −2.5), P(0, 4.2), Q(−7.8, 0), R(8.2, 2.2), S(8.7, −6.1), T(−9.4, −5.7), U(−0.8, 6.3), V(1.7, −6.6), W(−1.2, −7.1), X(−8.8, 0.9), Y(9.3, 1.4), Z(3.6, 1.7)

Solutions to Section 8.2 Exercises

1. 8.9 m **2.** 11.3 m **3.** 13.5 m
4. 9.4 m **5.** 9.0 m **6.** No
7. AEDHCGBFA

Solution to Section 8.3 Exercises

1. BA, BF, BG
3. See Figure 8-31.

2. $m_{BA} = 1/3$
m_{BC} undefined (cannot divide by 0)
$m_{BD} = -1$
$m_{BE} = -3/8$
$m_{BF} = 1/5$
$m_{BG} = 3$
$m_{BH} = -1$

FIGURE 8-31
Section 8.3 Exercises, Problem 3

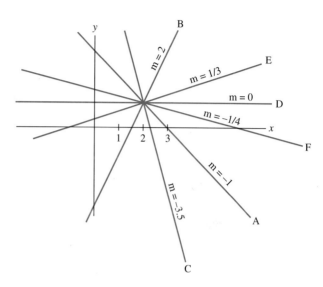

Solutions to Section 8.4 Exercises

1.–4. See Figure 8-32.

FIGURE 8-32
Solution to Section 8.4 Exercises, Problems 1 through 4

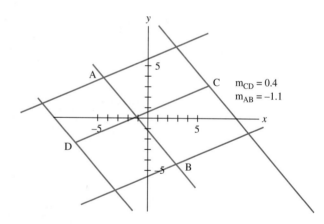

Solutions to Section 8.5 Exercises

1.–4. See Figure 8-33.

FIGURE 8-33
Solution to Section 8.5 Exercises, Problems 1 through 4

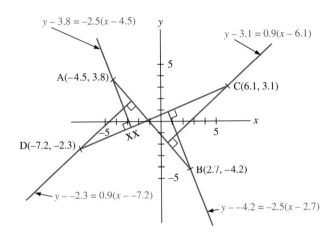

SOLUTIONS TO SECTION 8.6 EXERCISES

1. 105 ft

2. 2.34%

Solutions to Section 8.7 Exercises

1. (2, 2) **2.** (−1, −1) **3.** (−1/2, 1)
4. (−5/2, 0) **5.** (−3/2, 0)

Solutions to Section 8.8 Exercises

1.–8. See Figure 8-34.

FIGURE 8-34
Solutions to Section 8.8 Exercises, Problems 1 through 8

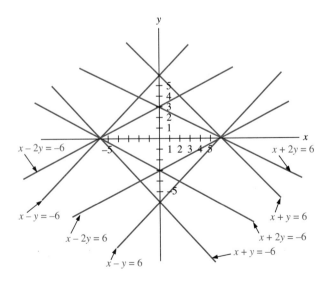

Solution to Section 8.9 Exercises

1.–3. See Figure 8-35.

FIGURE 8-35
Solution to Section 8.9 Exercises, Problems 1 through 3

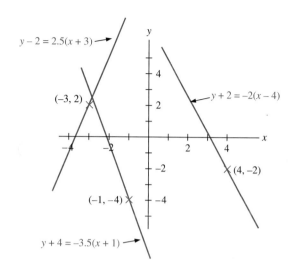

Solutions to Section 8.10 Exercises

1. A $y = -x + 5$
   B $y = -x - 5$
   C $y = 1x + 0 = x$
   D $y = -x$
   E $y = x/2 - 7$
2. See Figure 8-36.

FIGURE 8-36
Solution to Section 8.10 Exercises, Problem 2

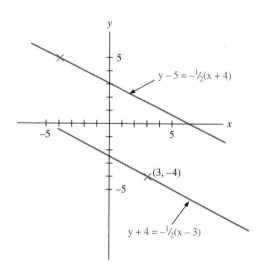

3. See Figure 8-37.

FIGURE 8-37
Solution to Section 8.10 Exercises, Problem 3

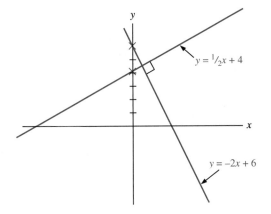

Solutions to Section 8.11 Exercises

A–D. See Figure 8-38.

FIGURE 8-38
Solutions to Section 8.11 Exercises, A through D

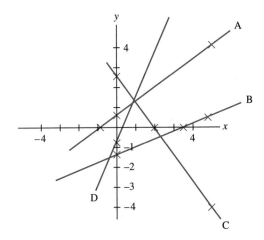

E–H. See Figure 8-39.

FIGURE 8-39
Solution to Section 8.11 Exercises, E through G

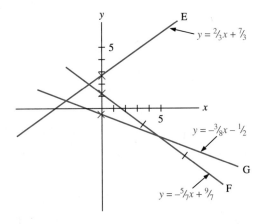

Solutions to Section 8.12 Exercises

1. a. See Figure 8-40.
   b. 8 V
   c. 28 pF
   d. 2 V

FIGURE 8-40
Solution to Section 8.12 Exercises, Problem 1

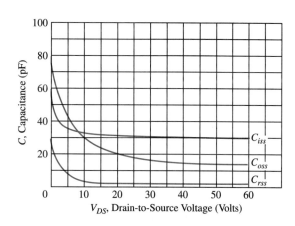

2. See Figure 8-41.

FIGURE 8-41
Solution to Section 8.12 Exercises, Problem 2

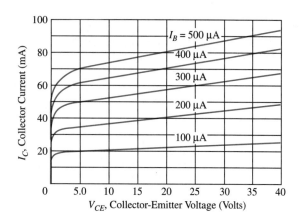

3. a. See Figure 8-42.
   b. 8.2 MHz
   c. $-17°C$

4. a. See Figure 8-43.
   b. 3.9×10^3 psi
   c. 100 psi
   d. 16 days

FIGURE 8-42
Solution to Section 8.12 Exercises, Problem 3

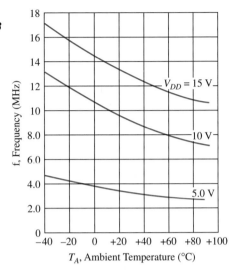

FIGURE 8-43
Solution to Section 8.12 Exercises, Problem 4

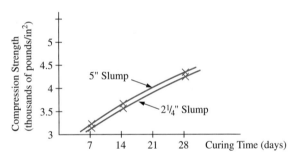

5. a. See Figure 8-44.
   b. 3.21 kHz
   c. 3.16 and 3.26 kHz
   d. 10 dB

FIGURE 8-44
Solution to Section 8.12 Exercises, Problem 5

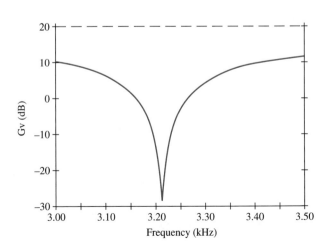

Solution to Section 8.13 Exercises, Part 1

1. See Figure 8-45.

FIGURE 8-45
Solution to Section 8.13 Exercises, Problem 1

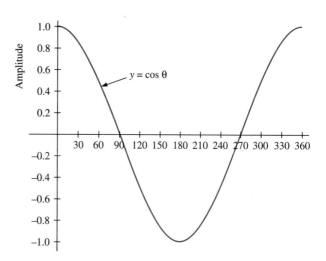

2. See Figure 8-46.

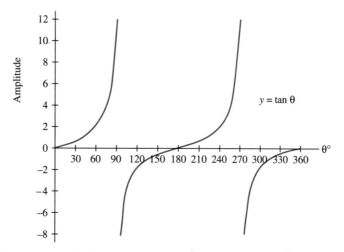

FIGURE 8-46
Solution to Section 8.13 Exercises, Problem 2

Solutions to Section 8.13 Exercises, Part 2

3. See Figure 8-47.

4. See Figure 8-48.

5. See Figure 8-49.

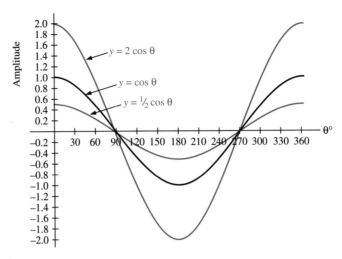

FIGURE 8-47
Solution to Section 8.13 Exercises, Problem 3

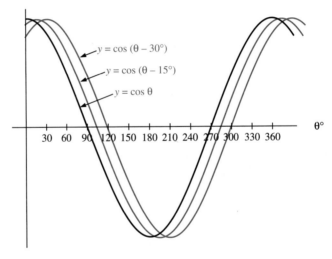

FIGURE 8-48
Solution to Section 8.13 Exercises, Problem 4

FIGURE 8-49
Solution to Section 8.13 Exercises, Problem 5

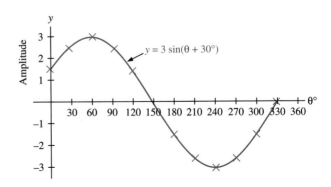

Solution to Section 8.14 Exercises

1. See Figure 8-50.

FIGURE 8-50
Solution to Section 8.14 Exercises, Problem 1

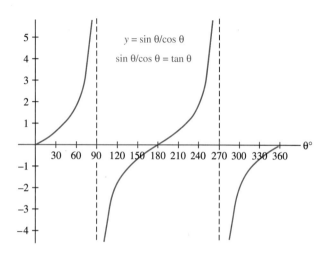

2. See Figure 8-51.

FIGURE 8-51
Solution to Section 8.14 Exercises, Problem 2

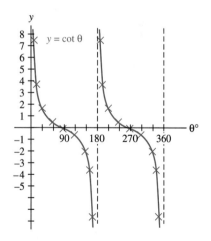

3. See Figure 8-52.

4. See Figure 8-53.

FIGURE 8-52
Solution to Section 8.14 Exercises, Problem 3

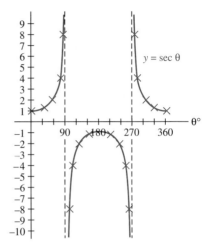

FIGURE 8-53
Solution to Section 8.14 Exercises, Problem 4

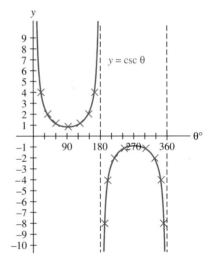

CHAPTER 8 PROBLEMS, PART 1

Figure 8-54 shows the drill pattern for the holes to be drilled in a printed circuit board. U_1 and U_3 are 14 pin dual in-line package (DIP) integrated circuits, and U_2 is a 16 pin DIP. The DIPs will be inserted into holes that will be drilled at each x. The rectangular outline of each DIP is shown. The holes are spaced on 0.1" centers with 0.3" between the two rows of holes.

$R_1 - R_4$ are resistors, $C_1 - C_3$ are capacitors, and D_1 is a diode. Each of these components requires two holes 0.4" apart. The center of Index A is the reference from which the centers of all other holes are measured. Its coordinates are (0, 0). The coordinates of all other holes are measured from Index A as an abscissa (to the right) and an ordinate (up). These holes are to be drilled on a numerical control machine that is precise to the thousandth of an inch, so all coordinates should be specified to the nearest thousandth of an inch. Index A and Index B will be used to align the artwork when traces are etched into the copper. The coordinates of Index B are (2.100, 1.300).

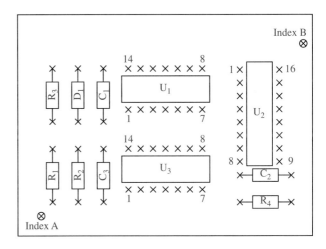

FIGURE 8-54
Printed Circuit Board, Chapter 8 Problems, Part 1

1. Calculate the coordinates of the center of the board.
2. Complete the drill chart.

	Abscissa x	Ordinate y
Index A	0	0
Index B	2.100	1.300
U_1 pin 1	0.700	0.800
U_1 pin 2	_____	_____
U_1 pin 3	_____	_____
U_1 pin 4	_____	_____
U_1 pin 5	_____	_____
U_1 pin 6	_____	_____
U_1 pin 7	_____	_____
U_1 pin 8	_____	_____
U_1 pin 9	_____	_____
U_1 pin 10	_____	_____
U_1 pin 11	_____	_____
U_1 pin 12	_____	_____
U_1 pin 13	_____	_____

U_1 pin 14	——	——
U_2 pin 1	1.600	1.100
U_2 pin 2	——	——
U_2 pin 3	——	——
U_2 pin 4	——	——
U_2 pin 5	——	——
U_2 pin 6	——	——
U_2 pin 7	——	——
U_2 pin 8	——	——
U_2 pin 9	——	——
U_2 pin 10	——	——
U_2 pin 11	——	——
U_2 pin 12	——	——
U_2 pin 13	——	——
U_2 pin 14	——	——
U_2 pin 15	——	——
U_2 pin 16	——	——
U_3 pin 1	0.700	0.200
U_3 pin 2	——	——
U_3 pin 3	——	——
U_3 pin 4	——	——
U_3 pin 5	——	——
U_3 pin 6	——	——
U_3 pin 7	——	——
U_3 pin 8	——	——
U_3 pin 9	——	——
U_3 pin 10	——	——
U_3 pin 11	——	——
U_3 pin 12	——	——
U_3 pin 13		——
U_3 pin 14	——	——
R_1 top	——	——
R_1 bottom	0.100	0.100
R_2 top	——	——
R_2 bottom	0.300	0.100

R_3 top	_____	_____
R_3 bottom	0.100	0.700
R_4 left	1.600	0.100
R_4 right	_____	_____
C_1 top	_____	_____
C_1 bottom	0.500	0.700
C_2 left	1.600	0.300
C_2 right	_____	_____
C_3 top	_____	_____
C_3 bottom	0.500	0.100
D_1 top	_____	_____
D_1 bottom	0.300	0.700

Calculate the distance between

3. U_1 pin 14, U_1 pin 7
4. U_3 pin 1, U_3 pin 8
5. C_2 left, R_3 top
6. U_3 pin 1, U_1 pin 8
7. D_1 top, R_4 right
8. U_2 pin 16, U_2 pin 9
9. Index A, U_3 pin 8
10. Index A, center of board

■ CHAPTER 8 PROBLEMS, PART 2

1. Graph these points: A $(-7, 4)$, B $(8, 2)$, C $(-5, 7)$, and D $(4, -3)$.
2. Using the points in Problem 1, calculate the slope of lines AB, AC, AD, BC, BD, and CD.
3. Are any of the lines in Problem 2 parallel to each other?
4. Are any of the lines in Problem 2 perpendicular to each other?
5. Using the points from Problem 1, find the midpoint of lines AB, AC, AD, BC, BD, and CD.
6. Using the points from Problem 1, find the distance between points A and B, A and C, A and D, B and C, B and D, and C and D.
7. Write the equation of a line that passes through (6, 4) with a slope of -3.
8. Write the equation of a line that passes through $(-4, -2)$ with a slope of 2/3.
9. Write the equation of a line that intercepts the *y*-axis at -5 and has a slope of 3.

10. Write the equation of a line that intercepts the y-axis at 4 and has a slope of −5.
11. Write the equation of a line that passes through (−3, 4) and is parallel to the line $y = -2x + 3$.
12. Write the equation of a line that passes through (5, −5) and is parallel to the line $y = 8x - 7$.
13. Write the equation of a line that passes through (3, −2) and is perpendicular to the line $y = -3x - 4$.
14. Write the equation of a line that passes through (2, 2) and is perpendicular to the line $y = -x + 4$.
15. Write the equation of each line in Figure 8-55.

FIGURE 8-55
Chapter 8 Problems, Part 2, Problem 15

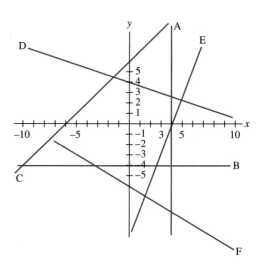

16. Graph by selecting points.
    A. $3y + 7x = 4$
    B. $-4x - 3y = 6$
    C. $6x - 6y = 1$
    D. $8x - 2y = 15$
17. Graph by converting to slope-intercept form.
    A. $3y + 7x = 4$
    B. $-4x - 3y = 6$
    C. $6x - 6y = 1$
    D. $8x - 2y = 15$
18. Graph on the same axes: $y = 10 \sin \theta$, $y = 10 \cos \theta$, $y = 10 \tan \theta$.
19. Graph on the same axes: $y = 10 \cos \theta$, $y = 10 \cos(\theta - 22.5)$, $y = 10 \sin(\theta + 22.5)$.
20. Graph: $y = 5 \sin(x + 45°)$.
21. Graph on the same axes: $y = \sin x$, $y = 2 \sin x$, $y = 3 \sin x$.

9
Systems of Linear Equations

CHAPTER OUTLINE

- **9.1** Graphical Solution
- **9.2** Solution by the Substitution Method
- **9.3** Solution by the Addition Method
- **9.4** General Solution of Two Equations, Two Unknowns
- **9.5** Matrices and Determinants
- **9.6** Cramer's Rule
- **9.7** Loop Equations

OBJECTIVES

After completing this chapter, you should be able to

1. solve systems of equations graphically.
2. identify systems as independent, inconsistent, or dependent.
3. solve systems of equations by the substitution method.
4. solve systems of equations by the addition method.
5. use the general solution to solve a system.
6. define matrix and determinant.
7. solve systems of equations using Cramer's rule.
8. apply Cramer's rule to DC circuits.

NEW TERMS TO WATCH FOR

system of equations
independent
inconsistent
dependent
substitution method
addition method

general solution
matrix
determinant
Cramer's rule
loop equation

9.1 ■ GRAPHICAL SOLUTION

$12x - 60 = 0$ is a linear equation with one unknown, x. $y = 12x + 7$ is a linear equation with two unknowns, x and y. There is one solution to the first equation, $x = 5$. The second equation has an infinite number of solutions. Each point on the graph of $y = 12x + 7$ is a solution to the equation.

Systems of equations will be studied in this chapter. A system is a set of equations being considered as a group. For example, here is a system of linear equations that contains two equations with two unknowns.

$$(1)\ y = 2x - 3$$
$$(2)\ y = -1/2\ x + 1$$

Each equation in the system imposes its own set of conditions on the solution to the system. The solution or solutions are those values that satisfy each of the equations in the system. An approximate solution to a system of equations can be found graphically by graphing each of the equations. The solution or solutions are those points at which the graphs intersect. This graphical solution is an approximation of the actual solution. To check the solution, substitute the coordinates into the original equations to see if they are satisfied.

EXAMPLE 9.1 Solve this system of equations graphically.

$$(1)\ y = 2x - 3$$
$$(2)\ y = -1/2\ x + 1$$

Solution Equation (1) has a slope of 2 and an intercept of −3. See Figure 9-1. Equation (2) has a slope of −1/2 and an intercept of 1. Their point of intersection is approximately (1.6, 0.2). Since these two lines intersect at one point, they are said to be **independent** and the system has a solution.

Check: (1) $y = 2x - 3$
$0.2 = 2 \cdot 1.6 - 3$
$0.2 = 3.2 - 3$
$0.2 = 0.2$

FIGURE 9-1
Example 9.1

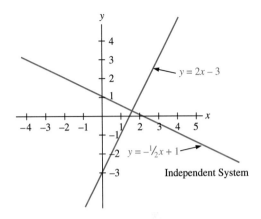

Independent System

(2) $y = -1/2\, x + 1$
$0.2 = -1/2 \cdot 1.6 + 1$
$0.2 = -0.8 + 1$
$0.2 = 0.2$

This graphical solution exactly satisfies both equations.

EXAMPLE 9.2 Solve this system of equations graphically.

$$y = -3x + 2$$
$$y = -3x - 3$$

FIGURE 9-2
Example 9.2

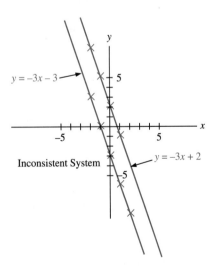

Inconsistent System

Solution These lines both have a slope of −3. There is no point of intersection because the lines are parallel. Since these two lines do not intersect, they are said to be **inconsistent,** and the system has no solution. See Figure 9-2.

EXAMPLE 9.3

Solve this system of equations graphically.

$$y = -3x + 4$$
$$2y = -6x + 8$$

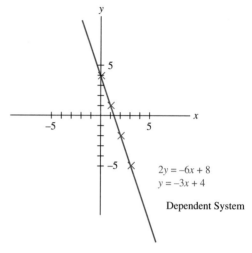

FIGURE 9-3
Example 9.3

Solution Divide both sides of the second equation by 2 and it reduces to the first equation. The graphs of the two are identical. See Figure 9-3. Since the two graphs lie on top of each other, the system is said to be **dependent**.

To summarize:

1. The graphs of the equations of an independent system of linear equations intersect at one point, which is the solution.

2. The graphs of the equations of an inconsistent system of linear equations are parallel lines. There is no solution to an inconsistent system.

3. The graphs of the equations of a dependent system of linear equations are identical. The equations are the same.

SECTION 9.1 EXERCISES

Solve each system graphically.

1. $y = 1/2x + 6$
   $y = -2x - 1$

2. $2y + 3x = 4$
   $3y - 4x = 3$

3. $y - 2 = -2(x + 3)$
   $y - 6 = -1/2(x - 2)$

4. $y = 2x - 2$
   $y = 2x + 4$

9.2 ■ SOLUTION BY THE SUBSTITUTION METHOD

If one of the equations is easy to solve for one of the variables, then the **substitution method** is a good choice for solving the system. Solve for one of the variables and then substitute into the other equation.

9.2 SOLUTION BY THE SUBSTITUTION METHOD

EXAMPLE 9.4 Solve by substitution.

$$(1)\ y - 2 = 4(x - 12)$$
$$(2)\ y + 8 = 3(x + 4)$$

Solve the first equation for y by adding 2 to each side.

$$(1)\ y = 4(x - 12) + 2$$

Substitute $4(x - 12) + 2$ for y in equation (2).

$$(2)\ 4(x - 12) + 2 + 8 = 3(x + 4)$$

Now we have an equation with just one unknown, x. It is a bit of a mess, but y has been eliminated.

$$4x - 48 + 10 = 3x + 12$$
$$x = 12 + 38$$
$$x = 50$$

Substitute 50 for x back in equation (1).

$$(1)\ y - 2 = 4(50 - 12)$$
$$y = 4 \cdot 38 + 2$$
$$y = 154$$

(50,154) is the solution to this system. It should satisfy each of the equations.

Check: The solution can be checked by substituting 50 for x and 154 for y in each of the original equations.

$$(1)\quad y - 2 = 4(x - 12)$$
$$154 - 2 = 4(50 - 12)$$
$$152 = 4\,(38)$$
$$152 = 152$$

$$(2)\quad y + 8 = 3(x + 4)$$
$$154 + 8 = 3(50 + 4)$$
$$162 = 162$$

(50,154) satisfies both equations.

EXAMPLE 9.5 Solve by substitution.

$$(1)\ 3x - 4y = 6$$
$$(2)\ 5x + 2y = 7$$

Solution Solve equation (1) for x.

$$3x - 4y = 6$$
$$3x = 6 + 4y$$
$$x = (6 + 4y)/3$$

Substitute $(6 + 4y)/3$ for x in equation (2).

$$5(6 + 4y)/3 + 2y = 7$$

Solve for y.

$$5(6 + 4y)/3 + 2y = 7$$
$$5(6 + 4y) + 6y = 21$$
$$30 + 20y + 6y = 21$$
$$20y + 6y = 21 - 30$$
$$26y = -9$$
$$y = -9/26$$

Substitute $-9/26$ for y in equation (2) and solve for x.

$$5x + 2y = 7$$
$$5x + 2(-9/26) = 7$$
$$5x + -9/13 = 7$$
$$5x = 7 + 9/13$$
$$5x = 91/13 + 9/13$$
$$5x = 100/13$$
$$x = 100/(13 \cdot 5)$$
$$x = 20/13$$

$(20/13, -9/26)$ should satisfy both equations.

Check:
$$(1) 3x - 4y = 6$$
$$3(20/13) - 4(-9/26) = 6$$
$$60/13 + 18/13 = 6$$
$$78/13 = 6$$
$$6 = 6$$
$$(2)\ 5x + 2y = 7$$
$$5(20/13) + 2(-9/26) = 7$$
$$100/13 - 9/13 = 7$$
$$91/13 = 7$$
$$7 = 7$$

It checks!

SECTION 9.2 EXERCISES

Solve these systems by substitution. They are the same systems you graphed in the last exercise. Compare the solutions from the two methods.

1. $y = 1/2x + 6$
   $y = -2x - 1$
2. $2y + 3x = 4$
   $3y - 4x = 3$
3. $y - 2 = -2(x + 3)$
   $y - 6 = -1/2(x - 2)$
4. $y = 2x - 2$
   $y = 2x + 4$

9.3 ■ SOLUTION BY THE ADDITION METHOD

To solve a system of equations by the **addition method**, manipulate the equations using the multiplication property of equality so that the coefficients of one variable in each equation are additive inverses, such as $5x$ and $-5x$. Then add the two equations together to eliminate that variable. Solve for the other variable, and substitute back into one of the original equations to solve for the variable that was eliminated. This method is also called the *elimination method*.

EXAMPLE 9.6   Solve by the addition method.

$$(1)\ 3x + 5y = 7$$
$$(2)\ -2x - y = 3$$

Solution   Multiply both sides of equation (2) by 5 so that the coefficients of y are $+5$ and -5.

$(1)\quad 3x + 5y = 7$
$(2)\ -10x - 5y = 15$

Add equation (1) to equation (2).

$(1)\quad 3x + 5y = 7$
$(2)\ \dfrac{-10x - 5y = 15}{-7x = 22}$
$x = -22/7$

Substitute $-22/7$ for x in one of the original equations, the one easier to work with.

$(2)\ -2(-22/7) - y = 3$
$44/7 - 3 = y$
$y = 44/7 - 21/7$
$y = 23/7$

$(-22/7, 23/7)$ is the solution if it checks.

Check:
$(1)\quad 3x + 5y = 7$
$3(-22/7) + 5(23/7) = 7$
$-66/7 + 115/7 = 7$
$49/7 = 7$
$7 = 7$

288 ■ CHAPTER 9 SYSTEMS OF LINEAR EQUATIONS

$$(2) \quad -2x - y = 3$$
$$-2(-22/7) - (23/7) = 3$$
$$44/7 - 23/7 = 3$$
$$21/7 = 3$$
$$3 = 3$$

It checks!

EXAMPLE 9.7 Solve by the addition method.

$$(1) \ 3x - 4y = 6$$
$$(2) \ 5x - 3y = 7$$

Solution Multiply equation (1) by 5 and equation (2) by −3 to make the x coefficients 15 and −15.

$$(1) \ 15x - 20y = 30$$
$$(2) \ -15x + 9y = -21$$

Add equation (1) to equation (2) to eliminate x.

$$15x + -15x - 20y + 9y = 30 + -21$$
$$-11y = 9$$

Solve for y.

$$y = -9/11$$

Substitute −9/11 for y in equation (2) and solve for x.

$$5x - 3(-9/11) = 7$$
$$5x + 27/11 = 7$$
$$5x = 7 - 27/11$$
$$5x = 77/11 - 27/11$$
$$5x = 50/11$$
$$x = 50/(11 \cdot 5)$$
$$x = 10/11$$

(10/11, −9/11) should satisfy the original equations.

Check: (1) $3x - 4y = 6$
$$3(10/11) - 4(-9/11) = 6$$
$$30/11 + 36/11 = 6$$
$$66/11 = 6$$
$$6 = 6$$

$$(2) \quad 5x - 3y = 7$$
$$5(10/11) - 3(-9/11) = 7$$
$$50/11 + 27/11 = 7$$
$$77/11 = 7$$
$$7 = 7$$

It checks!

SECTION 9.3 EXERCISES

Solve by the addition method. These are the same systems you graphed and solved by substitution.

1. $y = 1/2x + 6$
   $y = -2x - 1$
2. $2y + 3x = 4$
   $3y - 4x = 3$
3. $y - 2 = -2(x + 3)$
   $y - 6 = -1/2(x - 2)$
4. $y = 2x - 2$
   $y = 2x + 4$

9.4 ■ GENERAL SOLUTION OF TWO EQUATIONS, TWO UNKNOWNS

$ax + by = c$ and $dx + ey = f$, where a, b, c, d, e, and f are constants, are two linear equations in two unknowns, x and y. Let's use the addition method to solve the system for y in terms of a, b, c, d, e, and f, a **general solution**.

To solve for y, manipulate the coefficients of the x terms until they are additive inverses.

$$(1) \; ax + by = c$$
$$(2) \; dx + ey = f$$

Multiply both sides of equation (1) by d, and multiply both sides of equation (2) by $-a$.

$$(1) \quad adx + bdy = cd$$
$$(2) \; -adx + -aey = -af$$

The first terms of each equation are additive inverses. Add equation (2) to equation (1).

$$adx + -adx + bdy + -aey = cd + -af$$
$$bdy - aey = cd - af$$

Factor out y.

$$y(bd - ae) = cd - af$$

Divide both sides by $(bd-ae)$.

$$y = (cd - af)/(bd - ae)$$

Repeat the process for x. To solve for x, manipulate the coefficients of the y terms until they are additive inverses.

Multiply both sides of equation (1) by e, and multiply both sides of equation (2) by –b.

(1) $aex + bey = ce$
(2) $-bdx + -bey = -bf$

The second terms of each equation are additive inverses. Add equation (2) to equation (1).

$$aex + -bdx + bey + -bey = ce + -bf$$
$$aex - bdx = ce - bf$$

Factor out x.

$$x(ae - bd) = ce - bf$$

Divide both sides by (ae−bd).

$$x = (ce - bf)/(ae - bd)$$
$$y = (cd - af)/(bd - ae)$$

In these general solutions, the denominators have the same terms, but the signs are different. The signs can be swapped by factoring −1 out of the numerator and the denominator.

$$y = -1(-cd + af)/-1(-bd + ae)$$

The −1's cancel. Rearrange the terms in the numerator and denominator.

$$y = (af - cd)/(ae - bd)$$

Now the solutions for x and y have the same denominators.

$$x = (ce - bf)/(ae - bd)$$
$$y = (af - cd)/(ae - bd)$$

where $ax + by = c$
$dx + ey = f$

These solutions can be used to solve systems of two equations, two unknowns.

EXAMPLE 9.8 Solve this system by using the general solutions.

(1) $2x - 7y = 12$
(2) $15x + 6y = 8$

Solution Compare $2x - 7y = 12$ with $ax + by = c$ and $15x + 6y = 8$ with $dx + ey = f$

$a = 2$ $b = -7$ $c = 12$ $d = 15$ $e = 6$ $f = 8$

Substitute these values into the general solutions.

$$x = (ce - bf)/(ae - bd)$$
$$x = (12 \cdot 6 - -7 \cdot 8)/(2 \cdot 6 - -7 \cdot 15)$$

Here is a good place to use parentheses on your calculator. Enter the expression just as it is written, including the parentheses.

$$x = 1.1$$
$$y = (cd - af)/(bd - ae)$$
$$y = (12 \cdot 15 - 2 \cdot 8)/(-7 \cdot 15 - 2 \cdot 6)$$
$$y = -1.4$$

(1.1,−1.4) should be the answer. Let's see if it checks.

Check: (1) $2x - 7y = 12$, $2 \cdot 1.1 - 7 \cdot -1.4 = 12$, $12 = 12$
(2) $15x + 6y = 8$, $15 \cdot 1.1 + 6 \cdot -1.4 = 8$, $8.1 = 8$

There is a bit of discrepancy in the check due to round-off error. (1.1,−1.4) is close to the true solution.

The general solutions are not easy to remember, but fortunately this process can be automated by the use of matrices and determinants.

9.5 ■ MATRICES AND DETERMINANTS

A **matrix** is an array of coefficients.

$$\begin{matrix} a & b \\ d & e \end{matrix}$$

is a matrix of two rows (horizontal) and two columns (vertical). It is called a 2×2 (two by two) matrix.

$$\begin{matrix} a & b \\ d & e \end{matrix}$$

is a matrix consisting of the coefficients of the x and y terms of the system of equations

FIGURE 9-4
Evaluating a Determinant

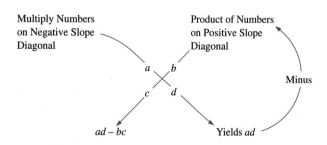

$$\begin{vmatrix} a & b \\ c & d \end{vmatrix} = ad - bc$$

$$(1)\ ax + by = c$$
$$(2)\ dx + ey = f.$$

A **determinant** is the value of a matrix when it is evaluated according to the following rule. Multiply the numbers on the negative slope diagonal and subtract the product of the numbers on the positive slope diagonal, as shown in Figure 9-4. Vertical lines are drawn to the left and right of the matrix to represent its determinant.

$$\begin{vmatrix} a & b \\ c & d \end{vmatrix} = ad - bc$$

EXAMPLE 9.9 Evaluate:

$$\begin{vmatrix} 3 & -2 \\ 4 & 9 \end{vmatrix}$$

Solution

$$\begin{vmatrix} 3 & -2 \\ 4 & 9 \end{vmatrix} = 3 \cdot 9 - (-2 \cdot 4) = 27 - -8 = 35$$

EXAMPLE 9.10 Evaluate:

$$\begin{vmatrix} -1 & -2 \\ -3 & 4 \end{vmatrix}$$

Solution

$$\begin{vmatrix} -1 & -2 \\ -3 & 4 \end{vmatrix} = -1 \cdot 4 - (-2 \cdot -3) = -4 - 6 = -10$$

SECTION 9.5 EXERCISES

Evaluate each determinant.

1. $\begin{vmatrix} -4 & 3 \\ -5 & 4 \end{vmatrix}$

2. $\begin{vmatrix} 3 & -3 \\ -4 & -5 \end{vmatrix}$

3. $\begin{vmatrix} 8 & 1/2 \\ 3/4 & 3/2 \end{vmatrix}$

4. $\begin{vmatrix} 6.2 & 4.7 \\ -2.4 & -3.2 \end{vmatrix}$

9.6 ■ CRAMER'S RULE

Now let's return to our sytem of equations.

$$(1)\ ax + by = c$$
$$(2)\ dx + ey = f$$

Compare this determinant

$$\begin{vmatrix} a & b \\ d & e \end{vmatrix} = ae - bd$$

with the denominator of the general solutions for x and y.

$$x = (ce - bf)/(ae - bd)$$
$$y = (af - cd)/(ae - bd)$$

They are the same. Now watch what happens when the coefficients on the right of the equal sign are substituted in the matrix for the coefficients of the x terms.

$$\begin{matrix} c & b \\ f & e \end{matrix}$$

The determinant of this matrix is $ce - bf$, the same as the numerator in the solution for x. Likewise, when c and f are substituted for the coefficients of y in the original matrix, the corresponding determinant is the numerator in the solution for y.

$$\begin{matrix} a & c \\ d & f \end{matrix}$$

has as its determinant $af - cd$. Putting all this together produces a procedure for solving systems of equations called **Cramer's rule**.

The solution to the system of equations

$$(1) \ ax + by = c$$
$$(2) \ dx + ey = f$$

is

$$x = \frac{\begin{vmatrix} c & b \\ f & e \end{vmatrix}}{\begin{vmatrix} a & b \\ d & e \end{vmatrix}} \text{ and } y = \frac{\begin{vmatrix} a & c \\ d & f \end{vmatrix}}{\begin{vmatrix} a & b \\ d & e \end{vmatrix}} \text{ Cramer's rule}$$

These determinants give the solution for x and y.

$$x = (ce - bf)/(ae - bd)$$
$$y = (af - cd)/(ae - bd)$$

just as we determined in the last section.

EXAMPLE 9.11 Solve using Cramer's rule.

$$(1) \ -31x + 56y = 123$$
$$(2) \ -74x + 91y = \ \ 83$$

Solution

$$x = \frac{\begin{vmatrix} 123 & 56 \\ 83 & 91 \end{vmatrix}}{D} \quad y = \frac{\begin{vmatrix} -31 & 123 \\ -74 & 83 \end{vmatrix}}{D} \text{ where } D = \begin{vmatrix} -31 & 56 \\ -74 & 91 \end{vmatrix}$$

$D = -31 \cdot 91 - 56 \cdot -74 = 1323$

$x = (123 \cdot 91 - 56 \cdot 83) / 1323 = 4.9470899$

$y = (-31 \cdot 83 - 123 \cdot -74) / 1323 = 4.9349962$

Rounded to two significant digits, the solution is (4.9, 4.9).

Check:
$$-31x + 56y = 123$$
$$-31 \cdot 4.9 + 56 \cdot 4.9 = 123$$
$$122.5 = 123$$
$$-74x + 91y = 83$$
$$-74 \cdot 4.9 + 91 \cdot 4.9 = 83$$
$$83.3 = 83$$

EXAMPLE 9.12 Solve using Cramer's rule.

$$(1)\ y = -3/8\ x - 7$$
$$(2)\ y = 5x + 3$$

Solution Rearrange the equations into an $ax + by = c$ form.

(1) $3/8\ x + y = -7$

(1) $3x + 8y = -56$

(2) $5x - y = -3$

$$x = \frac{\begin{vmatrix} -56 & 8 \\ -3 & -1 \end{vmatrix}}{D} \quad y = \frac{\begin{vmatrix} 3 & -56 \\ 5 & -3 \end{vmatrix}}{D} \quad D = \begin{vmatrix} 3 & 5 \\ 8 & -1 \end{vmatrix}$$

$D = 3 \cdot -1 - 5 \cdot 8 = -43$

$x = (-56 \cdot -1 - 8 \cdot -3)/-43 = -1.8604651$

$y = (3 \cdot -3 - -56 \cdot 5)/-43 = -6.3023256$

Rounded to three significant digits (arbitrarily), the solution is (−1.86, −6.30).

Check: Substitute back into the original equations.

(1) $\quad y = -3/8\ x - 7$
$$-6.30 = -3/8 \cdot -1.86 - 7$$
$$-6.30 = -6.30$$

(2) $\quad y = 5x + 3$
$$-6.30 = 5 \cdot -1.86 + 3$$
$$-6.30 = -6.30$$

SECTION 9.6 EXERCISES

Solve by Cramer's rule. These are the same systems you graphed and solved by substitution and elimination.

1. $y = 1/2\, x + 6$
   $y = -2\, x - 1$
2. $2y + 3x = 4$
   $3y - 4x = 3$
3. $y - 2 = -2(x + 3)$
   $y - 6 = -1/2\,(x - 2)$
4. $y = 2x - 2$
   $y = 2x + 4$

We will apply what we know about two equations, two unknowns to a DC circuit with two voltage sources.

FIGURE 9-5
Circuit 1

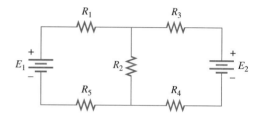

9.7 ■ LOOP EQUATIONS

The circuit shown in Figure 9-5 has two voltage sources, batteries, named E_1 and E_2. It has five resistors, R_1 through R_5. The following procedure solves for the current flowing through each resistor by writing an equation for each loop in the circuit, called **loop equations**. We will develop a system of general equations that apply to the circuit in Figure 9-5. Then we will use those equations and Cramer's rule to solve some specific circuits.

Step 1. Draw arrows around the loops to indicate the direction of current flow and label each of the currents. In Figure 9-6, I_1 and I_2 represent electrons flowing out of the negative terminal of the battery around the circuit to the positive terminal of the battery. This system of representing current is called electron flow. If a current actually flows in the opposite direction from the chosen direction, its value will come out negative.

FIGURE 9-6
Currents and Polarities

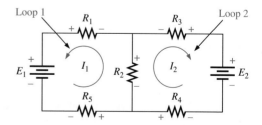

Step 2. Mark the − and + polarities of the voltage drops across each of the resistors. As electrons flow through a resistor, they cause a voltage drop across the resistor. The end the electrons flow into is negative and the other end is positive. The voltage drop across a resistor is equal to the current flowing through the resistor in amps times resistance of the resistor in ohms, $V = I \cdot R$. Notice that both currents I_1 and I_2 flow through R_2. Its voltage drop is $R_2(I_1 + I_2)$.

Step 3. Use Kirchhoff's voltage law to write an equation for each loop. Kirchhoff's voltage law states that the sum of the voltage drops around any closed loop equals the voltage applied to that loop. To write the equation for loop 1, start at the positive terminal of E_1 and travel clockwise around the loop. Since we reach the positive end of R_1 first, the drop across R_1 will be considered positive. The current through R_1 is I_1, and the voltage drop across R_1 is $I_1 R_1$. So, for R_1, write $+ I_1 R_1$. Both I_1 and I_2 flow through R_2, so the voltage drop across R_2 is $(I_1 + I_2)R_2$. For R_2, write $+ (I_1 + I_2)R_2$. For R_5, write $+ I_1 R_5$.

For loop 1:
$$E_1 = I_1 R_1 + (I_1 + I_2)R_2 + I_1 R_5$$

Combine the terms containing R_1 and the terms containing R_2.
$$E_1 = I_1 R_1 + I_1 R_2 + I_2 R_2 + I_1 R_5$$
$$E_1 = I_1(R_1 + R_2 + R_5) + I_2 R_2$$

Follow the same procedure for loop 2.

For loop 2:
$$E_2 = I_2 R_3 + (I_1 + I_2)R_2 + I_2 R_4$$
$$E_2 = I_2 R_3 + I_1 R_2 + I_2 R_2 + I_2 R_4$$
$$E_2 = I_1 R_2 + I_2(R_2 + R_3 + R_4)$$

Here are two equations with two unknowns, I_1 and I_2, that offer a general solution to the circuit in Figure 9-5.

$$E_1 = I_1(R_1 + R_2 + R_5) + I_2(R_2)$$
$$E_2 = I_1(R_2) \qquad\qquad + I_2(R_2 + R_3 + R_4)$$

Let's use these equations in a specific example to solve for the current through each resistor.

EXAMPLE 9.13 Find the current through each resistor in Figure 9-7.

Solution **Given:**
$E_1 = 12$ V, $E_2 = 20$ V
$R_1 = 220 \, \Omega$, $R_2 = 100 \, \Omega$, $R_3 = 470 \, \Omega$,
$R_4 = 330 \, \Omega$, $R_5 = 560 \, \Omega$

Find: I_1 and I_2

$$E_1 = I_1(R_1 + R_2 + R_5) + I_2(R_2)$$
$$E_2 = I_1(R_2) \qquad\qquad + I_2(R_2 + R_3 + R_4)$$
$$12 = I_1\,(220 + 100 + 560) + I_2\,(100)$$
$$20 = I_1\,(100) + I_2\,(100 + 470 + 330)$$
$$12 = 880\,I_1 + 100\,I_2$$
$$20 = 100\,I_1 + 900\,I_2$$

FIGURE 9-7
Example 9.13

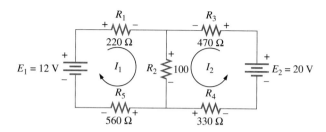

Swap sides.

$$880 I_1 + 100 I_2 = 12$$
$$100 I_1 + 900 I_2 = 20$$

To reduce the size of the numbers involved, divide the first equation by 4 and the second equation by 20.

$$220 I_1 + 25 I_2 = 3$$
$$5 I_1 + 45 I_2 = 1$$

$$I_1 = \frac{\begin{vmatrix} 3 & 25 \\ 1 & 45 \end{vmatrix}}{\begin{vmatrix} 220 & 25 \\ 5 & 45 \end{vmatrix}} = \frac{3 \cdot 45 - 25 \cdot 1}{220 \cdot 45 - 25 \cdot 5}$$

$$I_1 = 0.0113 \text{ A} = 11.3 \text{ ma}$$

$$I_2 = \frac{\begin{vmatrix} 220 & 3 \\ 5 & 1 \end{vmatrix}}{\begin{vmatrix} 220 & 25 \\ 5 & 45 \end{vmatrix}} = \frac{220 \cdot 1 - 3 \cdot 5}{220 \cdot 45 - 25 \cdot 5}$$

$$I_2 = 0.0210 \text{ A}$$
$$= 21.0 \text{ mA}$$

The current through R_1 and R_5 is I_1 or 11.3 mA. The current through R_3 and R_4 is I_2 or 21.0 mA. Both I_1 and I_2 flow through R_2. Since they flow in the same direction, the current through R_2 equals I_1 plus I_2 or 32.3 mA.

Check: The currents can be used to find the voltage drops across each resistor. The voltage drops around each loop must equal the applied voltage in that loop.

$$V_1 = I_1 R_1 = 11.3 \text{ mA} \cdot 220 \text{ } \Omega = 2.49 \text{ V}$$
$$V_2 = (I_1 + I_2) R_2 = (11.3 \text{ m} + 21.0 \text{ m}) 100 = 3.23 \text{ V}$$
$$V_5 = I_1 R_5 = 11.3 \text{ mA} \cdot 560 \text{ } \Omega = 6.32 \text{ V}$$

The sum of these three voltage drops should equal E_1 or 12 V.

$$V_1 + V_2 + V_5 = E_1$$
$$2.49 \text{ V} + 3.23 \text{ V} + 6.32 \text{ V} = 12 \text{ V}$$
$$12.04 \text{ V} = 12 \text{ V}$$
$$V_3 = I_2 \cdot R_3 = 21.0 \text{ mA} \cdot 470 \text{ } \Omega = 9.87 \text{ V}$$
$$V_4 = I_2 \cdot R_4 = 21.0 \text{ mA} \cdot 330 \text{ } \Omega = 6.93 \text{ V}$$

The sum of the voltages around loop 2 should equal 20 volts.

$$V_3 + V_2 + V_4 = E_2$$
$$9.87\text{ V} + 3.2\text{ V} + 6.93\text{ V} = 20\text{ V}$$
$$20\text{ V} = 20\text{ V}$$

It checks!

SECTION 9.7 EXERCISES, PART 1

1. Find the current flowing through each resistor in Figure 9-8.

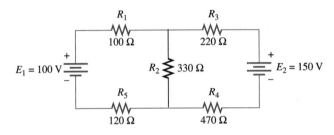

FIGURE 9-8
Section 9.7 Exercises, Part 1, Problem 1

Let's repeat the loop equation process for a circuit with a different configuration. See Figure 9-9.

FIGURE 9-9
Circuit 2

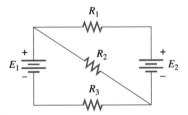

First assume a direction of current flow and draw arrows to indicate the current. See Figure 9-10. Assign polarities to the voltage drop across each resistor.

FIGURE 9-10
Currents and Polarities

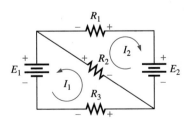

Write the loop equations.

$$E_1 = (I_1 + I_2) R_2 + I_1 R_3$$
$$E_2 = I_2 R_1 + (I_1 + I_2) R_2$$
$$E_1 = I_1 R_2 + I_2 R_2 + I_1 R_3$$
$$E_2 = I_2 R_1 + I_1 R_2 + I_2 R_2$$
$$E_1 = I_1 (R_2 + R_3) + I_2 R_2$$
$$E_2 = I_1 R_2 \qquad + I_2 (R_1 + R_2)$$

These are general equations for the circuit in Figure 9-9. Let's use the equations to solve a specific circuit.

EXAMPLE 9.14 Find the voltage across and the current through each resistor in the circuit in Figure 9-11.

FIGURE 9-11
Example 9.14

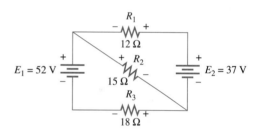

Solution

$$E_1 = I_1 (R_2 + R_3) + I_2 R_2$$
$$E_2 = I_1 R_2 \qquad + I_2 (R_1 + R_2)$$

$$52 = I_1 (15 + 18) + I_2 \cdot 15$$
$$37 = I_1 \cdot 15 + I_2 (12 + 15)$$

$$52 = 33 I_1 + 15 I_2$$
$$37 = 15 I_1 + 27 I_2$$

According to Cramer's rule:

$$I_1 = \frac{\begin{vmatrix} 52 & 15 \\ 37 & 27 \end{vmatrix}}{D} \quad I_2 = \frac{\begin{vmatrix} 33 & 52 \\ 15 & 37 \end{vmatrix}}{D} \quad D = \begin{vmatrix} 33 & 15 \\ 15 & 27 \end{vmatrix} = 666$$

$$I_1 = (52 \cdot 27 - 15 \cdot 37)/666 = 1.27 \text{ A}$$
$$I_2 = (33 \cdot 37 - 52 \cdot 15)/666 = 0.66 \text{ A}$$
$$I_{R_1} = 0.66 \text{ A}$$
$$I_{R_2} = I_1 + I_2 = 1.27 \text{ A} + 0.66 \text{ A} = 1.93 \text{ A}$$
$$I_{R_3} = 1.27 \text{ A}$$

Voltage across R_1: $V_{R_1} = I_2 \cdot R_1$
$$V_{R_1} = 0.66 \text{ A} \cdot 12 \text{ }\Omega = 7.9 \text{ V}$$
Voltage across R_2: $V_{R_2} = (I_1 + I_2)R_2$
$$V_{R_2} = 1.93 \text{ A} \cdot 15 \text{ }\Omega = 29 \text{ V}$$
Voltage across R_3: $V_{R_3} = I_1 \cdot R_3$
$$V_{R_3} = 1.27 \text{ A} \cdot 18 \text{ }\Omega = 23 \text{ V}$$

Check: Kirchhoff's voltage law says that the sum of the voltage drops around any loop is equal to the supply voltage.

$$E_1 = V_2 + V_3 \qquad\qquad E_2 = V_1 + V_2$$
$$52 \text{ V} = 29 \text{ V} + 23 \text{ V} \qquad 37 \text{ V} = 7.9 \text{ V} + 29 \text{ V}$$
$$52 \text{ V} = 52 \text{ V} \qquad\qquad 37 \text{ V} = 36.9 \text{ V}$$

Close enough!

SECTION 9.7 EXERCISES, PART 2

2. Calculate the voltage across and the current through each resistor in the circuit in Figure 9-12.

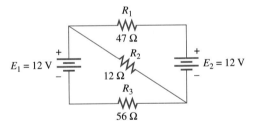

FIGURE 9-12
Section 9.7 Exercises, Part 2, Problem 2

3. The circuit in Figure 9-12 is equivalent to the circuit in Figure 9-5 with two of the resistors reduced to 0 Ω. Rework Problem 2 using the loop equations written for Figure 9-5. Hint: Redraw the circuit as shown in Figure 9-13.

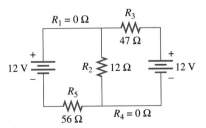

FIGURE 9-13
Section 9.7 Exercises, Part 2, Problem 3

4. For the circuit in Figure 9-14, write the loop equations. Hint: In the loop for I_1, after the voltage drops across R_1 and R_2 are written, the positive terminal of E_2 is encountered. Write $+ E_2$ to represent that source. Do the same for loop 2.

FIGURE 9-14
Section 9.7 Exercises, Part 2, Problem 4

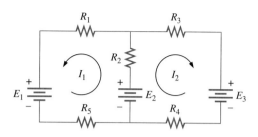

5. Use your loop equations and Cramer's rule to solve for the currents and voltages in Figure 9-15.

FIGURE 9-15
Section 9.7 Exercises, Part 2, Problem 5

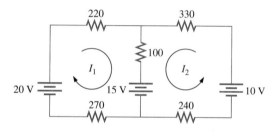

■ SUMMARY

1. A system of linear equations is a set of equations being considered as a group. Each equation in the system imposes its own set of conditions on the solution to the system.

2. An approximate solution to a system of equations can be found graphically by graphing each of the equations. The solution or solutions are those points at which the graphs intersect.

3. The graphs of the equations of an independent system of linear equations intersect at one point, which is the solution. The graphs of the equations of an inconsistent system of linear equations are parallel lines. There is no solution to an inconsistent system. The graphs of the equations of a dependent system of linear equations are identical. The equations are the same.

4. To solve a system of equations by the substitution method, solve one of the equations for one of the variables and then substitute into the other equation.

5. To solve a system of equations by the addition method, manipulate the equations so that the coefficients of one variable in each equation are additive inverses, such as $5x$ and $-5x$. Then add the two equations together to eliminate that variable. Solve for the other variable, and substitute back into one of the original equations to solve for the variable that was eliminated. This method is also called the elimination method.

6. The general solution to the system of equations

$$(1)\ ax + by = c$$
$$(2)\ dx + ey = f$$

is

$$x = \frac{ce - bf}{ae - bd}$$
$$y = \frac{cd - af}{bd - ae}$$

7. A matrix is an array of coefficients.

$$\begin{matrix} a & b \\ d & e \end{matrix}$$

is a matrix of two rows (horizontal) and two columns (vertical). It is called a 2×2 (two by two) matrix.

8. A determinant

$$\begin{vmatrix} a & b \\ d & e \end{vmatrix}$$

is the value of a matrix when it is evaluated as $ae - bd$.

9. A procedure called Cramer's rule uses determinants to solve a system of equations. The solution to

$$(1)\ ax + by = c$$
$$(2)\ dx + ey = f$$

is

$$x = \frac{\begin{vmatrix} c & b \\ f & e \end{vmatrix}}{\begin{vmatrix} a & b \\ d & e \end{vmatrix}} \text{ and } y = \frac{\begin{vmatrix} a & c \\ d & f \end{vmatrix}}{\begin{vmatrix} a & b \\ d & e \end{vmatrix}} \text{ Cramer's rule}$$

10. To solve a circuit that contains more than one battery, loop equations can be written. Cramer's rule or one of the other methods of solving a system of equations is used to solve for the currents that flow in each loop.

SOLUTIONS TO CHAPTER 9 SECTION EXERCISES

Solutions to Section 9.1 exercises

1. Figure 9-16.
2. Figure 9-17.
3. Figure 9-18.
4. Figure 9-19.

SOLUTIONS TO CHAPTER 9 SECTION EXERCISES ▪ 303

FIGURE 9-16
Solution to Section 9.1 Exercises, Problem 1

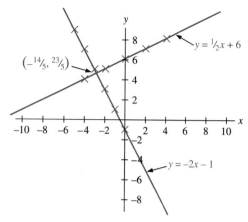

FIGURE 9-17
Solution to Section 9.1 Exercises, Problem 2

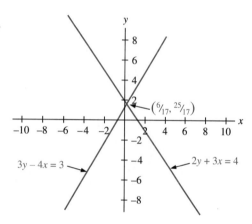

FIGURE 9-18
Solution to Section 9.1 Exercises, Problem 3

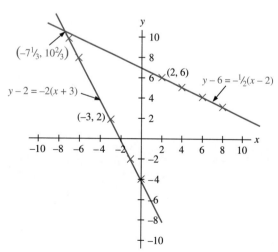

FIGURE 9-19
Solution to Section 9.1 Exercises, Problem 4

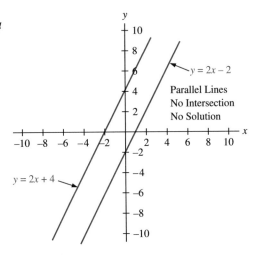

Solutions to Sections 9.2 and 9.3 Exercises

1. $(-14/5, 23/5)$
2. $(6/17, 25/17)$
3. $(-22/3, 32/3)$
4. Parallel lines. No intersection, no solution.

Solutions to Section 9.5 Exercises

1. -1
2. -27
3. $11\ 5/8$
4. -8.56

Solutions to Section 9.6 Exercises

1. $(-14/5, 23/5)$
2. $(6/17, 25/17)$
3. $(-22/3, 32/3)$
4. Parallel lines. No intersection, no solution.

Solution to Section 9.7 Exercises, Part 1

1. $I_{R1} = I_{R5} = 120$ mA, $I_{R3} = I_{R4} = 110$ mA, the current through R_2 is the sum of I_1 and I_2 or 230 mA.

Solutions to Section 9.7 Exercises, Part 2

2. and 3.

	R_1	R_2	R_3
voltage	8.2 V	3.8 V	8.2 V
current	174 mA	320 mA	146 mA

4. $E_1 - E_2 = I_1(R_1 + R_2 + R_5) + I_2 R_2$
   $E_3 - E_2 = I_1(R_2) + I_2(R_2 + R_3 + R_4)$
5. $IR_1 = IR_5 = 10$ mA $IR_3 = IR_4 = 9.0$mA
   The current through R_2 is the difference between the two loop currents or 1 mA.

CHAPTER 9 PROBLEMS

Solve each system by (a) graphing, (b) the addition method, (c) the substitution method, and (d) Cramer's rule.

1. $10y = 2x - 15$
   $10y = -2x - 35$
2. $y = -15/7x - 3$
   $y = 11/7x - 3$
3. $y = 7/2x$
   $y + 7/2 = 1/2(x + 1)$
4. $34y + 20x + 19 = 0$
   $2y - 2x + 17 = 0$

Use loop equations and Cramer's rule to find the voltage across and current through each resistor in the following circuits.

5. Figure 9-20.

FIGURE 9-20
Chapter 9, Problem 5

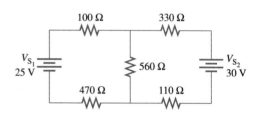

6. Figure 9-21.

FIGURE 9-21
Chapter 9, Problem 6

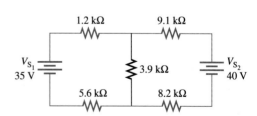

7. Figure 9-22.
8. Figure 9-23.

FIGURE 9-22
Chapter 9, Problem 7

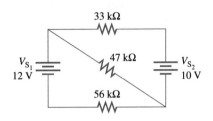

FIGURE 9-23
Chapter 9, Problem 8

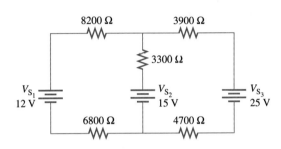

10
Trigonometry

CHAPTER OUTLINE

- **10.1** Trig Functions
- **10.2** Inverse Trig Functions
- **10.3** Snell's Law
- **10.4** Angles of Elevation and Depression
- **10.5** Surveying

OBJECTIVES

After completing this chapter, you should be able to

1. use the Pythagorean theorem to find unknown sides of right triangles.
2. use trig functions to find unknown sides of right triangles.
3. use inverse trig functions to solve for unknown angles.
4. determine the accuracy or precision of the result of a problem involving trig or inverse trig functions.
5. apply trig to problems involving refraction.
6. apply trig to problems involving angle of elevation and angle of depression.
7. apply trig to surveying problems.

NEW TERMS TO WATCH FOR

side opposite	angle of incidence, θ_i
side adjacent	angle of transmission, θ_t
sine	normal
cosine	refraction
tangent	critical angle
cosecant	angle of elevation
secant	angle of depression
cotangent	transversal
degree mode	alternate interior angles
inverse trig function	latitude
Snell's law	departure
refracted	bearing

Right triangles and the Pythagorean theorem were studied in Chapter 7. Review them as needed before continuing. In this chapter, trigonometric functions and the Pythagorean theorem will be used to solve numerous problems. The skills learned will be applied in several of the remaining chapters. The examples are solved using the eight-step approach modeled in Chapter 6. If one of the eight steps requires no action, it is omitted and the individual steps are not numbered. Comments about a step appear *after* the step has been taken so that you have a chance to discover what has happened before seeing the comment. Perhaps you will not need to rely on the comments. The most lengthy comments are written about determining the proper number of significant digits in the result. Remember that in step 6 the raw result is taken from the calculator and that result is rounded to the proper number of significant digits in step 8.

10.1 ■ TRIG FUNCTIONS

The hypotenuse of the right triangle shown in Figure 10-1 is labeled H. In Figure 10-1A, one of the acute angles is labeled A. The side across from angle A is labeled SO for **side opposite**. The side of the triangle that forms angle A with the hypotenuse is labeled SA for **side adjacent**. Figure 10-1B shows the same triangle with the other acute angle labeled B. The side opposite angle B is labeled SO, and the side adjacent to it is labeled SA. H is always across from the right angle. SA and SO are relative to the angle being studied.

FIGURE 10-1
Side Opposite, Side Adjacent

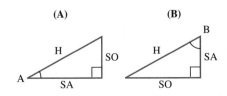
(A)   (B)

10.1 TRIG FUNCTIONS ▪ 309

EXAMPLE 10.1 Relative to acute ∠B in each triangle in Figure 10-2, label the sides SO, SA, and H.

FIGURE 10-2
Example 10.1

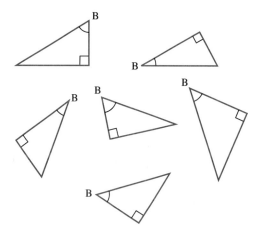

Solution In each triangle, locate the right angle and label the side opposite it H. Locate ∠B and label the side opposite it SO. The remaining side should form ∠B with the hypotenuse. Label it SA. The results are shown in Figure 10-3.

FIGURE 10-3
Solution to Example 10.1

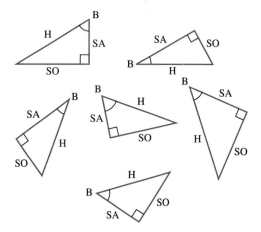

In all right triangles with a given acute angle, the ratio of the lengths of any two sides is a constant. For example, in all right triangles with an acute angle of 35°, the ratio SO to H is the same. Likewise, the ratio of SA to H or SO to SA is the same. (This is equivalent to saying that the quotients SO/H, SA/H, and SO/SA are the same.) These ratios are so useful that each of the six possibilities has a name. Here is a list of each of the ratios.

Ratio	Name	Abbreviation
Side Opposite/Hypotenuse (SO/H)	**Sine**	sin
Side Adjacent/Hypotenuse (SA/H)	**Cosine**	cos
Side Opposite/Side Adjacent (SO/SA)	**Tangent**	tan
Hypotenuse/Side Opposite (H/SO)	**Cosecant**	csc
Hypotenuse/Side Adjacent (H/SA)	**Secant**	sec
Side Adjacent/Side Opposite (SA/SO)	**Cotangent**	cot

The first three of these ratios are included on scientific calculators. The last three are reciprocals of the first three and are not found on calculators. Note the relationships: Cosecant is the reciprocal of sine; secant is the reciprocal of cosine; and cotangent is the reciprocal of tangent. The first three ratios, sin, cos, and tan, should be memorized right now.

EXAMPLE 10.2 For the triangle in Figure 10-4, calculate sin A, cos A, tan A, sin B, cos B, and tan B.

FIGURE 10-4
Example 10.2

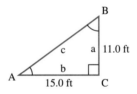

Solution

Given: right triangle ABC, a = 11.0 ft, b = 15.0 ft
Find: sin A, cos A, tan A, sin B, cos B, and tan B

The hypotenuse is needed to calculate sin and cos. Use the Pythagorean theorem.

$$c = (a^2 + b^2)^{1/2}$$
$$c = [(11.0 \text{ ft})^2 + (15.0 \text{ ft})^2]^{1/2}$$
$$c = 18.6 \text{ ft}$$

$$\sin A = \text{SO/H} = 11.0/18.6 = 0.591$$
$$\cos A = \text{SA/H} = 15.0/18.6 = 0.806$$
$$\tan A = \text{SO/SA} = 11.0/15.0 = 0.733$$
$$\sin B = \text{SO/H} = 15.0/18.6 = 0.806$$
$$\cos B = \text{SA/H} = 11.0/18.6 = 0.591$$
$$\tan B = \text{SO/SA} = 15.0/11.0 = 1.36$$

Note that sin A = cos B = 0.591 and cos A = sin B = 0.806. In a right triangle with acute angles A and B, sin A will always equal cos B and cos A will always equal sin B. Since ∠A and ∠B are complementary (add to 90°), B = 90° − A. These relationships can be expressed formally as follows:

$$\sin A = \cos B \qquad \cos A = \sin B$$
$$\sin A = \cos(90° - A) \qquad \cos A = \sin(90° - A)$$

EXAMPLE 10.3   For the triangle in Figure 10-5, find sin R, cos R, tan R, sin P, cos P, and tan P.

FIGURE 10-5
Example 10.3

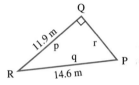

Solution   **Given:**   right triangle PQR, p = 11.9 meters, q = 14.6 meters
Find: sin R, cos R, tan R, sin P, cos P, and tan P

$$\cos R = p/q = 11.9/14.6 = 0.815$$
$$\sin P = \cos R = 0.815$$

Use the Pythagorean theorem to find the length of side r.

$$q^2 = p^2 + r^2$$
$$r^2 = q^2 - p^2$$

Transpose p^2.

$$r = (q^2 - p^2)^{1/2}$$

Take the square root of each side.

$$r = [(14.6 \text{ m})^2 - (11.9 \text{ m})^2]^{1/2}$$
$$r = 8.4587 \text{ m}$$
Round r to 8.46 m.

Significant digits: The given two sides have an accuracy of three, round R to three significant digits.

$$\sin R = r/q = 8.46/14.6 = 0.579$$
$$\cos P = \sin R = 0.579$$
$$\tan R = r/p = 8.46/11.9 = 0.711$$
$$\tan P = p/r = 11.9/8.46 = 1.41$$

Significant digits: In each case, the numerator and denominator have an accuracy of three. Round the sines, cosines, and tangents to three significant digits.

In the last example, the values of the trig functions were found by using the lengths of the sides of the triangles. In some applications, the angle itself is known and the value of the trig function is found from the angle. To do this on your calculator, first ensure that your calculator is set to the **degree mode** (as opposed to another system of angular measure like radians). Some Casios are set to degree mode by pressing the MODE key and then 4. Some TIs are set by pressing the DGR key, which causes the calculator to alternate between degrees, radians, and grads. Watch for DEG to appear in small letters in the display. Once your calculator is in degree mode, enter the angle and then press sin, cos, or tan. For example, to find the tangent of 65°, enter 65 and then press tan. 2.144 506 921 should appear in the display.

EXAMPLE 10.4 Use your calculator to complete the chart. Round to the thousandth's place.

Angle°	sin	cos	tan
0			
15			
30			
45			
60			
75			
80			
85			
87			
89			
90			

Solution

Angle°	sin	cos	tan
0	0.000	1.000	0.000
15	0.259	0.966	0.268
30	0.500	0.866	0.577
45	0.707	0.707	1.000
60	0.866	0.500	1.732
75	0.966	0.259	3.732
80	0.985	0.174	5.671
85	0.996	0.087	11.430
87	0.999	0.052	19.081
89	0.999	0.017	57.290
90	1.000	0.000	Undefined

Note that as the angle increased from 0 to 90°, sin A increased from 0 to 1 and cos A decreased from 1 to 0. As the angle increased from 0 to 45°, tangent increased from 0 to 1.

As the angle increased from 45° to 90°, tangent increased from 1 to large without bound.

Figure 10-6 will help explain the behavior of tangent. In the sequence of triangles from left to right, ∠B gets progressively larger while the side opposite ∠B is drawn the same length each time. Side adjacent to ∠B gets progressively smaller. Since tangent is the quotient of SO/SA, as we proceed from left to right we are dividing by a smaller number and the quotient gets larger. When we hit 90°, SA has been reduced to zero, and division by zero is not defined. Your calculator will indicate an error when you enter tan 90°.

Also, sin 15° = cos 75°   cos 15° = sin 75°
sin 30° = cos 60°   cos 30° = sin 60°
sin 45° = cos 45°
sin 0° = cos 90°   cos 0° = sin 90°

FIGURE 10-6
Tangent B

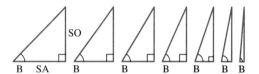

The sines, cosines, and tangents in the last example were arbitrarily rounded to three significant digits. How can we decide the proper number of significant digits for a trig function? Let's investigate sin 63°. The angle is precise to the nearest degree. It is closer to 63° than it is to 62° or 64°. It can range from 62.5° to 63.5°. Let's compare the sine of these two extremes.

sin 62.5° = 0.887 010 83
sin 63° = 0.891 006 52
sin 63.5° = 0.894 934 36

These results are the same in the tenth's place and begin to differ in the hundredth's place. Round the result to the hundredth's place. sin 63° = 0.89

EXAMPLE 10.5 Calculate tan 73.4° to the proper number of significant digits.

Solution The angle is precise to the tenth degree. It is closer to 73.4° than it is to 73.3° or 73.5°. Its value can range from 73.35° to 73.45°.

tan 73.35° = 3.343 772 43
tan 73.4° = 3.354 433 30
tan 73.45° = 3.365 156 78

These results are the same in the tenth's place and begin to differ in the hundredth's place. Round the result to the hundredth's place. tan 73.4° = 3.35

This process is tedius, but do not despair. Here is a shortcut. As a rule of thumb, **for an angle precise to the nearest degree, round the sine, cosine, or tangent of that angle to two significant digits**. Also, in every example so far, for an angle precise to the nearest tenth of degree, the sine, cosine, or tangent of that angle has been shown to have an accuracy of three. So, as a rule of thumb, **for an angle precise to the nearest tenth of degree, round the sine, cosine, or tangent of that angle to three digits**. These rules of thumb are not accurate for small angles of 0° to 10°, or for large angles of 80° to 90°. For those extremes, use the process that we have used up to this point.

EXAMPLE 10.6 Use the rule of thumb to round these results to the proper number of significant digits.

a. tan 15.7°

b. cos 63°

Solution **a.** tan 15.7° = 0.281 087 32. The angle is precise to the tenth of degree; round the result to three significant digits. Round tan 15.7° to 0.281.

b. cos 63° = 0.453 990 50. The angle is precise to the degree; round the result to two significant digits. Round 63° to 0.45.

SECTION 10.1 EXERCISES, PART 1

1. Relative to acute ∠D in each triangle in Figure 10-7, label the sides SO, SA, and H.

FIGURE 10-7
Section 10.1 Exercise, Problem 1

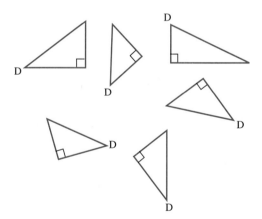

2–5. For each triangle in Figure 10-8, find sin A, cos A, tan A, sin B, cos B, and tan B.

6. For each of these angles, find sin, cos, and tan: 0°, 12.6°, 30°, 40.9°, 60°, 64.3°, 87.2°, and 90°. Round the result to the proper number of significant digits.

7. Find the largest value of tangent that your calculator will display.

Armed with these basics, we are ready to solve a few right triangles. In the next three examples, the length of one side of a right triangle and an acute angle are given and

FIGURE 10-8
Section 10.1 Exercise, Problems 2 through 5

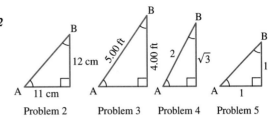

Problem 2   Problem 3   Problem 4   Problem 5

Note: In Problems 4 and 5, 1 means exactly 1 and 2 means exactly 2 (unlimited significant digits).

another side is solved for. To solve each example, determine which trig function is appropriate, that is, which trig function involves the given information and the side being solved for. Write the equation for that trig function and solve it for the unknown side. Each example requires the use of a different trig function.

EXAMPLE 10.7   If the side opposite a 30° angle in a right triangle measures 14 m, calculate the length of the hypotenuse.

Solution   **Given:** right triangle ABC, A = 30°, BC = 14 m
Find: H. See Figure 10-9.

FIGURE 10-9
Example 10.7

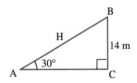

Relative to the 30° angle, the 14 m side is the side opposite and the hypotenuse is the unknown. Which trig function relates SO to H? Sine does.

$$\sin 30° = SO/H$$

Solve for H.

$$H \cdot \sin 30° = SO$$

Multiply both sides by H.

$$H = SO/\sin 30°$$

Divide both sides by sin 30°.

$$H = 14 \text{ m}/\sin 30°$$

$$H = 28 \text{ m}$$

Calculator sequence used: 14 / 30 sin =
Significant digits: sin 30° = 0.50, H = 0.50, H = 14/0.50. Round H to two significant digits.

316 ■ CHAPTER 10 TRIGONOMETRY

EXAMPLE 10.8 In a right triangle, if the side adjacent to a 40° angle measures 19 m, calculate the length of the other side (not the hypotenuse).

Solution **Given:** right triangle ABC, A = 40°, AC = 19 m
Find: BC. See Figure 10-10.

FIGURE 10-10
Example 10.8

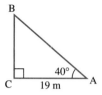

Relative to ∠A, side adjacent is known and side opposite is being solved for. Which trig function relates side opposite and side adjacent? Tangent does.

$$\tan 40° = CB/AC$$

Solve for CB.

$$AC \cdot \tan 40° = CB$$

Multiply both sides by AC.

$$CB = AC \cdot \tan 40°$$
$$CB = 19 \text{ m} \cdot \tan 40°$$
$$CB = 15.942\,892\,99 \text{ m}$$

Calculator sequence used: 19 × 40 tan =

Round CB to 16 m.

Significant digits: tan 40° = 0.84, CB = 19 · 0.84. Round CB to two significant digits.

EXAMPLE 10.9 If the hypotenuse of a right triangle measures 64.5 m, and one acute angle measures 52.2°, calculate the side adjacent to the 52.2° angle.

Solution **Given:** right triangle ABC, A = 52.2°, AB = 64.5 m
Find: AC. See Figure 10-11.

FIGURE 10-11
Example 10.9

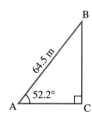

Relative to the 52.2° angle, the unknown side AC is side adjacent and the hypotenuse is known. Which trig function relates SA to H? Cosine does.

$$\cos A = AC/AB$$

Solve for AC.

$$AB \cdot \cos A = AC$$

Multiply both sides by AB.

$$AC = AB \cdot \cos A$$
$$AC = 64.5 \text{ m} \cdot \cos 52.2°$$
$$AC = 39.532\ 504\ 9 \text{ m}$$

Calculator sequence used: 64.5 × 52.2 cos =
Round AC to 39.5 m.
Significant digits: cos 52.2° = 0.613 Round 64.5 · cos 52.2° to three significant digits.

SECTION 10.1 EXERCISES, PART 2

8–16. For each triangle shown in Figure 10-12, calculate the length of side b.

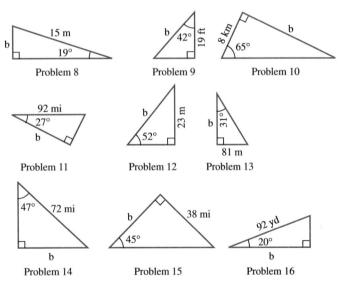

FIGURE 10-12
Section 10.1 Exercise, Problems 8 through 16

10.2 ■ INVERSE TRIG FUNCTIONS

If sin A = SO/H, then A is *the angle whose* sin is SO/H. "The angle whose" is denoted in two different ways. $A = \sin^{-1}$ SO/H and A = arcsin SO/H are two ways of stating "A is the angle whose sin is SO/H." These expressions are called **inverse trig functions**.

Both notations will be used in this book. In the next three examples, an acute angle of a right triangle will be written in inverse trig form.

EXAMPLE 10.10 For the right triangle in Figure 10-13, write ∠A in terms of the given sides.

Solution In relation to ∠A, 12.6 mm is the side opposite and 22.1 mm is the hypotenuse. Which trig function uses SO and H? Sine does. See Figure 10-13.

FIGURE 10-13
Example 10.10

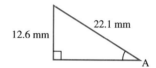

$$\sin A = SO/H$$
$$A = \sin^{-1}(SO/H)$$
$$A = \sin^{-1}(12.6/22.1) \text{ or } A = \arcsin(12.6/22.1)$$

Both measurements are in mm, so the units cancel.

EXAMPLE 10.11 For the right triangle in Figure 10-14, write ∠B in terms of the given sides.

FIGURE 10-14
Example 10.11

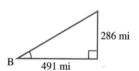

Solution In relation to ∠B, 286 mi is side opposite and 491 mi is side adjacent. Which trig function uses SO and SA? Tangent does. See Figure 10-14.

$$\tan B = SO/SA$$
$$B = \tan^{-1}(SO/SA) \text{ or } B = \arctan(SO/SA)$$
$$B = \tan^{-1}(286/491) \text{ or } B = \arctan(286/491)$$

Both measurements are in mi, so the units cancel.

EXAMPLE 10.12 For the right triangle in Figure 10-15, write ∠R in terms of the given sides.

Solution In relation to ∠R, 38.2 km is side adjacent and 63.4 km is the hypotenuse. Which trig function uses SA and H? Cosine does. See Figure 10-15.

FIGURE 10-15
Example 10.12

$$\cos R = SA/H$$
$$R = \cos^{-1}(SA/H) \text{ or } R = \arccos(SA/H)$$
$$R = \cos^{-1}(38.2/63.4) \text{ or } R = \arccos(38.2/63.4)$$

Both measurements are in km, so the units cancel.

On many calculators, inverse trig functions are written above the corresponding trig keys using the −1 notation. For example, over the sin key is written "$\sin^{-1}$." Other calculators use the "arcsin" notation but shorten it to "asin." To get to the $\sin^{-1}$ function on a Casio, first press INV and then sin. On a TI, press 2nd and then sin.

EXAMPLE 10.13 Use a calculator to find the angles in the last three examples.

Solution

Example 10.10 $A = \sin^{-1}(12.6/22.1)$ or $A = \arcsin(12.6/22.1)$
$A = 34.8°$
Calculator sequence used: 12.6/22.1 = INV sin

Example 10.11 $B = \tan^{-1}(286/491)$ or $B = \arctan(286/491)$
$B = 30.2°$
Calculator sequence used: 286/491 = INV tan

Example 10.12 $R = \cos^{-1}(38.2/64.3)$ or $R = \arccos(38.2/64.3)$
$R = 53.6°$
Calculator sequence used: 38.2/64.3 = INV cos

In each case, the result was rounded to the nearest tenth of a degree. The following discussion shows why. In Example 10.10, $A = \sin^{-1}(12.6/22.1) = \sin^{-1} 0.570$, where 0.570 has an accuracy of three because 12.6 and 22.1 each have an accuracy of three. This means the sine of $\angle A$ is closer to 0.570 than it is to 0.569 or 0.571. The sine of $\angle A$ can range from 0.569 5 to 0.570 5. Calculate angle A for the two extremes.

$$A = \sin^{-1} 0.569\ 5 = 34.715\ 367°$$
$$A = \sin^{-1} 0.570 = 34.750\ 226°$$
$$A = \sin^{-1} 0.570\ 5 = 34.785\ 1°$$

In each case, $\angle A$ is the same in the tenth's place, but varies greatly in the hundredth's place. Round $\angle A$ to the tenth's place. Round 34.750 226° to 34.8°.

Let's try another one. In Example 10.11, B = arctan (286/491) = arctan 0.582, where 0.582 has an accuracy of three because 286 and 491 each have an accuracy of three. This means the tangent of $\angle B$ is closer to 0.582 than it is to 0.581 or 0.583. The tangent of $\angle B$ can range from 0.5815 to 0.5825. Calculate $\angle B$ for the two extremes.

$$B = \arctan 0.581\ 5 = 30.178\ 001°$$
$$B = \arctan 0.582 = 30.199\ 405°$$
$$B = \arctan 0.582\ 5 = 30.220\ 8°$$

$\angle B$ is the same in each case to the unit's place, but begins to vary in the tenth's place. Round $\angle B$ to the tenth's place. Round 30.199 405° to 30.2°.

Here is a shortcut for determining the precision of an angle. **For a number with an accuracy of two, round the arcsin, arccos, or arctan of that number to the nearest degree. For a number with an accuracy of three, round the arcsin, arccos, or arctan to the nearest tenth of degree.** These rules of thumb are not accurate for small angles of 0° to 10° and large angles of 80° to 90°.

EXAMPLE 10.14   Use the rule of thumb to round these results to the proper number of significant digits.

   a. arctan 3.3

   b. arcsin (32/47)

   c. arccos [1.45 · sin 35°/1.55]

Solution   **a.** arctan 3.3 = 73.141 601 23°. 3.3 has an accuracy of two; round the result to the nearest degree. Round arctan 3.3 to 73°. Calculator sequence used: 3.3 INV tan

   b. arcsin (324/477) = 42.784 694 91°. 324/477 has an accuracy of two; round the result to the tenth of degree. Round arcsin (324/477) to 42.8°. Calculator sequence used: 324 ÷ 477 = INV sin

   c. arccos [1.45 · sin 35°/1.55] = 57.549 450 10°. sin 35° has an accuracy of two and limits [1.45 · sin 35°/1.55] to an accuracy of two. Round arccos [1.45 · sin 35°/1.55] to the nearest degree. Calculator sequence used: 1.45 × 35 sin ÷ 1.55 = INV cos. arccos [1.45 · sin 35°/1.55] = 58°

SECTION 10.2 Exercises, PART 1

1. Write the six inverse trig functions in both notations. For example, A = $\sin^{-1}$(SO/H)

Use the rules of thumb to round these results to the proper number of significant digits.

2. arcsin 0.555 =

3. arccos 0.45 =

4. arctan 2.02 =

5. arcsin [2.4 · tan 34°/3.3] =
6. arctan [2.55 · cos 65.5°/3.78] =
7. Find ∠B in Problem 2 of Figure 10-8.
8. Find ∠A in Problem 3 of Figure 10-8.
9. Find ∠B in Problem 4 of Figure 10-8.

In the next example, the hypotenuse and one side of a right triangle are given and the angles are found.

EXAMPLE 10.15 The hypotenuse of a right triangle measures 12.5 ft and one side measures 9.5 ft. Calculate the angles of the triangle.

Solution **Given:** H = 12.5 ft, SA = 9.5 ft
Find: ∠A and ∠B. See Figure 10-16.

$$\cos A = SA/H$$

FIGURE 10-16
Example 10.14

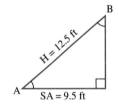

Which trig function relates SA and H? Cosine.

$$A = \arccos (SA/H)$$

If cos A = SA/H, then A is the angle whose cos is SA/H. "The angle whose cos" is written arccos or $\cos^{-1}$.

$$A = \arccos (9.5\text{ft}/12.5\text{ft})$$
$$A = 40.535\ 802°$$

Calculator sequence used: 9.5 / 12.5 = INV cos

$$\text{Round A to } 41°.$$

Significant digits: 9.5/12.5 = 0.76. 0.76 has an accuracy of two. Round to the nearest degree.

$$A + B = 90°$$

The problem requested both A and B. They are complementary angles.

$$B = 90° - A$$

Transpose A.

$$B = 90° - 41°$$
$$B = 49°$$

In the next example, two sides of a right triangle are given and the hypotenuse is found.

EXAMPLE 10.16 Two sides of a right triangle measure 72.8 m and 48.2 m. Find the hypotenuse.

Solution 1 **Given:** SA = 72.8 m, SO = 48.2 m
Find: H. See Figure 10-17.

$$\tan A = SO/SA$$

First find angle A. SA and SO are given. Which trig function contains SO and SA? Tangent.

FIGURE 10-17
Example 10.16

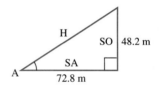

$$A = \tan^{-1}(SO/SA)$$

If tan A = SO/SA, then A is the angle whose tan is SO/SA.

$$A = \tan^{-1}(48.2/72.8)$$
$$A = 33.508\ 061°$$
Round A to 33.5°.

48.2/72.8 = 0.662 (accuracy of three). Round A to 33.5°.
Significant digits: Now that we know angle A, we can use a trig function to find H.

$$\cos A = SA/H$$

sin A = SO/H can be used instead.

$$H = SA/\cos A$$

Multiply both sides by H to get H out of the denominator. Multiply both sides by 1/cos A to isolate H.

$$H = 72.8/\cos 33.5°$$
$$H = 87.3021\ m$$
Round H to 87.3 m.

Significant digits: 33.5° is precise to the nearest degree. Round cos 33.5° and H to three significant digits.

This problem can also be solved by using the Pythagorean theorem.

SECTION 10.2 EXERCISES, PART 2

Draw a figure and solve.

10. If the hypotenuse of a right triangle measures 2.0 miles and one side measures 1.7 miles, find the angles.
11. If the hypotenuse of a right triangle measures 178 cm and one side measures 101 cm, find the angles.
12. If the hypotenuse of a right triangle measures 8.1 km and one angle measures 36°, find the lengths of the two sides.
13. If the hypotenuse of a right triangle measures 4.7 ft and one angle measures 52.6°, find the lengths of the two sides.
14. If two sides of a right triangle measure 56.2 m and 82.6 m, find the hypotenuse by two different methods.
15. If two sides of a right triangle measure 9.4 km and 4.8 km, find the hypotenuse by two different methods.
16. If the hypotenuse of a right triangle measures 47 in and one side measures 39 in, find the length of the other side and the two angles.
17. If the hypotenuse of a right triangle measures 2.4 km and one side measures 1.6 km, find the length of the other side and the two angles.

10.3 ■ SNELL'S LAW

Armed with these basics, let's look at some actual applications in which trig is used to find the solution. The following examples use **Snell's law** (described by Willebrord Snell in 1621) to determine the path of light through a fiber optic cable. Snell's law describes how light is **refracted** (bent) as it passes from one medium into another. The angle at which light approaches the junction between the two mediums is called the **angle of incidence**, θ_i. See Figure 10-18. The angle at which light travels through the second medium is called the **angle of transmission, θ_t**. θ_i and θ_t are both measured from the **normal** (perpendicular) to the boundary between the two mediums. Index of refraction was introduced in an earlier chapter. It is the property of a material that affects the velocity and wavelength of light passing through it. This change in velocity causes the light to change direction of travel. The change in direction of travel is called **refraction**. n_i is the index of refraction of the material light is passing through before it hits the junction, and n_t is the index of refraction of the material that light passes into. The index of refraction of a vacuum is 1, and all other materials have an index of refraction greater than 1. Air is only slightly greater at 1.0003. Snell's law states that

FIGURE 10-18
Snell's Law

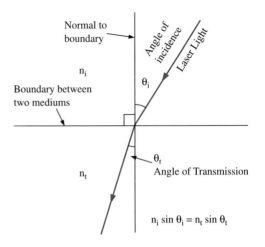

$$n_i \sin \theta_i = n_t \sin \theta_t \quad \text{Snell's law}$$

where n_i is the index of refraction of the medium incident light is passing through,

n_t is the index of refraction of the medium refracted light is passing through,

θ_i is the angle of incidence, and

θ_t is the angle of transmission.

EXAMPLE 10.17 A beam of laser light is traveling from air (index of refraction = 1.00) into water, whose index of refraction is 1.33. If the light approaches the junction at an angle of 35° to the normal, calculate the angle of the light as it passes through the water.

Solution **Given:** $n_i = 1.00$, $n_t = 1.33$, $\theta_i = 35°$
Find: θ_t. See Figure 10-19.

$$n_i \sin \theta_i = n_t \sin \theta_t \quad \text{Snell's Law}$$

where n_i is the index of refraction of the medium incident light is passing through,

n_t is the index of refraction of the medium refracted light is passing through,

θ_i is the angle of incidence,

θ_t is the angle of transmission.

$$n_i \sin \theta_i / n_t = \sin \theta_t$$

Transpose n_t.

$$\sin \theta_t = n_i \sin \theta_i / n_t$$
$$\theta_t = \arcsin [n_i \sin \theta_i / n_t]$$

FIGURE 10-19
Example 10.17

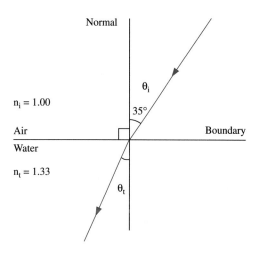

Write in inverse trig form to isolate θ_t.

$$\theta_t = \arcsin[1.00 \sin 35°/1.33]$$
$$\theta_t = 25.547\ 58°$$

Sequence used: 35 sin/1.33 = INV sin

Round θ_t to 26°.

Significant digits: 35° is precise to the nearest degree so sin 35° has an accuracy of two. The quotient sin 35°/1.33 is limited to two significant digits, 0.43. Arcsin 0.43 is precise to the nearest degree.

When light passes from a material of lower index of refraction into a material of higher index of refraction ($n_i < n_t$), light is bent toward the normal. In the next example, light passes from a material of higher index of refraction to lower index of refraction ($n_i > n_t$).

EXAMPLE 10.18 Light is passing from glass ($n_i = 1.50$) into air ($n_t = 1.00$) at an angle of incidence of 30°. Calculate the angle of transmission.

Solution **Given:** $n_i = 1.50$, $n_t = 1.00$, $\theta_i = 30°$
Find: θ_t. See Figure 10-20.
θ_t was solved for in the last example.

$$\theta_t = \arcsin[n_i \sin \theta_i / n_t]$$
$$\theta_t = \arcsin[1.50 \sin 30° / 1.00]$$
$$\theta_t = 48.590\ 377\ 89°$$
Round θ_t to 49°.

FIGURE 10-20
Example 10.18

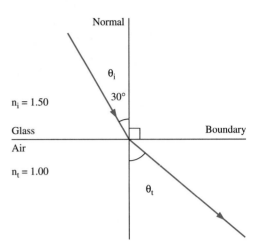

Significant digits: Assuming 30° is precise to the nearest degree, sin 30° has an accuracy of two. The arcsin of a quantity with an accuracy of two is precise to the nearest degree. Round θ to the nearest degree.

Light approached the junction at 30° and was transmitted into the air at an angle of 49°. **When light passes from a material of higher index of refraction into a material of lower index of refraction ($n_i > n_t$), light is bent away from the normal.** Figure 10-21 summarizes the relationships between angle of transmission and index of refraction.

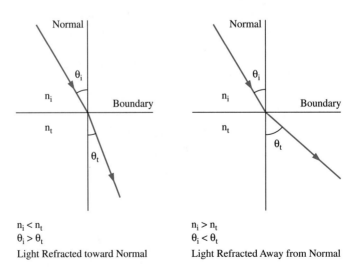

FIGURE 10-21
Refraction

10.3 SNELL'S LAW ■ 327

When $n_i > n_t$, as θ_i increases, θ_t also increases and soon will approach 90°. Let's calculate the angle of incidence that will cause an angle of transmission of 90°. See Figure 10-22.

FIGURE 10-22
Critical Angle, θ_C

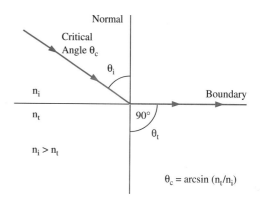

$$n_i \sin \theta_i = n_t \sin \theta_t \quad \text{Snell's law}$$

where n_i is the index of refraction of the medium incident light is passing through,
n_t is the index of refraction of the medium refracted light is passing through,
θ_i is the angle of incidence, and
θ_t is the angle of transmission.

$$\sin \theta_i = n_t \sin \theta_t / n_i$$

Multiply both sides by $1/n_i$ to isolate $\sin \theta_i$.

$$\theta_i = \arcsin [n_t \sin \theta_t / n_i]$$

Write in inverse trig form to isolate θ_i.

$$\theta_i = \arcsin [n_t \sin 90°/n_i]$$

Substitute 90° for θ_t.

$$\theta_i = \arcsin [n_t/n_i]$$

$\sin 90° = 1$

When $\theta_i = \arcsin [n_t/n_i]$, light does not enter the second medium but travels along the junction or boundary of the two mediums instead. This angle of incidence is called the **critical angle, θ_c**.

$$\theta_c = \arcsin [n_t/n_i]$$

EXAMPLE 10.19 Calculate the critical angle for the situation described in the last example.

Solution **Given:** $n_i = 1.50$, $n_t = 1.00$
Find: θ_c. See Figure 10-23.

FIGURE 10-23
Example 10.19

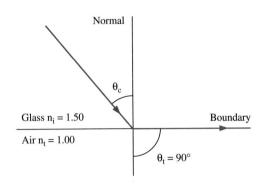

$$\theta_c = \arcsin [n_t / n_i]$$
$$\theta_c = \arcsin [1.00/1.50]$$
$$\theta_c = 41.81031490°$$
Round θ_c to 41.8°.

Significant digits: 1.00/1.50 has an accuracy of three, so arcsin [1.00/1.50] has a precision of tenth's of a degree. Round θ_c to 41.8°.

The following example investigates what happens if the angle of incidence is greater than the critical angle.

EXAMPLE 10.20 Suppose the angle of incidence in the situation described in the last example increases to 50°. Calculate the angle of transmission.

Solution **Given:** $n_i = 1.50$, $n_t = 1.00$, $\theta_i = 50°$
Find: θ_t. See Figure 10-24.

$$\theta_t = \arcsin [n_i \sin \theta_i / n_t]$$
$$\theta_t = \arcsin [1.50 \sin 50° / 1.00]$$
$$\theta_t = \arcsin 1.15$$

FIGURE 10-24
Example 10.20

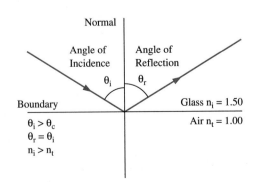

Here, Snell's law falls apart because the maximum value of sin is 1. Above the critical angle all the light is reflected at the boundary; none of the light passes into the air. The angle of reflection, θ_r is equal to the angle of incidence. See Figure 10-24. Note that this total reflection only happens when light is passing from a higher index of refraction to a lower index of refraction.

SECTION 10.3 EXERCISES

1. A beam of light is passing from water (index of refraction = 1.35) into air (index of refraction = 1.00) at an angle of 35° to the normal. Calculate the angle of the light as it passes through the air.

2. A beam of light is passing from air (index of refraction = 1.00) into water. If the light approaches the boundary at an angle of 37° and travels through the water at an angle of 25°, calculate the index of refraction of the water.

3. A ray of laser light passes from air into a block of glass at an angle of incidence of 33.3°. The block of glass is 3.88-in thick. Calculate the angle at which the light will reemerge from the back of the glass into the air. The index of refraction of the air is 1.00 and of the glass is 1.50.

10.4 ■ ANGLES OF ELEVATION AND DEPRESSION

To solve the following problems, draw a figure and form a right triangle that involves the unknown quantity. Use right angle trig to solve the triangle. Two new terms will be used in these problems. An **angle of elevation** is an angle formed by the horizontal and the line of sight above the horizontal. An **angle of depression** is an angle formed by the horizontal and the line of sight below the horizontal. See Figure 10-25.

FIGURE 10-25
Angle of Elevation, Angle of Depression

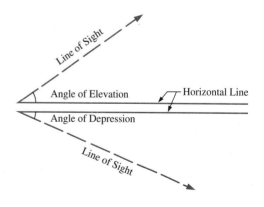

EXAMPLE 10.21 At a distance of 200 ft (to the nearest foot) from a water tower, the angle of elevation to the top of the tower is 24.6°. Find the height of the tower.

Solution Assume that the observation point is level (at the same height) with the base of the tower.

FIGURE 10-26
Example 10.21

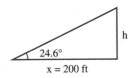

Given: x = 200 ft, θ = 24.6°
Find: h. See Figure 10-26.

$$\tan \theta = h/x$$
$$h = x \tan \theta$$

Transpose x and swap sides.

$$h = 200 \text{ ft} \cdot \tan 24.6°$$
$$h = 91.567149 \text{ ft}$$
Round h to 91.6 ft.

Significant digits: 24.6° is precise to the tenth's place, so tan 24.6° has an accuracy of three. 200 has an accuracy of three. Round the product to three significant digits.

When working with an angle of depression, a theorem from geometry becomes useful. In Figure 10-27, the horizontal and ground are parallel lines. Because the line of sight intersects both lines, it is called a **transversal**. The acute angles formed by the intersection are called **alternate interior angles**. The theorem states that when parallel lines are cut by a transversal, alternate interior angles are equal. The right triangle formed to solve an angle of depression problem often contains the lower angle, angle B in Figure 10-27. This theorem states that its measure is the same as the angle of depression. The next example uses this theorem twice.

FIGURE 10-27
Transversal, Alternate Interior Angles

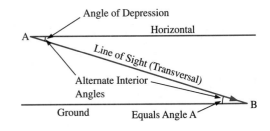

EXAMPLE 10.26 See Figure 10-28. From the top of Lookout Point, a 284-ft cliff, a surveyor measures the angle of depression to a point on the shore on the near side of Lake Lancer as 64.7°. The angle of depression to a point on the opposite shore is 21.0°. How wide is Lake Lancer at this point? Note: This problem assumes that the two measurements were taken in the same plane.

10.4 ANGLES OF ELEVATION AND DEPRESSION ■ 331

FIGURE 10-28
Example 10.27

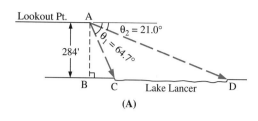

(A)

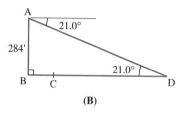

(B)

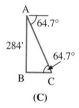

(C)

Solution **Given:** $\theta_1 = 64.7°$, $\theta_2 = 21.0°$
Find: CD. See Figure 10-28A.

Using the large right triangle ABD, find the length of side BD. See Figure 10-28B.

$$\tan D = AB/BD$$
$$BD = AB/\tan D$$
$$BD = 284 \text{ ft}/\tan 21.0°$$
$$BD = 739.84529 \text{ ft}$$

tan 21.0° is a ratio of lengths of sides and has no units.

Round BD to 739.8 ft.

Significant digits: 21.0° is precise to the tenth's place, so tan 21.0° has an accuracy of three. One extra significant digit is kept for this intermediate result.
Using the small triangle ABC, find the length of side BC. See Figure 10-28C.

$$\tan C = AB/BC$$
$$BC = AB/\tan C$$

Two steps. Multiply each side by BC and divide each side by tan C.

$$BC = 284/\tan 64.7°$$
$$BC = 134.24618 \text{ ft}$$

Round BC to 134.2 ft

One extra significant digit is retained for this intermediate result. Find CD by subtracting BC from BD.

$$CD = BD - BC$$
$$= 739.8 - 134.2$$
$$= 605.6 \text{ ft}$$

Round CD to 606 ft.

Significant digits: BD and BC each had an extra significant digit. Lake Lancer is 606-ft wide.

SECTION 10.4 EXERCISES

1. A 56-ft pole casts a 125-ft shadow. What is the angle of elevation of the sun?
2. If the angle of elevation of the sun is 27°, how long a shadow does a 62-ft flag pole cast?
3. From a plane 2450 ft high, the angle of depression to the far end of the runway is 42°, and to the near end is 62°. Calculate the length of the runway.
4. An antenna is mounted on top of a tall building. From a point 235 ft from the base of the building, the angle of elevation to the top of the building is 37°. The angle of elevation to the top of the antenna is 43°. How high is the antenna?
5. A guy wire for a 47-ft power pole has an angle of elevation of 52°. How far is the ground anchor from the base of the pole?

10.5 ■ SURVEYING

In surveying, the distance traveled north or south from a starting point is called **latitude**. The distance traveled east or west is called **departure**. See Figure 10-29. Latitude and departure form a right angle. A right triangle can be formed with the hypotenuse representing the distance of travel and the direction of travel (**bearing**). The bearing angle B is calculated using the tangent function.

$$B = \arctan(\text{departure}/\text{latitude})$$

FIGURE 10-29
Latitude, Departure

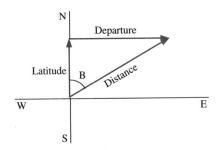

To specify the bearing, first state whether the latitude was north or south in direction, then the bearing, angle B, and finally whether the departure was east or west; for example, N56°E or S33°W. Several examples are shown in Figure 10-30. Note that the bearing angle is always measured from the north–south line. The distance is calculated using trig or the Pythagorean theorem.

$$\text{distance} = (\text{latitude}^2 + \text{departure}^2)^{1/2}$$

FIGURE 10-30
Bearing

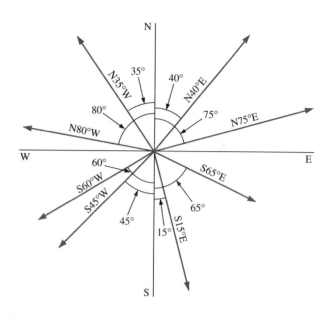

EXAMPLE 10.23 Calculate the bearing and distance for the latitude and departure shown in Figure 10-31.

FIGURE 10-31
Example 10.23

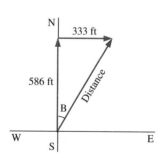

Solution **Given:** latitude = 586 feet north, departure = 333 ft east.
Find: bearing and distance

$$B = \arctan(\text{departure}/\text{latitude})$$
$$B = \arctan(333/586)$$
$$B = 29.6° \quad \text{bearing} = \text{N}29.6°\text{E}$$
$$\text{distance} = (\text{latitude}^2 + \text{departure}^2)^{1/2}$$
$$\text{distance} = (586^2 + 333^2)^{1/2}$$
$$\text{distance} = 674 \text{ ft}$$

334 ■ CHAPTER 10 TRIGONOMETRY

EXAMPLE 10.24 Calculate the bearing and distance for the latitude and departure shown in Figure 10-32.

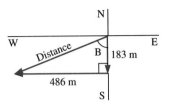

FIGURE 10-32
Example 10.24

Solution Given: latitude = 183 m south, departure = 486 m west
Find: Find: bearing and distance

$$B = \arctan(\text{departure}/\text{latitude})$$
$$B = \arctan 486/183$$
$$B = 69.4°$$
$$\text{bearing} = S69.4°W$$
$$\text{distance} = (\text{latitude}^2 + \text{departure}^2)^{1/2}$$
$$\text{distance} = ((183 \text{ m})^2 + (486 \text{ m})^2)^{1/2}$$
$$\text{distance} = 519 \text{ m}$$

EXAMPLE 10.25 Given a bearing of S70.3°W and a distance of 46.3 m, calculate the latitude and departure.

Solution Given: bearing S70.3°W, distance 46.3 m
Find: latitude, departure. See Figure 10-33.

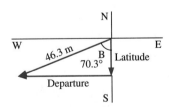

FIGURE 10-33
Example 10.25

$\cos B = \text{latitude}/\text{distance}$ $\sin B = \text{departure}/\text{distance}$
latitude = distance $\cos B$ departure = distance $\sin B$
latitude = 46.3 $\cos 70.3°$ departure = 46.3 $\sin 70.3°$
latitude = 15.60751047 m   departure = 43.590 086 22 m
latitude = 15.6 m   departure = 43.6 m

Significant digits: 70.3° is precise to the tenth of a degree, its cosine can be written to three significant digits. Round the result to three significant digits.

SECTION 10.5 EXERCISES

Calculate the bearing and distance.

	Latitude	Departure
1.	2.4 mi N	3.7 mi E
2.	24.6 m S	57.3 m E
3.	222 ft S	345 ft W
4.	3.02 km N	2.66 km W

Calculate latitude and departure.

	Bearing	Distance
5.	N24.6°W	657 ft
6.	S45°W	1.7 mi
7.	S68.4°E	34.5 km
8.	N12.3°E	1230 m

■ SUMMARY

1. Here are the six trig functions and their definitions.

Ratio	Name	Abbreviation
Side Opposite/Hypotenuse (SO/H)	Sine	sin
Side Adjacent/Hypotenuse (SA/H)	Cosine	cos
Side Opposite/Side Adjacent (SO/SA)	Tangent	tan
Hypotenuse/Side Opposite (H/SO)	Cosecant	csc
Hypotenuse/Side Adjacent (H/SA)	Secant	sec
Side Adjacent/Side Opposite (SA/SO)	Cotangent	cot

2. "The angle whose" is denoted in two different ways. $A = \sin^{-1} SO/H$ and $A = \arcsin SO/H$ are two ways of stating "A is the angle whose sin is SO/H." These expressions are called inverse trig functions.

3. For an angle precise to the nearest degree, round the sine, cosine, or tangent of that angle to two significant digits. Also, for an angle precise to the nearest tenth of degree, round the sine, cosine, or tangent of that angle to three significant digits. These rules of thumb are not accurate for small angles of 0° to 10°, or for large angles of 80° to 90°.

4. For a number with an accuracy of two, round the arcsin, arccos, or arctan of that number to the nearest degree. For a number with an accuracy of three, round the arcsin, arccos, or arctan to the nearest tenth of degree. These rules of thumb are also not accurate for small angles of 0° to 10° and large angles of 80° to 90°.

5. Snell's law describes how light is refracted (bent) as it passes from one medium into another.

6. Index of refraction or refractive index is the property of a material that affects the velocity and wavelength of light passing through it. This change in velocity causes the light to change direction of travel. The index of refraction of a vacuum is 1, and all other materials have a greater index of refraction.

7. Snell's law states that

$$n_i \sin \theta_i = n_t \sin \theta_t \quad \text{Snell's law}$$

where n_i is the index of refraction of the medium incident light is passing through,

n_t is the index of refraction of the medium refracted light is passing through,

θ_i is the angle of incidence,

θ_t is the angle of transmission.

8. When light passes from a material of lower index of refraction into a material of higher index of refraction ($n_i < n_t$) such as air into glass or water, light is bent toward the normal. When light passes from a material of higher index of refraction into a material of lower index of refraction ($n_i > n_t$), light is bent away from the normal. Figure 10-21 summarizes the relationships between the angle of transmission and the index of refraction.

9. The angle of incidence that causes a beam of light to be refracted along the junction between the two mediums is called the critical angle, θ_c.

$$\theta_c = \arcsin [n_t/n_i]$$

10. An angle of elevation is an angle formed by the horizontal and the line of sight above the horizontal. An angle of depression is an angle formed by the horizontal and the line of sight below the horizontal.

11. In surveying, the distance traveled north or south from a starting point is called latitude. The distance traveled east or west is called departure. See Figure 10-29.

12. The bearing, angle B, is calculated from latitude and departure using the tangent function.

$$B = \arctan (\text{departure}/\text{latitude})$$

13. The distance is calculated from latitude and departure using trig or the Pythagorean theorem.

$$\text{distance} = (\text{latitude}^2 + \text{departure}^2)^{1/2}$$

14. To specify the bearing, first state whether the latitude was north or south in direction, then the bearing, angle B, and finally whether the departure was east or west, for example, N56°E or S33°W.

■ SOLUTIONS TO CHAPTER 10 SECTION EXERCISES

Solutions to Section 10.1 Exercises

1. See Figure 10-34.

FIGURE 10-34
Section 10.1 Exercises, Problem 1

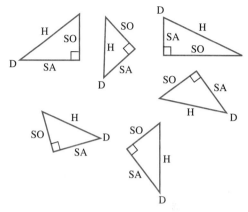

	sin A	cos A	tan A	sin B	cos B	tan B
2.	12/16	11/16	12/11	11/16	12/16	11/12
3.	4/5	3/5	4/3	3/5	4/5	3/4
4.	$3^{1/2}/2$	1/2	$3^{1/2}$	1/2	$3^{1/2}/2$	$3^{1/2}/3$
5.	$2^{1/2}/2$	$2^{1/2}/2$	1	$2^{1/2}/2$	$2^{1/2}/2$	1

6.

angle °	sin	cos	tan
0	0	1	0
12.6	0.218	0.976	0.224
30	0.50	0.87	0.58
40.9	0.655	0.756	0.866
60	0.87	0.50	1.7
64.3	0.901	0.434	2.08
87.2	0.9988	0.049	20
90	1.0	0.0	undefined

7. tan 89.999 999 999 9° = 572 957 795 131
8. 4.9 m
9. 26′
10. 17 km
11. 82 mi
12. 29 m
13. 135 m
14. 53 mi
15. 38 mi
16. 86 yd

Solutions to Section 10.2 Exercises

1.

$A = \arcsin \text{SO/H}$ $A = \sin^{-1} \text{SO/H}$
$A = \arccos \text{SA/H}$ $A = \cos^{-1} \text{SA/H}$
$A = \arctan \text{SO/SA}$ $A = \tan^{-1} \text{SO/SA}$
$A = \text{arccsc H/SO}$ $A = \csc^{-1} \text{H/SO}$
$A = \text{arcsec H/SA}$ $A = \sec^{-1} \text{H/SA}$
$A = \text{arccot SA/SO}$ $A = \cot^{-1} \text{SA/SO}$

2. 33.7°

3. 63°

4. 63.7°

5. 29°

6. 67.7°

7. 42.5°

8. 53°

9. 30.0°

10. A = 58° See Figure 10-35.
B = 32°

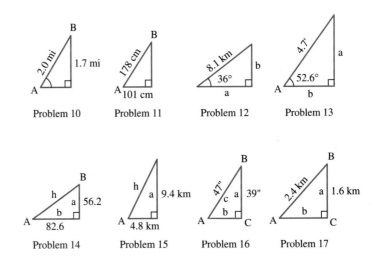

FIGURE 10-35
Solutions to Section 10.2 Exercises, Problems 10 through 17

11. A = 55° See Figure 10-35.
    B = 35°
12. b = 4.8 km See Figure 10-35.
    a = 6.6 km
13. a = 3.7 ft See Figure 10-35.
    b = 2.9 ft
14. See Figure 10-35.
    h = 99.9 m
15. See Figure 10-35.
    h = 11 km
16. A = 56° See Figure 10-35.
    B = 34°
    b = 26 in
17. A = 42° See Figure 10-35.
    B = 48°
    b = 1.8 km

Solutions to Section 10.3 Exercises

1. 51° See Figure 10-36.   2. 1.42 See Figure 10-37.   3. 33.3° See Figure 10-38.

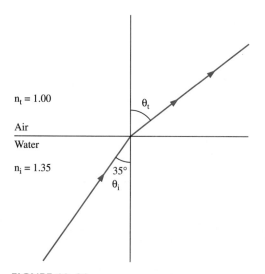

FIGURE 10-36
Solution to Section 10.3 Exercises, Problem 1

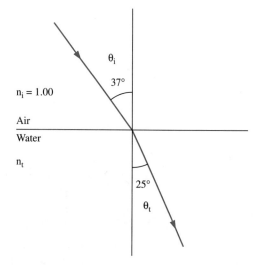

FIGURE 10-37
Solution to Section 10.3 Exercises, Problem 2

FIGURE 10-38
Solution to Section 10.3 Exercises, Problem 3

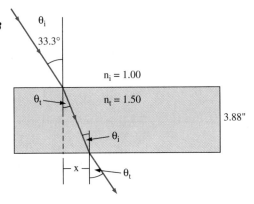

Solutions to Section 10.4 Exercises

1. 24°
2. 120 ft
3. 1400 ft
4. 40 ft
5. 37 ft

Solutions to Section 10.5 Exercises

	Bearing	Distance
1.	N57°E	4.4 mi
2.	S66.8°E	62.4 m
3.	S57.2°W	410 ft
4.	N41.4°W	4.02 km

	Latitude	Departure
5.	597 ft N	273 ft W
6.	1.2 mi S	1.2 mi W
7.	12.7 km S	32.1 km E
8.	1200 m N	262 m E

CHAPTER 10 PROBLEMS

1. Sketch a right triangle. Label the hypotenuse c and the two sides a and b. Write the Pythagorean theorem in terms of a, b, and c. Solve the theorem for b.
2. In Figure 10-39, calculate the length of the third side of the triangle.

FIGURE 10-39
Chapter 10 Problems 2, 5, and 7

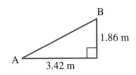

FIGURE 10-40
Chapter 10 Problems 3, 6, and 8

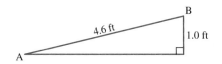

3. In Figure 10-40, calculate the length of the third side of the triangle.

Complete the chart.

	sin A	cos A	tan A	csc A	sec A	cot A
Ratio of	____	____	____	____	____	____

4. *Complete the chart.*

	sin A	cos A	tan A	sin B	cos B	tan B
5. Figure 10-39	____	____	____	____	____	____
6. Figure 10-40	____	____	____	____	____	____

Complete the chart.

	Angle A	Angle B
7. Figure 10-39	____	____
8. Figure 10-40	____	____

9. If the hypotenuse of a right triangle measures 3.56 miles and one side measures 2.89 miles, find the angles.

10. If the hypotenuse of a right triangle measures 84 cm and one side measures 68 cm, find the angles.

11. If the hypotenuse of a right triangle measures 87.6 km and one angle measures 22°, find the lengths of the two sides.

12. If the hypotenuse of a right triangle measures 78.2 ft and one angle measures 54.6°, find the lengths of the two sides.

13. If two sides of a right triangle measure 23.7 m and 38.2 m, find the hypotenuse by two different methods.

14. If two sides of a right triangle measure 10.6 km and 7.26 km, find the hypotenuse by two different methods.

15. If the hypotenuse of a right triangle measures 33.7 in and one side measures 47.3 in, find the length of the other side and the two angles.

16. If the hypotenuse of a right triangle measures 8.5×10^3 m and one side measures 6.4×10^3 m, find the length of the other side and the two angles.

Use the rules of thumb to round these results to the proper number of significant digits.

17. sin 32° =
18. cos 14.2° =
19. tan 58.7° =
20. sin 15.6° =
21. cos 47° =
22. tan 13° =
23. arcsin 0.326 =
24. arccos 0.73 =
25. arctan 1.32 =
26. arcsin (3.62/4.98) =
27. arccos (106/133) =
28. arctan (13/10) =
29. arcsin [cos 34°] =
30. arccos [tan 23.4°] =
31. arctan [sin 57.9°] =
32. A beam of light is passing from water (index of refraction = 1.30) into air (index of refraction = 1.00) at an angle of 18° to the normal. Calculate the angle of the light as it passes through the air.
33. A beam of light is passing from air (index of refraction = 1.00) into a liquid. If the light approaches the boundary at an angle of 29.6° and travels through the liquid at an angle of 19.0°, calculate the index of refraction of the liquid.
34. A beam of light is passing from water (index of refraction = 1.35) into air (index of refraction = 1.00) at an angle of 35° to the normal. Calculate the angle of the light as it passes through the air.
35. For the situation in Problem 34, what angle of incidence will cause the light ray to be totally reflected?
36. A beam of light is passing from air (index of refraction = 1.00) into glass. If the light approaches the boundary at an angle of 27.0° and travels through the glass at an angle of 18.5°, calculate the index of refraction of the glass.
37. A ray of light enters a cube of glass at an angle of incidence of 25°. The cube is 2.5 in thick and has an index of refraction of 1.43. Calculate the angle at which the light ray will emerge from the glass and the point at which it will emerge.
38. A 35-ft pole casts a 20-ft shadow. What is the angle of elevation of the sun?
39. If the angle of elevation of the sun is 65.2°, and a pole casts a 10.5' shadow, how high is the pole?
40. From a plane 1550-ft high, the angle of depression to the far end of the runway is 32°, and to the near end is 52°. Calculate the length of the runway.

41. An antenna is mounted on top of a tall building. From a point 122 ft from the base of the building, the angle of elevation to the top of the antenna is 67° and the angle of elevation to the top of the building is 55°. Calculate the height of the antenna.
42. From the top of a building, the angle of depression to a point on the ground 235 m from the base of the building is 37°. Calculate the height of the building.
43. From an altitude of 1240 ft above the shore, a Coast Guard plane spots a boat that is one mile offshore. Calculate the angle of depression to the boat from the plane.
44. A guy wire for a 58-ft power pole is anchored into the ground 47 ft from the pole. Calculate the angle of elevation.
45. From a lighthouse 265 ft above the surface of the ocean, the angle of depression to a ship is 13.7°. How far is the ship from the base of the lighthouse?
46. From atop a 157-ft cliff, the angle of depression to the far side of a lake is 23°. The angle of depression to the near side of the lake is 64°. How wide is the lake?

Sketch a figure and calculate the bearing and distance:

	Latitude	Departure
47.	328 ft S	462 ft E
48.	58 m S	76 m W
49.	2.83 mi N	1.43 mi W
50.	876 yd N	345 yd E
51.	69.4 km S	73.5 km W
52.	59 m N	73 m E

Sketch a figure and calculate the latitude and departure:

	Bearing	Distance
53.	N32°W	17 km
54.	S62.4°W	14.3 mi
55.	S27°E	87 ft
56.	N55.5°E	346 ft
57.	S25°W	58.4 yd
58.	N49.1°E	91 m

Unscramble these letters:

I R T G
I S E N
O S E I C N
A T S C E N

Now unscramble the circled letters to solve the puzzle.

SAME   UNDER   APART

_ _ _ _ _ _ _

11
Vectors and Phasors

CHAPTER OUTLINE

- **11.1** Adding Vectors—Graphical Solution Daisy Chain Method
- **11.2** Adding Vectors—Graphical Solution Parallelogram Method
- **11.3** Subtracting Vectors—Graphical Solution
- **11.4** Resolving a Vector into Horizontal and Vertical Components
- **11.5** Combining Horizontal and Vertical Components into a Resultant Vector
- **11.6** Adding Vectors Algebraically
- **11.7** Subtracting Vectors Algebraically
- **11.8** Phasors

OBJECTIVES

After completing this chapter, you should be able to

1. define scalar quantity and vector quantity.
2. add vectors graphically using the daisy chain method.
3. add vectors graphically using the parallelogram method.
4. subtract vectors graphically.
5. resolve vectors into their horizontal and vertical components.
6. combine horizontal and vertical components into a resultant.
7. add vectors algebraically.
8. subtract vectors algebraically.
9. apply vectors to technical problems.

NEW TERMS TO WATCH FOR

scalar	instantaneous value
vector	cycle
resultant	peak voltage
daisy chain	peak current
parallelogram	leads
resolving	lags
horizontal component	phase shift
vertical component	phasor diagram
vector sum	frequency
vector difference	cycle
phasors	Hertz
unit phasor	

Mass, area, and volume are examples of quantities that have a magnitude associated with them but not a direction. They are called **scalar** quantities. Scalar quantities are combined using the rules for signed numbers. Force, velocity, and acceleration have a magnitude and direction. They are called **vectors**. Vectors are combined using the rules for vector addition and vector subtraction presented in this chapter.

In applications, vectors are represented by arrows. Each arrow has a *tip* and a *tail*. The tip of the arrow points in the specified direction of the vector. The length of the vector is proportional to the magnitude of the vector. Vectors are identified in this book with boldface uppercase letters, such as vector **A** or vector **B**. Sometimes vectors are named by two letters, a letter at the tail followed by a letter at the tip, such as vector **OA**. In many applications of vectors, the associated angle is measured from the positive *x*-axis. Counterclockwise is considered positive, and clockwise is considered negative.

Figure 11-1 shows two vectors **A** and **B** that represent forces acting on a common point O. The vectors can also be named **OA** and **OB**. **A** has a magnitude of 2 kips (a kip is 1000 pounds) acting at +60°: it is written 2 ∠60° kips. **B** has a magnitude of three kips acting at −30°, 3 ∠−30° kips. The sum of these two vectors is called the **resultant**. **A resultant is a single vector that has the same effect as two or more individual vectors.** The right angle trigonometry learned in the last chapter will be used in this chapter to add and subtract vectors.

11.1 ■ ADDING VECTORS—GRAPHICAL SOLUTION: DAISY CHAIN METHOD

Forces acting on a body are vector quantities and can be added using graphical methods. One method of adding vectors graphically is to plot the vectors tip to tail to form a

FIGURE 11-1
Vector Quantities

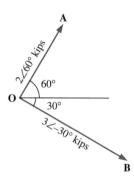

daisy chain. That is, redraw them to scale so that the tail of the second vector lies on the tip of the first vector, and the tail of the third lies on the tip of the second. The process can be continued for any number of vectors. The sum of the vectors, the resultant, is the vector that runs from the tail to the first vector to the tip of the last vector. A suitable scale is used to plot the magnitude of each vector and to measure the magnitude of the resultant. A protractor can be used to plot the angles of the vectors and to measure the angle of the resultant.

The vectors that are translated must remain parallel to their original positions. One way to maintain the proper direction is to measure the original angle with a protractor and then redraw the vector using the protractor.

In Figure 11-2, vector **OA** is being translated so that its tail lies at the tip of vector **OB**. The angle of vector **OA** measures 38°. The length of **OA** is measured so that **OA** can be drawn the same length in its new position. The resultant **R** is drawn from the tail of **B** to the tip of **A**.

FIGURE 11-2
Daisy Chain Using Protractor

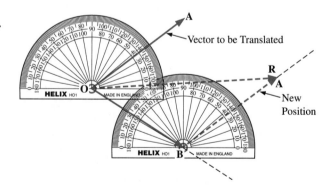

A second method is shown in Figure 11-3. Vector **OB** is to be added to **OA** by translating **OB** so that its tail lies on the tip of **OA**. Two triangles or a straight edge and a triangle are required. Arrange the triangles so that the hypotenuse of triangle 2 lies against one leg of triangle 1. Hold triangle 2 stationary and slide triangle 1 to the desired position. Some trial and error might be required.

11.2 ADDING VECTORS—GRAPHICAL SOLUTION: PARALLELOGRAM METHOD ▪ 347

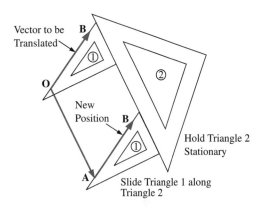

FIGURE 11-3
Drawing Parallel Lines

EXAMPLE 11.1 Figure 11-1 shows two forces **A** and **B** acting on a stationary body. Find the resultant force **R** graphically.

Solution See Figure 11-4. Leave one of the vectors in place and redraw the other to form the tip–to–tail chain. In the solution of Figure 11-4, the 3 kip force **B** was left in place. The 2 kip force **A** was redrawn so that its tail rests on the tip of vector **B**. The resultant vector **R** has its tail at the tail of vector **B** and its tip at the tip of vector **A**. The length and direction of vector **A** is maintained. The resultant measures about 3.7 kips at an angle of about 5°. **R** = 3.7 ∠5° kips.

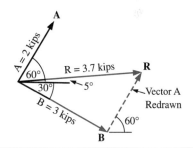

FIGURE 11-4
Example 11.1

11.2 ▪ ADDING VECTORS—GRAPHICAL SOLUTION: PARALLELOGRAM METHOD

In Chapter 7, we defined a **parallelogram** as a four-sided figure whose opposite sides are parallel and equal. Figure 11-5 shows several parallelograms.

To add vectors graphically using the parallelogram method, form a parallelogram using two of the vectors as sides. The resultant vector is the diagonal of the parallelogram. If a third vector is to be added, form a second parallelogram using the third vector and the diagonal of the first parallelogram as sides. The diagonal of the second parallelogram is the overall resultant vector. The process can be continued for any number of vectors.

FIGURE 11-5
Parallelograms

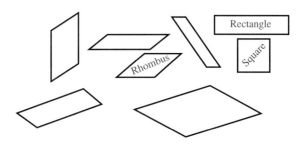

EXAMPLE 11.2   Use the parallelogram method to add vectors **A** and **B** in Figure 11-1.

Solution   See Figure 11-6. Use **A** and **B** as sides and form a parallelogram. Draw a line parallel to **A** through the tip of **B**. Draw a second line parallel to **B** through the tip of **A**. Their point of intersection is the tip of the resultant **R**. Its tail lies on the tail of **A** and **B**. In other words, the resultant **R** is the diagonal of the parallelogram. **R** measures about 3.7 kips at an angle of 5°.

$$\mathbf{R} = 3.7 \angle 5° \text{ kips}$$

FIGURE 11-6
Example 11.2

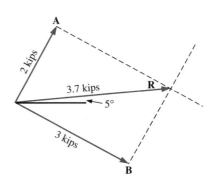

EXAMPLE 11.3   Figure 11-7 shows three forces acting on a body. Find the resultant force graphically.

FIGURE 11-7
Example 11.3

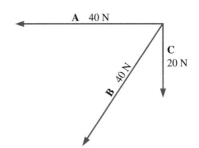

Solution 1 —Daisy Chain Method

See Figure 11-8. Translate two of the vectors to form a daisy chain. It does not matter which two. The resultant runs from the tail of the first to the tip of the last. In Figure 11-8, **B** and **C** have been translated to form a daisy chain. The resultant **R** runs from the tail of **A** to the tip of **C**. It measures about 83 newtons at an angle of $40° + 180° = 220°$.

$$\mathbf{R} = 83 \angle 220° \text{ N}$$

Solution 2 —Parallelogram method

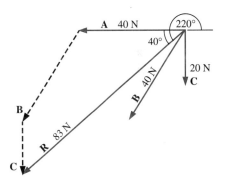

FIGURE 11-8
Example 11.3, Daisy Chain Method

See Figure 11-9. Form a parallelogram using any two of the vectors as sides. Use the resultant of the first parallelogram along with the third vector to form a second parallelogram. The diagonal of the second parallelogram is the overall resultant. In Figure 11-9, **A** and **C** are used to form the first parallelogram. The resultant is $\mathbf{R}_1$. A second parallelogram is formed with $\mathbf{R}_1$ and **B** as sides. The diagonal of the second parallelogram is $\mathbf{R}_2$. It is the solution to the problem. $\mathbf{R}_2$ measures about 82 N at an angle of about $180° + 42° = 222°$.

$$\mathbf{R}_2 = 82 \angle 222° \text{ N}$$

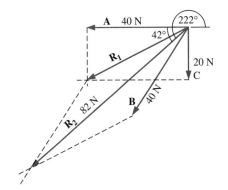

FIGURE 11-9
Example 11.3, Parallelogram Method

Solution 1 and Solution 2 do not agree exactly. This is a disadvantage of graphical solutions. Their precision depends on the skill of the draftsperson and the equipment used. They are not as precise as the algebraic solutions presented later in the chapter.

In the next example, a vector and a resultant are given and the second vector is found.

EXAMPLE 11.4 Two tractors are pulling a truck out of loose sand. One is pulling at an angle of 30° with a force of 800 lb, as shown in Figure 11-10A. Graphically find the direction and magnitude of the force exerted by the second tractor to produce a resultant of 1200 lb due east.

FIGURE 11-10
Example 11.4

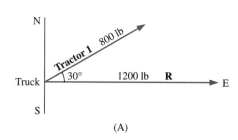

(A)

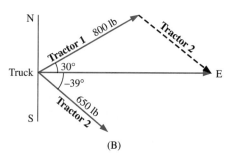

(B)

Solution The resultant and one of the vectors are given. Complete the daisy chain by drawing the unknown force. In Figure 11-10B, the required force is labeled Tractor 2, a vector whose tail lies at the tip of the 800-lb vector and whose tip lies at the tip of the resultant. It is drawn with a dashed line. Two triangles are used to translate the force provided by Tractor 2 over to the location of the truck where it will be applied. Tractor 2 must pull with a force of 650 lb at −39°.

Tractor 2 = 650 ∠−39° lb

11.3 ■ SUBTRACTING VECTORS—GRAPHICAL SOLUTION

When, as in the last example, two vectors are being added to form a resultant and one of the vectors and the resultant are given, the unknown vector can be found by subtracting the known vector from the resultant. If $\mathbf{A} + \mathbf{B} = \mathbf{R}$, then $\mathbf{B} = \mathbf{R} + (-\mathbf{A})$. See Figure 11-11. So, to subtract one vector from another graphically, first reverse the direction of the vector to be subtracted, and then use the daisy chain or parallelogram method to add the original resultant to the reversed vector. The resultant formed is the unknown vector. This subtraction method yields a vector whose tail is common to the tails of the two given vectors.

FIGURE 11-11
Vector Subtraction, $\mathbf{R} - \mathbf{A} = \mathbf{B}$

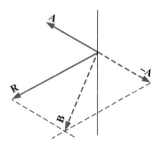

In Figure 11-11, **A** is being added to an unknown **B** to form the resultant **R**. To find **B**, draw **−A** the same magnitude as **A** but in the opposite direction. Form a parallelogram with **R** and **−A** as sides. The diagonal is the required vector **B**.

EXAMPLE 11.5 Rework the truck in the sand example, Figure 11-10, using vector subtraction.

Solution The unknown vector, **Tractor 2**, can be found by subtracting the vector representing the first tractor, **Tractor 1**, from the resultant force **R**. See Figure 11-12. Draw **−Tractor 1** 180° from **Tractor 1**. Then draw a parallelogram using **−Tractor 1** and **R** as sides. The diagonal **Tractor 2** is the required vector. The second tractor needs to pull at an angle of −39° with a force of 650 lb.

FIGURE 11-12
Example 11.5

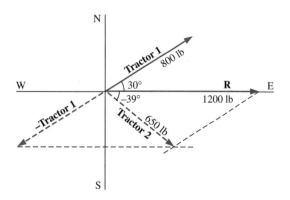

Tractor 2 = 650 ∠ −39° lb

SECTION 11.1 THROUGH 11.3 EXERCISES

See Figure 11-13. Use the daisy chain method to solve Problems 1, 2, 5, and 6 and the parallelogram method for problems 3, 4, 7, and 8.

FIGURE 11-13
Section 11.3 Exercises

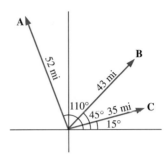

1. A + B
2. A + C
3. B + C
4. A + B + C
5. A − B
6. B − A
7. C − A
8. B − C

Vectors can be added and subtracted more precisely by using mathematical methods. The first step in adding vectors mathematically is **resolving**, or breaking down, the vectors into their horizontal and vertical components. The last step is combining horizontal and vertical components back into a resultant. After learning these two steps, we will be ready to add and subtract vectors mathematically.

11.4 ■ RESOLVING A VECTOR INTO HORIZONTAL AND VERTICAL COMPONENTS

Figure 11-14 shows **Z** in the first quadrant of an *x–y* coordinate axes. In Figure 11-14A, a line has been drawn perpendicular to the horizontal axis through the tip of the vector. That line, the *x*-axis, and the vector form a right triangle, with the vector as the hypotenuse. The **horizontal component** of **Z** is labeled Z_x and the **vertical component** is labeled Z_y. Z_x and Z_y are vectors themselves since they have both magnitude and direction. Z_x and Z_y are drawn in daisy chain form in Figure 11-14A to represent **Z**. **Z** is the vector sum of Z_x and Z_y.

$$Z = Z_x + Z_y.$$

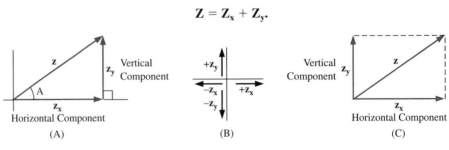

FIGURE 11-14
Horizontal and Vertical Components

11.4 RESOLVING A VECTOR INTO HORIZONTAL AND VERTICAL COMPONENTS

As shown in Figure 11-14B, **vertical components that point up are usually considered positive, and vertical components that point down are negative. Horizontal components that point to the right are usually taken as positive, and those that point to the left are negative.**

In Figure 11-14C, **Z** is again broken down into horizontal and vertical components, but this time the components are drawn in parallelogram form. $\mathbf{Z_y}$ can be drawn along the y-axis as shown in Figure 11-14B or as side opposite in Figure 11-14A, whichever is most convenient.

Cosine can be used to solve for $\mathbf{Z_x}$, and sine can be used to solve for $\mathbf{Z_y}$.

$$\cos A = Z_x/Z \qquad \sin A = Z_y/Z$$
$$Z \cos A = Z_x \qquad Z \sin A = Z_y$$
$$Z_x = Z \cos A \qquad Z_y = Z \sin A$$

where Z is the magnitude of vector **Z**, A is the angle that **Z** forms with the horizontal (reference angle), and Z_x and Z_y are the magnitudes of the horizontal and vertical components of **Z**.

Note that when the magnitude of a vector is being referred to, it is not printed in bold type. When the total vector is being referred to, it is printed in bold type.

EXAMPLE 11.6 Figure 11-15 shows a 2.18 kip force acting upward at an angle of +35°. Resolve the vector into its horizontal and vertical components.

FIGURE 11-15
Example 11.6

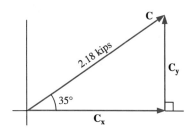

Solution   Given: **C**
   Find: C_x and C_y

$$C_x = C \cos A \qquad\qquad C_y = C \sin A$$
$$C_x = 2180 \text{ lb} \cdot \cos 35° \qquad C_y = 2180 \text{ lb} \cdot \sin 35°$$
$$C_x = 1785.751 \text{ lb} \qquad\qquad C_y = 1250.396 \text{ lb}$$
Round C_x to 1.8 kips.   Round C_y to 1.3 kips.

Significant digits: 35° is precise to the nearest degree, so sin 35° and cos 35° have an accuracy of two. The product is limited to two significant digits.

EXAMPLE 11.7 Figure 11-16 shows a vector of 38.6 N acting at an angle of 140°. Resolve the vector into its horizontal and vertical components.

FIGURE 11-16
Example 11.7

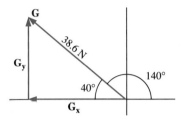

Solution 1 Using the reference angle:

Given: $G = 38.6 \angle 140°$ N
Find: G_x and G_y. See Figure 11-16.

$$\cos \theta = G_x/G \qquad \sin \theta = G_y/G$$
$$G_x = G \cos \theta \qquad G_y = G \sin \theta$$
$$G_x = 38.6 \text{ N} \cdot \cos 40° \qquad G_y = 38.6 \text{ N} \cdot \sin 40°$$

$180° - 140° = 40°$ 40° is the reference angle.

$$G_x = -29.569 \text{ N} \qquad G_y = 24.811 \text{ N}$$

From Figure 11-16, it can be seen that the horizontal component is the left. Make G_x negative. The vertical component is up. Make G_y positive.

$$G_x = -30 \text{ N} \qquad G_y = 25 \text{ N}$$

Significant digits: Assuming 140° is precise to the nearest degree, 40° is precise to the nearest degree. Sin 40° and cos 40° have an accuracy of two. 38.6 has an accuracy of three. The product is limited to two significant digits.

In the second solution, the full angle is used. Scientific calculators can calculate trig functions of angles larger than 90° and assign the proper signs to the results. It is still recommended to draw a figure to see that the result is reasonable.

Solution 2 Using full 140° angle:

$$G_x = G \cos \theta \qquad G_y = G \sin \theta$$
$$G_x = 38.6 \text{ N} \cdot \cos 140° \qquad G_y = 38.6 \text{ N} \cdot \sin 140°$$
$$G_x = -29.569 \text{ N} \qquad G_y = 24.811 \text{ N}$$
$$G_x = -30 \text{ N} \qquad G_y = 25 \text{ N}$$

Significant digits: Assuming 140° is precise to the nearest degree, sin 140° and cos 140° have an accuracy of two. 38.6 has an accuracy of three. The product is limited to two significant digits.

SECTION 11.4 EXERCISES

Resolve these vectors into horizontal and vertical components. Be sure to draw a figure to see that your answers are reasonable.

1. $1.03 \times 10^4 \angle 78°$ lb
2. $4.8 \angle 45°$ kN
3. $14 \angle 105°$ kN
4. $12 \angle 135°$ kips
5. $9.4 \angle 212°$ kips
6. $175 \angle -107°$ lb
7. $138 \angle -49°$ N
8. $48 \angle 295°$ oz

Figure 11-17 summarizes the results of the exercises. The signs of A_x and A_y are determined by the quadrant that the vector **A** lies in.

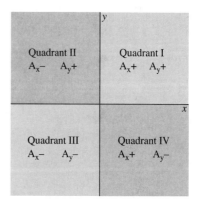

FIGURE 11-17
Signs of Components of Vector **A**

11.5 ■ COMBINING HORIZONTAL AND VERTICAL COMPONENTS INTO A RESULTANT VECTOR

Figure 11-18 shows a horizontal component $\mathbf{R}_x$ and a vertical component $\mathbf{R}_y$. We want to use mathematical vector addition to obtain the resultant **R**.

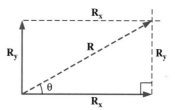

FIGURE 11-18
Combining Horizontal and Vertical Components

$$\mathbf{R} = \mathbf{R}_x + \mathbf{R}_y$$

Following the parallelogram method for vector addition, construct a rectangle as shown in Figure 11-18. The diagonal is the hypotenuse of a right triangle and is the resultant **R**. The Pythagorean theorem can be used to calculate the magnitude of the resultant.

$$R^2 = R_x^2 + R_y^2 \quad \text{or} \quad R = (R_x^2 + R_y^2)^{1/2}$$

where R_x is the magnitude of the horizontal component of vector **R**, R_y is the magnitude of the vertical component of **R**, and R is the magnitude of vector **R**.

The angle associated with the resultant can be calculated by using R_y and R_x as side opposite and side adjacent. Which trig function uses SO and SA? Tangent does.

$$\tan \theta = SO/SA = R_y/R_x$$
$$\theta = \arctan R_y/R_x$$

EXAMPLE 11.8 Given a horizontal component of +13 N (to the right) and a vertical component of +20 N (up), find the resultant vector.

Solution **Given:** $R_x = 13N$, $R_y = 20N$
Find: **R**. See Figure 11-19.

FIGURE 11-19
Example 11.8

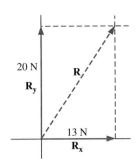

$$R^2 = R_x^2 + R_y^2 \qquad \tan \theta = R_y/R_x$$

where R_x is the magnitude of the horizontal component, R_y is the magnitude of the vertical component, R is the magnitude of the resultant vector, and θ is the angle of the resultant vector.

$$R = (R_x^2 + R_y^2)^{1/2} \qquad \theta = \tan^{-1}(R_y/R_x)$$
$$R = (13^2 + 20^2)^{1/2} \qquad \theta = \tan^{-1}(20/13)$$
$$R = 23.853 \text{ N} \qquad \theta = 56.976°$$

Sequence used for R: 13 INV x^2 + 20 INV x^2 = $\sqrt{}$
Sequence used for θ: 20/13 = INV tan

Round R to 24 N.   Round θ to 57°.

11.5 COMBINING HORIZONTAL AND VERTICAL COMPONENTS INTO A RESULTANT VECTOR

Significant digits:
For the magnitude:

$13^2 + 20^2 = 570$

— accuracy of two
— Round to two significant digits, Precise to ten's place.

For the angle:

$20/13 = 1.5$ — accuracy of two

$\tan^{-1} 1.5 = 57°$

— accuracy of two
— nearest degree

What if the resultant does not lie in the first quadrant? The reference angle is formed with the horizontal axis and solved for. Then the total angle from the positive x-axis is found.

EXAMPLE 11.9 For a horizontal component of −75.4 kips and a vertical component of −106.4 kips, find the resultant.

Solution **Given:** $C_x = -75.4$ kips, $C_y = -106.4$ kips
Find: **C**. See Figure 11-20.

FIGURE 11-20
Example 11.9

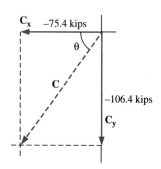

$$C = (C_x^2 + C_y^2)^{1/2} \qquad \theta = \arctan(C_y/C_x)$$
$$C = [(-75.4)^2 + (-106.4)^2]^{1/2} \qquad \theta = \arctan(106.4/75.4)$$

The signs of C_x and C_y are ignored in the calculation of θ because a reference angle less than 90° is being calculated. The total angle will be determined from Figure 11-20.

$$C = 130.407 \text{ kips} \qquad \theta = 54.6768°$$

Shortcut: Both forces are given in kips; the resultant is in kips. k is not entered into the equation.

Round C to 130.4 kips.   Round $\theta = 54.7°$.

Significant digits:
For the magnitude:

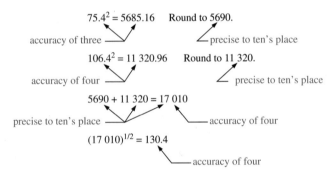

For the angle:

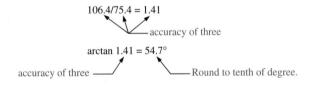

The total angle is 180° + 54.7° = 234.7°. It can also be written as a negative angle.

$$-(180° - 54.7°) = -125.3°$$

SECTION 11.5 EXERCISES

Combine the horizontal and vertical components in Problems 1 through 8 of Figure 11-21 into resultant vectors.

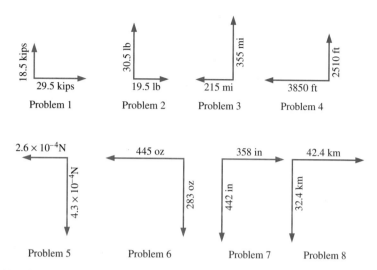

FIGURE 11-21
Section 11.5 Exercises

11.6 ■ ADDING VECTORS ALGEBRAICALLY

We have seen how to resolve or break vectors into horizontal and vertical components and the reverse operation of how to combine components into a resultant vector. Now we will use those skills to add two or more vectors that are acting at any angles.

Step 1. Resolve each vector into its horizontal and vertical components.

Step 2. Add the horizontal components, using the rules for signed numbers. Components pointing to the left are considered negative and those pointing to the right are considered positive.

Step 3. Add the vertical components using the rules for signed numbers. Components pointing down are considered negative and those pointing up are positive.

Step 4. Combine the equivalent horizontal component from step 2 and the equivalent vertical vector from step 3 into a final resultant vector. This vector is called the **vector sum** of the given vectors.

EXAMPLE 11.10 See Figure 11-22. Add vectors **R** and **S** algebraically. Call the resultant **T**.

Solution **Given:** $R = 86.5 \angle 35°N$, $S = 73.4 \angle -65°N$
Find: **T**

Step 1.

$R_x = R \cos A$ $R_y = R \sin A$
$R_x = 86.5 \text{ N} \cdot \cos 35°$ $R_y = 86.5 \text{ N} \cdot \sin 35°$
$R_x = 70.85 \text{ N}$ $R_y = 49.614 \text{ N}$
Round R_x to 70.9 N.   Round R_y to 49.6 N.

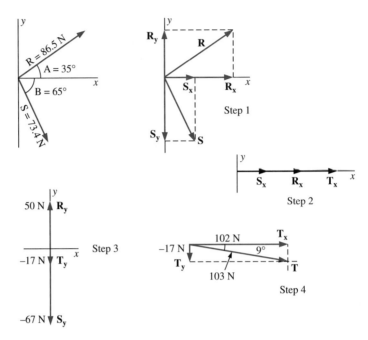

FIGURE 11-22
Example 11.10

Significant digits: 35° is precise to the nearest degree, so cos 35° and sin 35° have an accuracy of two. R_x and R_y are limited to two significant digits. As intermediate results, they are rounded to three significant digits.

$$S_x = S \cos B \qquad\qquad S_y = S \sin B$$
$$S_x = 73.4 \text{ N} \cdot \cos -65° \qquad S_y = 73.4 \text{ N} \cdot \sin -65°$$
$$S_x = 31.020 \text{ N} \qquad\qquad S_y = -66.522 \text{ N}$$
$$\text{Round } S_x \text{ to } 31.0 \text{ N}. \qquad \text{Round } S_y \text{ to } -66.5 \text{ N}.$$

Significant digits: −65° is precise to the nearest degree, so cos −65° and sin −65° have an accuracy of two. S_x and S_y are limited to an accuracy of two. As intermediate results, they are rounded to three significant digits.

Step 2.
$$T_x = R_x + S_x$$
$$T_x = 70.9 \text{ N} + 31.0 \text{ N}$$
$$T_x = 101.9 \text{ N}$$

Both horizontal components are to the right; both are positive.

Step 3.
$$T_y = R_y + S_y$$
$$T_y = 49.6 \text{ N} - 66.5 \text{ N}$$
$$T_y = -16.9 \text{ N}$$

Signs are unlike. Take the difference in magnitudes. Use the sign of the larger magnitude.

Step 4. $T = (T_x^2 + T_y^2)^{1/2}$ $\theta = \arctan(T_y/T_x)$
$T = [(101.9)^2 + (-16.9)^2]^{1/2}$ $\theta = \arctan(-16.9/101.9)$
$T = 103.291$ N $\theta = -9.416°$
Round T to 103 N.   Round θ to $-9.4°$.

Significant digits: 101.9 and −16.9 are carrying an extra significant digit. $102^2 + (-17)^2 = (10\,400 + 290)^{1/2} = (10\,700)^{1/2}$; 102 has an accuracy of three as does 102^2 or 400. 10 400 has a precision of hundreds. 17 has an accuracy of two as does 17^2 or 290. 290 has a precision of tens. 10 400 + 290 is limited to a precision of hundreds, 10 700. 10 700 has an accuracy of three as does $(10\,700)^{1/2}$. Round T to 103 N.

EXAMPLE 11.11 See Figure 11-23. Add vectors **F, G, H,** and **I** to produce vector **J**.

Solution **Step 1.**

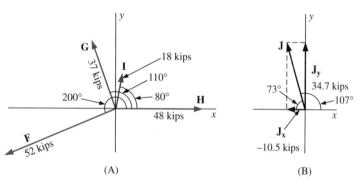

FIGURE 11-23
Example 11.11

$F_x = F \cos A$ $G_x = G \cos B$ $H_x = H \cos 0°$ $I_x = I \cos C$
$F_x = 52 \cos 200°$ $G_x = 37 \cos 110°$ $H_x = 48 \cdot 1$ $I_x = 18 \cos 80°$
$F_x = -48.864$ $G_x = -12.654$ $H_x = 48$ $I_x = 3.125$

all in kips

Do the negative signs agree with Figure 11-23? Note that H is horizontal and has no vertical component.
Round as shown:

$F_x = -48.9$ kips $G_x = -12.7$ kips $H_x = 48$ kips $I_x = 3.13$ kips

Significant digits: Assuming each angle is precise to the nearest degree, the cos of that angle has an accuracy of two. The horizontal components are limited to two significant digits. An extra digit will be carried on these intermediate results.

$$F_y = F \sin A \qquad G_y = G \sin B \qquad H_y = H \sin 0° \qquad I_y = I \sin C$$
$$F_y = 52 \sin 200° \qquad G_y = 37 \sin 110° \qquad H_y = 48 \cdot 0 \qquad I_y = 18 \sin 80°$$
$$F_y = -17.785 \qquad G_y = 34.768 \qquad H_y = 0 \qquad I_y = 17.726$$

all in kips
Round as shown:

$$F_y = -17.8 \text{ kips} \qquad G_y = 34.8 \text{ kips} \qquad H_y = 0 \text{ kips} \qquad I_y = 17.7 \text{ kips}$$

Significant digits: The magnitudes have an accuracy of two and limit the products to an accuracy of two. These intermediate results are carrying an extra significant digit.

Step 2.
$$J_x = F_x + G_x + H_x + I_x$$
$$J_x = -48.9 \text{ kips} + -12.7 \text{ kips} + 48 \text{ kips} + 3.13 \text{ kips}$$
$$J_x = -10.4 \text{ kips}$$
Round J_x to -10.5 kips.

Significant digits: 48 is precise to the unit's place and limits the precision of the sum to the unit's place. J_x is carrying an extra significant digit.

Step 3.
$$J_y = F_y + G_y + H_y + I_y$$
$$J_y = -17.8 \text{ kips} + 34.8 \text{ kips} + 0 + 17.7 \text{ kips}$$
$$J_y = 34.7 \text{ kips}$$

Step 4.
$$J = (J_x^2 + J_y^2)^{1/2} \qquad \theta = \arctan(J_y/J_x)$$
$$J = [(-10.5)^2 + (34.7)^2]^{1/2} \qquad \theta = \arctan(34.7/-10.5)$$
$$J = 36.253 \text{ kips} \qquad \theta = -73.164°$$
Round J to 36.3 kips. \qquad Round θ to 73°.
$$J = 36.3 \angle(180° - 73°) \text{ kips} = 36.3 \angle 107° \text{ kips}$$

Significant digits: For the magnitude, -10.5 and 34.7 are carrying an extra significant digit. $(-11^2 + 35^2)^{1/2} = (121 + 1230)^{1/2} = (1350)^{1/2}$. 1350 has an accuracy of three as does $(1350)^{1/2}$. Round J to 36.3 kips. For the angle, 34.7 has an accuracy of two, as does 34.7/–10.5. The angle is precise to the nearest degree.

The resultant J is shown in Figure 11-23B.

SECTION 11.6 EXERCISES

Find the resultant for each set of vectors in Problems 1 through 8 in Figure 11-24.

11.7 ■ SUBTRACTING VECTORS ALGEBRAICALLY

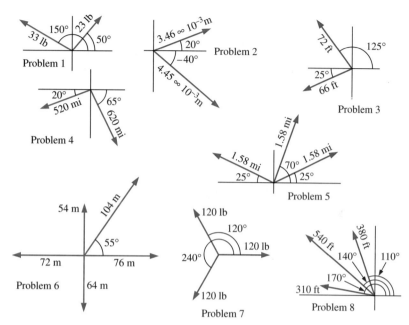

FIGURE 11-24
Section 11.6 Exercises

11.7 ■ SUBTRACTING VECTORS ALGEBRAICALLY

Suppose two vectors **A** and **B** are acting together to produce a required resultant vector **R**; **A**+**B**=**R**. If **A** and **R** are known, **B** can be found by subtracting **A** from **R**; **B**=**R**−**A**. To subtract vectors, follow these steps.

Step 1. Resolve each vector into its horizontal and vertical components.

Step 2. Subtract the horizontal components, using the rules for signed numbers. Components pointing to the left are considered negative and those pointing to the right are considered positive.

Step 3. Subtract the vertical components using the rules for signed numbers. Components pointing down are considered negative and those pointing up are positive.

Step 4. Combine the equivalent horizontal component from step 2 and the equivalent vertical vector from step 3 into a final vector. This vector is called the **vector difference** of the given vectors.

Let's rework the truck stuck in the sand problem using vector subtraction.

EXAMPLE 11.12 Two tractors are working together to pull a truck out of the sand. One is pulling with a force of 800 lb (nearest 100 lb) in a direction of N60°E as shown in Figure 11-25A.

Find the direction and magnitude of the force exerted by the second tractor if the required resultant force is 1200 lb (nearest 100 lb) due east.

Solution   See Figure 11-25.

FIGURE 11-25
Example 11.12

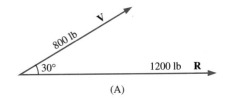
(A)

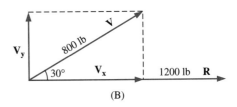
(B)

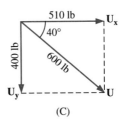
(C)

Step 1. Resolve **V** into its horizontal and vertical components. See Figure 11-25B. **V** makes an angle of 30° with the positive *x*-axis.

$$V_x = V \cos A \qquad V_y = V \sin A$$
$$V_x = 800 \text{ lb} \cdot \cos 30° \qquad V_y = 800 \text{ lb} \cdot \sin 30°$$
$$V_x = 692.820\ 32 \text{ lb} \qquad V_y = 400 \text{ lb}$$
$$\text{Round } V_x \text{ to 690 lb.} \qquad \text{Round } V_y \text{ to 400 lb.}$$

Significant digits: 800 has an accuracy of one. The product is limited to one significant digit. This is an intermediate result; retain an extra significant digit.

Step 2. Subtract the horizontal components.

$$U_x = R_x - V_x$$
$$U_x = 1200 \text{ lb} - 690 \text{ lb} = 510 \text{ lb}$$

11.7 SUBTRACTING VECTORS ALGEBRAICALLY ▪ 365

Step 3. Subtract the vertical components.

$$U_y = R_y - V_y$$
$$U_y = 0 - 400 \text{ lb}$$
$$U_y = -400 \text{ lb}$$

R has no vertical component.

Step 4. Combine U_x and U_y into **U**. See Figure 11-25C.

$U = (U_x^2 + U_y^2)^{1/2}$ $\theta = \arctan(U_y/U_x)$
$U = [(510)^2 + (-400)^2)]^{1/2}$ $\theta = \arctan(-400/510)$
$U = 648.1512 \text{ lb}$ $\theta = -38.1076°$
Round U to 600 lb.   Round θ to $-40°$.

Significant digits: For the magnitude, 510 is carrying an extra significant digit. It has an accuracy of one and so does its square. The square root of the sum of the squares will also have an accuracy of one. Round U to one significant digit. For the angle, since 510 is carrying an extra significant digit, −400/510 has an accuracy of one, 0.8. A rule of thumb for a single significant digit was not developed.

0.8 ranges from 0.75 to 0.85.

arctan 0.75 = 36.869 898°
arctan 0.8 = 38.659 808°
arctan 0.85 = 40.364 537°

These are close only to the ten's place. Round the angle to the nearest ten degrees. The second tractor must pull with a force of 600 ∠−40° lb as shown in Figure 11-25C.

EXAMPLE 11.13   Vector **A**, 286 ∠165.2° N is to act with vector **B** to produce a resultant **R** of 526 ∠−132.0° N. Find **B**. See Figure 11-26A.

Solution   **Given:**   **A** = 286 ∠165.2° N, **R** = 526 ∠−132.0° N
          Find: **B**. See Figure 11-26A.

$$B = R - A$$

Step 1. Break **A** into its component parts. See Figure 11-26B.

$A_x = V \cos A$ $A_y = V \sin A$
$A_x = 286 \text{ N} \times \cos 165.2°$ $A_y = 286 \text{ N} \times \sin 165.2°$
$A_x = -276.5115 \text{ N}$ $A_y = 73.0575 \text{ N}$
Round A_x to −276.5 N.   Round A_y to 73.06 N.

FIGURE 11-26
Example 11.13

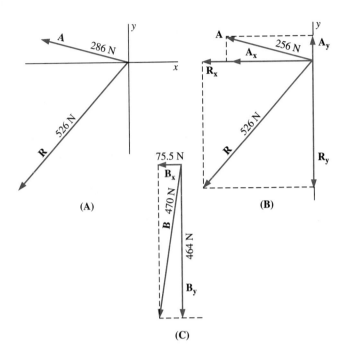

Significant digits: −165.2° is given to the nearest tenth of degree, so sin −165.2° and cos −165.2° both have an accuracy of three. A_x and A_y are limited to an accuracy of three, but as intermediate results are rounded to four.

Break **R** into its component parts. See Figure 11-26B.

$R_x = V \cos A$ $\qquad R_y = V \sin A$
$R_x = 526 \text{ N} \cdot \cos -132.0°$ $\qquad R_y = 526 \text{ N} \cdot \sin -132.0°$
$R_x = -351.9627 \text{ N}$ $\qquad R_y = -390.8942$
Round R_x to −352.0 N. $\qquad$ Round R_y to −390.9 N.

Significant digits: −132.0° is given to the nearest tenth of degree, so sin −132.0° and cos −132.0° both have an accuracy of three. R_x and R_y are limited to an accuracy of three, but as intermediate results are rounded to four.

Step 2. Subtract the horizontal components.

$B_x = R_x - A_x$
$B_x = -352.0 - -276.5$
$B_x = -352.0 + 276.5$
$B_x = -75.5 \text{ N}.$

Step 3. Subtract the vertical components.

$$B_y = R_y - A_y$$
$$B_y = -390.9 - 73.06$$
$$B_y = -463.96$$
Round B_y to -464.0 N.

Significant digits: R_y is least precise and limits B_y to a precision of tenths.

Step 4. Combine B_x and B_y into B. See Figure 11-26C.

$B = (B_x^2 + B_y^2)^{1/2}$ $\theta = \arctan(B_y/B_x)$
$B = [(-75.5)^2 + (-464.0)^2]^{1/2}$ $\theta = \arctan(-464/-75.5)$
$B = 470.1024$ N $\theta = 80.758°$
Round B to 470 N.   Round θ to 81°.

Significant digits: For the magnitude, −75.5 is carrying an extra significant digit. It has an accuracy of two and so does its square. The square root of the sum of the squares will also have an accuracy of two. Round B to two significant digits. For the angle, since −75.5 is carrying an extra significant digit, −464/−75.5 has an accuracy of two. The angle is precise to the nearest degree.

SECTION 11.7 EXERCISES

See Figure 11-27. Evaluate.

FIGURE 11-27
Section 11.7 Exercises

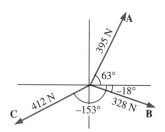

1. B − A   2. A − B   3. C − A
4. A − C   5. B − C   6. C − B

In electronic circuits that contain capacitance or inductance, phase shifts occur between voltage and current and between resistance and reactance. Vectors are used to represent these phase relationships. In this application, the vectors are called **phasors**. The techniques presented in this chapter for manipulation of vectors also apply to phasors. The remainder of this chapter is dedicated to the use of phasors to represent sine waves. The application of phasors to electronics will be covered in the following chapters.

11.8 ■ PHASORS

As a phasor rotates counterclockwise, its vertical component traces out the shape of a sine wave. In Figure 11-28, a phasor of magnitude 1 (**unit phasor**), is positioned with its tail at the origin of an $x-y$ coordinate axes system and its tip on the positive x-axis. In this position, its associated angle is 0° and the distance of its tip off the x-axis is also 0. As the phasor is rotated counterclockwise, the distance of its tip from the x-axis, the vertical component of the phasor, will be recorded. At each angle, this distance is called the **instantaneous value**. If the phasor is representing a voltage waveform, the instantaneous value is measured in volts. If the phasor is representing a current waveform, the instantaneous value is measured in amps.

FIGURE 11-28
Unit Phasor

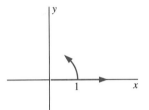

Figure 11-29A shows the phasor at an angle of 15°. The height of the tip of the phasor is translated over to the chart on the right and recorded at θ = 15°. Note that a height of 0 has been recorded for θ = 0°. The unit vector, the instantaneous value y, and a portion of the x-axis form a right triangle with the unit vector as the hypotenuse. This triangle is redrawn in Figure 11-29B. The instantaneous value of y at 15° can be calculated using the sine function. See Figure 11-29B.

FIGURE 11-29
Instantaneous Value

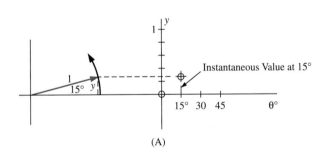

(A)

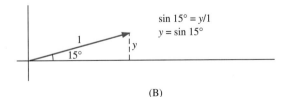

(B)

$$\sin\theta = y/1 \quad \text{or} \quad y = \sin\theta$$
$$y = \sin 15°$$
$$y = 0.2588$$

Since the instantaneous values can be calculated using the sine function, the curve being created in the chart is called a sine wave.

In Figure 11-30, the process is repeated for angles of 30°, 45°, 60°, 75°, and 90°, and the instantaneous values are connected with a smooth curve.

FIGURE 11-30
First Quadrant of Sine Wave

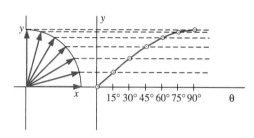

In Figure 11-31, the process of rotating the phasor and recording its instantaneous value at each angle is continued for a half rotation, 180°, of the phasor. Note the symmetry of the curve. At angles of 150° and 30°, the instantaneous values are the same. This is predicted by the trig identity:

$$\sin(180° - 30°) = \sin 30°$$
$$\sin 150° = \sin 30°$$

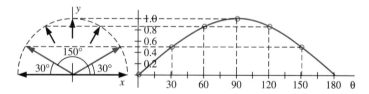

FIGURE 11-31
sin 150° = sin 30°

The curve traced from 90° to 180° is symmetric (mirror image) to the curve from 0° to 90°.

In Figure 11-32, the process is continued through three-fourths of a revolution, 270°. Once again, note the symmetry. As the unit phasor rotates through the third quadrant, the curve that is traced out is the same as the curve in the first quadrant, except it is negative. This is predicted by the trig identity:

$$\sin(180° + \theta) = -\sin\theta$$
$$\sin(180° + 30°) = -\sin 30°$$
$$\sin 210° = -\sin 30°$$

FIGURE 11-32
$\sin 210° = -\sin 30°$

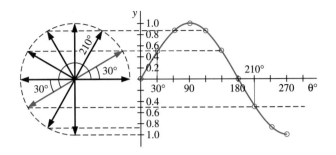

Also note that $\sin 270° = -1$.

In Figure 11-33, the process is continued through a full revolution, 360°. Once again, note the symmetry. As the phasor rotates through the fourth quadrant, the curve that is traced out is the mirror image of the curve in the first quadrant, except it is negative. This is predicted by the trig identity:

$$\sin(-\theta) = -\sin\theta$$
$$\sin 330° = \sin -30° = -\sin 30°$$

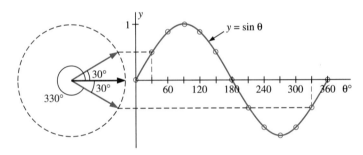

FIGURE 11-33
$\sin -30° = -\sin 30°$

Each revolution of the phasor sweeps out one complete sine wave called a **cycle**. The process is repetitive and produces a continuous sine wave. The waveform is labeled $y = \sin\theta$.

In Figure 11-33, the horizontal axis is marked off in degrees, with 360° completing a cycle. A second option is to measure the rotation of the phasor in radians, and mark the horizontal axis accordingly. One revolution of the phasor and one cycle of the waveform will be completed in 2π radians. Note the divisions of the horizontal axis in Figure 11-34. Tic marks are placed at intervals of $\pi/6$ radians (30°). 2/6 π has been reduced to $\pi/3$; 3/6 π has been reduced to $\pi/2$, and so on. Sin θ will reach a maximum value at $\pi/2$ radians, 0 at π radians, a negative maximum at $3\pi/2$ radians, and 0 again at 2π radians.

Now let's revolve two phasors, unit phasor **A** the same as before as a standard and a second phasor **B** of magnitude 2, and compare the sine waves generated. See Figure 11-34. The sine wave produced by phasor **A** is labeled y_A and the sine wave by phasor **B**,

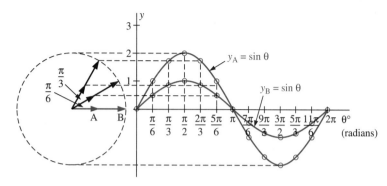

FIGURE 11-34
$y = 2 \sin \theta$

y_B. y_A and y_B cross zero at the same points because their phasors lie on the positive x-axis and the negative x-axis at the same time.
Figure 11-35 shows the two phasors frozen at $\pi/6$ radians.

FIGURE 11-35
$y_B = 2 y_A$

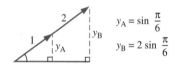

$y_A = \sin \dfrac{\pi}{6}$
$y_B = 2 \sin \dfrac{\pi}{6}$

$$\sin \pi/6 = y_A/1 \qquad \sin \pi/6 = y_B/2$$
$$y_A = \sin \pi/6 \qquad y_B = 2 \sin \pi/6$$
$$y_A = 0.5736 \qquad y_B = 1.1472$$

At $\pi/6$ radians, y_B is twice y_A. In fact, for any angle θ, the instantaneous value of y_B is twice the instantaneous value of the standard y_A.

$$\sin \theta = y_A/1 \qquad \sin \theta = y_B/2$$
$$y_A = \sin \theta \qquad y_B = 2 \sin \theta$$

Here is another way to approach this situation. In Figure 11-36, phasors **A** and **B** are shown frozen at any angle θ. Since each right triangle contains the same angle θ, the two triangles are similar.

$$y_B/y_A = 2/1$$
$$y_B = 2y_A$$

FIGURE 11-36
Similar Triangles

At any instant, the phasor of length 2 produces an amplitude of twice that of the unit phasor. In general, the instantaneous value of a sine wave $y = m \sin \theta$ will be m times greater than the instantaneous value of our standard sine wave $y = \sin \theta$.

If the phasor has a magnitude of 2, the instantaneous value of the sine wave at $\pi/2$ radians is 2 and the value at $3\pi/2$ radians is -2. This maximum instantaneous value is called the **peak voltage** or **peak current** of the sine wave. For example, $y = 170 \sin \theta$ has a peak voltage of 170 V. Its instantaneous value at any particular angle can be found by substituting the angle into the equation. If the angle is given in radians, don't forget to put your calculator into the radian mode.

At $\theta = \pi/6$:
$y = 170 \sin \pi/6$
$y = 170 \cdot 0.5$
$y = 85$ V

At $\theta = \pi/3$:
$y = 170 \sin \pi/3$
$y = 170 \cdot 0.866$
$y = 147$ V

At $\theta = \pi/2$:
$y = 170 \sin \pi/2$
$y = 170 \cdot 1$
$y = 170$ V positive peak voltage

At $\theta = 3\pi/2$:
$y = 170 \sin 3\pi/2$
$y = 170 \cdot -1$
$y = -170$ V negative peak voltage

What happens if one of the two phasors does not start on the *x*-axis?

Figure 11-37 shows phasor **B** of length 1 at an angle of $-\pi/6$ and phasor **A**, our unit phasor, on the *x*-axis. As they rotate at the same angular velocity, they remain separated by $\pi/6$ radians. Phasor **A** generates waveform $y_A = \sin \theta$, and phasor **B** generates $y_B = \sin(\theta - \pi/6)$. Phasor **A** will rotate $\pi/6$ radians before phasor **B** reaches the *x*-axis. the sine waves produced, waveform y_B starts up from zero $\pi/6$ after waveform y_A. Waveform y_A **leads** waveform y_B by $\pi/6$ radians. An alternate way to state this relationship is waveform y_B **lags** waveform y_A by $\pi/6$ radians. The $\pi/6$ **phase shift** is denoted in Figure 11-37. The sine waves on the right are as they would appear on an oscilloscope, with time passing as you proceed from left to right. The **phasor diagram** on the left contains the same information but is much easier and quicker to draw. You must learn to associate phasor diagrams with their corresponding sine waves and vice versa. Note that the phasor diagram shows **B** at $-\pi/6$, and its equation contains $(\theta - \pi/6)$, but waveform is shifted to the right in the waveform diagrams. This may seem backwards at first, but the oscilloscope sweeps (writes its pattern) from left to right. Time is passing as you move from left to right across the waveforms. Since **B** starts up from zero $\pi/6$ radians after **A**, **B** must be shifted to the right. Likewise, if a phasor is at a positive angle, more counterclockwise than another, it will appear shifted to the left.

FIGURE 11-37
$y_B = sin\ (\theta - \pi/6)$

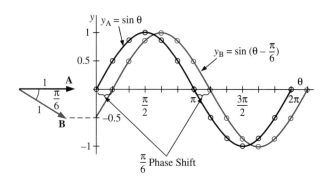

In Figure 11-38, the phasor diagram of Figure 11-37 has been rotated π/6 radians so that **B** is on the x-axis and **A** is at an angle of +π/6. As these phasors rotate, they produce the same waveforms shown in Figure 11-37. In Figure 11-38, **A** is the reference (starts at zero at 0) and **B** lags by π/6 radians. In Figure 11-38, **B** is the reference and **A** leads by π/6. Since these waveforms are repetitive, the result is the same; it's all relative.

FIGURE 11-38
$y_A = sin\ (\theta + \pi/6)$

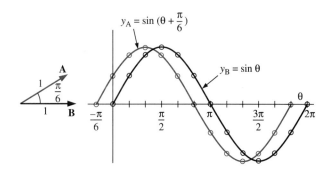

Let's use what we've learned about symmetry to graph two sine waves. Here are some suggestions to make the task easier.

1. Choose a vertical scale that will accommodate the peak values of the waves to be drawn. It is convenient to have a subdivision at one-half of the peak value. To graph the voltage waveform $y = 10\ sin\ \theta$, have divisions at 10 and 5. See Figure 11-39A. Spread your work out. Don't work microscopically.

2. On your graph or grid paper, choose a scale that will conveniently divide one cycle, 360° or 2π radians, into the four quadrants, and each of the four quadrants into convenient divisions. For example, if each division of the graph paper equals 15°, then there will be tic marks at 30°, 45°, 60°, 90°, 180°, 270°, and 360°. Label those angles. See Figure 11-39A.

3. Mark the zeros, the places where the graph crosses the θ axis. If the sine wave has not been shifted, it will cross at 0°, 180°, and 360°. Use a fine X or a small dot with a circle around it to mark the spot. A big blob is not precise.

374 ■ CHAPTER 11   VECTORS AND PHASORS

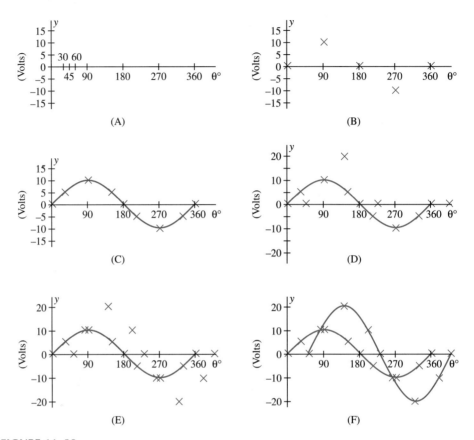

FIGURE 11-39
Graphing Sine Waves

4. Mark the peak values at 90° and 270°. See Figure 11-39B.

5. sin 30° = 0.5. The sine wave rises quickly and has a steep slope around the zero crossings. At 30° on either side of zero, the waveform has already reached half of its peak value. Mark these values as shown in Figure 11-39C.

6. These values give enough information for a fairly accurate sketch. Using a pencil, lightly draw a smooth curve through the points. When satisfied with the shape, draw the final curve in ink.

7. If more accuracy is required, add points at 30° on either side of the peak points. sin 60° = 0.866. Mark points at 0.866 of the peak value.

With this sine wave as the reference, let's graph a voltage sine wave that lags by 45° with double the amplitude, $y = 20 \sin(\theta - 45°)$.

1. Extend the vertical axis to 20 volts.

2. Since this waveform is shifted 45° to the right, mark the zero crossings three divisions to the right of the crossings on the reference waveform. See Figure 11-39D.

3. Mark the peak voltages. From the positive peak of the reference, shift three divisions to the right and up to 20 volts. From the negative peak of the reference, shift three divisions to the right and down to −20 volts. See Figure 11-39D.

4. At 30° either way from the zero crossings of the shifted waveform, mark an amplitude of 10 or −10 volts. See Figure 11-39E.

5. Smooth in the curve in pencil and then in ink. See Figure 11-39F.

Now draw a phasor diagram to represent these sine waves.

1. One of the sine waves crosses zero at 0°, so its phasor is drawn on the horizontal axis. It has an amplitude of 10 volts. See Figure 11-40.

FIGURE 11-40
Drawing Phasor Diagrams

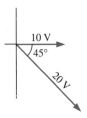

2. The second sine wave is shifted 45° to the right. It lags the reference by 45° and is drawn in the fourth quadrant. It has an amplitude of 20 volts. See Figure 11-40.

The next example uses a different approach to graphing the shifted waveforms. A calculator is used to find y values for angles from 0° to 90° in 15° intervals. Symmetry is used to complete the rest of the cycle.

EXAMPLE 11.14 Figure 11-41 shows three phasors of various magnitudes and phase shifts. Sketch the sine waves they represent.

FIGURE 11-41
Example 11.14

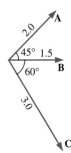

Solution Phasor **B** is the reference and will be graphed with an instantaneous value of zero at 0°. Since its magnitude is 1.5, it will have an amplitude of 1.5 at 90° and an amplitude of −1.5 at 270°. **A** leads **B** by 45° and has a peak amplitude of 2. The horizontal axis is extended to the left of 0° to graph **A** as it comes up from zero. Phasor **C** lags **B** by 60° and reaches a peak of 3. To graph these sine waves, write the equation of each and use your calculator to develop a table of values to be plotted for the first 90° of the waveform. Use what you have learned about the symmetry of the sine wave to complete the sketch.

Note the way the equation is "developed" in stages to reach the final equation.

$y_A = 2 \sin (\theta + 45)°$

θ	$\theta + 45$	$\sin (\theta + 45)°$	$y_A = 2 \sin (\theta + 45)°$
−45	0	0	0
−30	15	0.2588	0.5176
−15	30	0.5	1.0
0	45	0.707	1.414
15	60	0.866	1.732
30	75	0.9659	1.93
45	90	1	2

$y_B = 1.5 \sin \theta$

θ	$\sin \theta$	$y_B = 1.5 \sin \theta$
0	0	0
15	0.2588	0.3882
30	0.5	0.75
45	0.707	1.06
60	0.866	1.2990
75	0.9659	1.4489
90	1	1.5

$y_C = 3 \sin (\theta - 60°)$

θ	$\theta - 60$	$\sin (\theta - 60)°$	$y_A = 3 \sin (\theta - 60)°$
60	0	0	0
75	15	0.2588	0.7764
90	30	0.5	1.5
105	45	0.707	2.121
120	60	0.866	2.5980
135	75	0.9659	2.8977
150	90	1	3.0

In each case, plot the first column on the horizontal axis and the last column on the vertical axis. See Figure 11-42.

FIGURE 11-42
Solution to Example 11.14

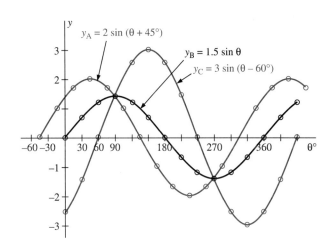

One more characteristic of the sine wave is determined by the revolving phasor. The **frequency**, f, of the sine wave is determined by the angular velocity of the phasor. For each revolution of the phasor, one **cycle** of the waveform is generated. The number of cycles generated each second is called the frequency of the sine wave. One cycle per second is given a special name, one **hertz**.

EXAMPLE 11.15   Calculate the frequency of a sine wave defined by a phasor rotating at 18 000 revolutions per minute. $\omega = 18$ k rpm

Solution
$$\frac{18 \text{ k revolutions}}{\text{minute}} \times \frac{1 \text{ minute}}{60 \text{ seconds}} = 300 \frac{\text{revolutions}}{\text{second}}$$

The phasor completes 300 revolutions each second, so the frequency is 300 hertz.

$$f = 300 \text{ Hz}$$

EXAMPLE 11.16   Calculate the frequency of a sine wave defined by a phasor rotating at 10 000 radians per second.

Solution
$$\frac{10\,000 \text{ radians}}{\text{second}} \times \frac{1 \text{ revolution}}{2\pi \text{ radians}} = 1592 \frac{\text{revolution}}{\text{second}}$$

The phasor completes 1592 revolutions each second, so the frequency is 1592 hertz.

378 ■ CHAPTER 11 VECTORS AND PHASORS

SECTION 11.8 EXERCISES

For Problems 1 through 6, refer to Figure 11-43. For Problems 1 through 3, answer True or False to each statement.

FIGURE 11-43
Section 11.8 Exercises, Problems 1 through 6

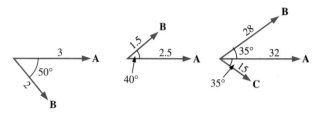

Problems 1 and 4   Problems 2 and 5   Problems 3 and 6

1. In Problem 1 of Figure 11-43.
   a. A is leading B by 50°.
   b. B is leading A by 50°.
   c. A is lagging B by 50°.
   d. B is lagging A by 50°.
2. In Problem 2 of Figure 11-43.
   a. A is leading B by 40°.
   b. B is leading A by 40°.
   c. A is lagging B by 40°.
   d. B is lagging A by 40°.
3. In Problem 3 of Figure 11-43.
   a. A is leading B by 35°.
   b. B is leading C by 70°.
   c. A is lagging C by 35°.
   d. C is lagging A by 35°.
4., 5., and 6. For each phasor diagram in Figure 11-43, sketch the corresponding sine waves.
7., 8., and 9. For each set of waveforms in Figure 11-44, sketch the corresponding phasor diagrams.

Calculate the frequency of sine waves generated by the phasors revolving at these angular frequencies.

10. 7000 revolutions per second
11. 12 000 revolutions per minute
12. 18 000 radians per second

■ **SUMMARY**

1. Quantities that have a magnitude associated with them but not a direction are called scalar quantities. Scalar quantities are combined using the rules for signed numbers.
2. Force, velocity, and acceleration have a magnitude and direction. They are called vectors. Vectors are combined using the rules for vector addition and vector subtraction.

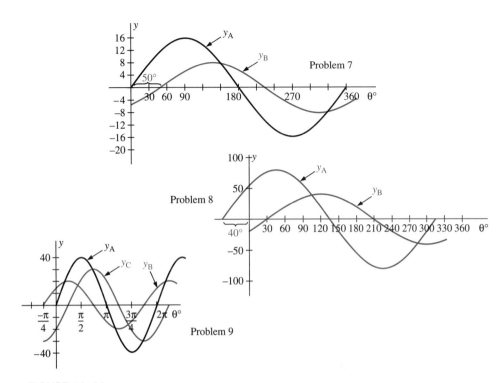

FIGURE 11-44
Section 11.8 Exercises, Problems 7 through 9

3. Vectors are represented by arrows. Each arrow has a tip and a tail. The tip of the arrow points in the specified direction of the vector. The length of the vector is proportional to the magnitude of the vector.
4. The sum of two or more vectors is called the resultant.
5. In the daisy chain method of adding vectors graphically, the vectors are plotted tip to tail to form a daisy chain. The sum of the vectors, the resultant, is the vector that runs from the tail of the first vector to the tip of the last vector.
6. In the parallelogram method of adding vectors graphically, form a parallelogram using two of the vectors as sides. The resultant vector is the diagonal of the parallelogram.
7. To subtract one vector from another graphically, first reverse the direction of the vector to be subtracted, and then use the daisy chain or parallelogram method to add the original resultant to the reversed vector.
8. To add vectors algebraically, follow these steps:
   Step 1. Resolve each vector into its horizontal and vertical components.
   Step 2. Add the horizontal components, using the rules for signed numbers. Components pointing to the left are considered negative, and those pointing to the right are considered positive.

Step 3. Add the vertical components, using the rules for signed numbers. Components pointing down are considered negative and those pointing up are considered positive.

Step 4. Combine the equivalent horizontal component from step 2 and the equivalent vertical vector from step 3 into a final resultant vector. This vector is called the vector sum of the given vectors.

9. To subtract vectors algebraically, follow these steps:
   Step 1. Resolve each vector into its horizontal and vertical components.
   Step 2. Subtract the horizontal components, using the rules for signed numbers. Components pointing to the left are considered negative and those pointing to the right are considered positive.
   Step 3. Subtract the vertical components, using the rules for signed numbers. Components pointing down are considered negative and those pointing up are considered positive.
   Step 4. Combine the equivalent horizontal component from step 2 and the equivalent vertical vector from step 3 into a final vector. This vector is called the vector difference of the given vectors.

10. Phasors are vectors used in electronics to represent phase relationships between quantities such as current and voltage, and resistance and reactance.

11. The length of the phasor is the peak or maximum amplitude of the sine wave it represents.

12. As the phasor is rotated counterclockwise, the distance of its tip from the x-axis, the vertical component of the phasor, is the instantaneous value of the quantity being represented. One revolution of the phasor traces out one cycle of the waveform.

13. Trig identities are true for any angle. Here are three trig identities.

$$\sin(180° - \theta°) = \sin \theta°$$
$$\sin(-\theta) = -\sin \theta$$
$$\sin(180° + \theta) = -\sin \theta$$

14. If phasor **A** is more counterclockwise than phasor **B**, then **A** leads **B**. When the sine waves are graphed, **A** will lie to the left of **B**.

15. The frequency, f, of the sine wave is determined by the angular velocity of the phasor. For each revolution of the phasor, one cycle of the waveform is generated. The number of cycles generated each second is called the frequency of the sine wave measured in hertz.

■ SOLUTIONS TO CHAPTER 11 EXERCISES

Solutions to Section 11.1 through 11.3 Exercises

1. $80 \angle 81°$ mi
2. $60 \angle 75°$ mi
3. $75 \angle 31.5°$ mi
4. $100 \angle 62°$ mi
5. $52 \angle 159°$ mi
6. $52 \angle -21°$ mi
7. $65 \angle -37°$ mi
8. $22 \angle 99°$ mi

SOLUTIONS TO CHAPTER 11 EXERCISES ■ 381

Solutions to Section 11.4 Exercises

	R_x	R_y
1.	2100 lb	10 000 lb
2.	3.4 kN	3.4 kN
3.	−3600 N	14 kN
4.	−8.5 kips	8.5 kips
5.	−8.0 kips	−5.0 kips
6.	−51 lb	−170 lb
7.	91 N	−100 N
8.	20 oz	−44 oz

Solutions to Section 11.5 Exercises

1. 34.8 ∠32.1° kips
2. 36.2 ∠57.4° lb
3. 415 ∠121.2° mi
4. 4600 ∠146.9° ft
5. 5.02×10^{-4} ∠239° N
6. 527 ∠212.5° oz
7. 569 ∠−51.0° in
8. 53 ∠−37.4° km

Solutions to Section 11.6 Exercises

1. 37 ∠112° lb
2. 6.87 ∠14.1° mm
3. 110 ∠163° ft
4. 774 ∠−107° mi
5. 2.9 ∠79° mi
6. 99 ∠50° m
7. 0
8. 1140 ∠138° ft

Solutions to Section 11.7 Exercises

1. 472 ∠−74° N
2. 472 ∠106° N
3. 770 ∠−135° N
4. 770 ∠45° N
5. 680 ∠7° N
6. 680 ∠−170° N

Solutions to Section 11.8 Exercises

1. **a.** True
   b. False
   c. False
   d. True
2. **a.** False
   b. True
   c. True
   d. False

3. a. False
   b. True
   c. False
   d. True

4., 5., and 6. See Figure 11-45.

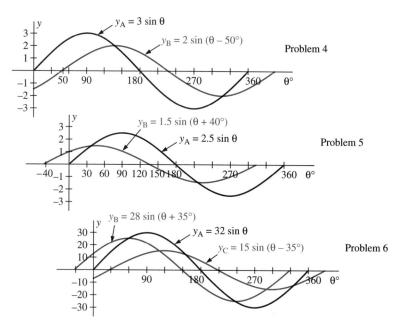

FIGURE 11-45
Solutions to Section 11.8 Exercises, Problems 4 through 6

7, 8, and 9. See Figure 11-46.

10. 7 kHz

11. 200 Hz

12. 286.48 Hz

CHAPTER 11 PROBLEMS

Answer True or False.

1. Velocity is a scalar quantity.
2. Acceleration is a vector quantity.
3. The length of a vector represents its magnitude.
4. Vectors can be added graphically using the daisy chain method.
5. The parallelogram method is a procedure for multiplying vectors.
6. To subtract vectors graphically, change the direction of the minuend and proceed as in addition.

FIGURE 11-46
Solutions to Section 11.8 Exercises, Problems 7 through 9

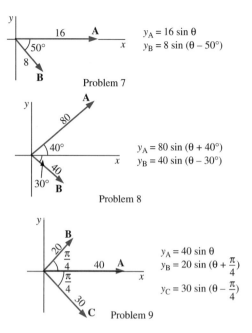

7. A resultant is a vector that produces the same result as two or more individual vectors.
8. Resolving a vector means breaking it down into its horizontal and vertical components.
9. A horizontal component to the left is usually considered negative.
10. A vertical component pointing up is usually considered negative.
11. The angle of a vector in the first quadrant ranges between 270° and 360°.
12. The angle of a vector in the third quadrant ranges between 90° and 270°.
13. The angle of a vector in the third quadrant ranges between −90° and −180°.
14. Two vectors can be added by first breaking them down into their horizontal and vertical components.
15. A phasor is similar to a pulsar.
16. Phasors rotate clockwise.
17. If phasor **A** is more clockwise than phasor **B**, then **A** leads **B**.
18. If phasor **A** leads **B**, then **B** lags **A**.
19. The length of the vector represents the instantaneous value of the vector.
20. The instantaneous value of the waveform is found using the tangent function.
21. If waveform A crosses zero to the left of waveform B, then A leads B.
22. In a phasor diagram, the phasor with the greatest magnitude will have the largest peak value.

23. The magnitude of the phasor determines the frequency of the waveform.
24. Each revolution of the phasor creates one cycle of the waveform.
25. sin (180° − θ) = sin θ.
26. sin (180° + θ) = sin θ.
27. sin (−θ) = sin θ.
28. y = 2 sin θ crosses the horizontal axis at the same places as y = sin θ.
29. y = 3 sin (θ − 30°) leads y = sin θ.

Combine these vectors from Figure 11-47 graphically.

FIGURE 11-47
Chapter 11 Problems 30 through 42, 49 through 59

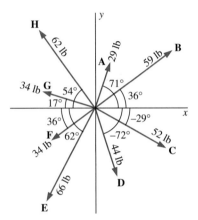

30. A + B
31. A + C
32. B + C
33. A + B + C
34. A − B
35. B − A
36. C − A
37. B − C

Resolve these vectors from Figure 11-47 into horizontal and vertical components. Check to see that your answers are reasonable.

38. D
39. E
40. F
41. G
42. H

Combine these horizontal and vertical components from Figure 11-48 into a resultant vector.

FIGURE 11-48
Chapter 11 Problems 43 through 48

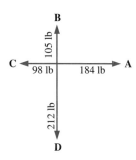

43. **A** and **B**
44. **A** and **B** and **C**
45. **B** and **C**
46. **A** and **D**
47. **B** and **C** and **D**
48. **A** and **B** and **C** and **D**

Combine these vectors from Figure 11-47 algebraically.

49. **A** + **B**
50. **A** + **C**
51. **B** + **C**
52. **A** + **B** + **C**
53. **A** − **B**
54. **B** − **A**
55. **C** − **A**
56. **B** − **C**
57. **D** + **E** + **F**
58. **G** + **H**
59. **D** + **E** + **F** + **G** + **H**

In Problems 60–63, answer True or False to each statement.

60. In Problem 60 of Figure 11-49:
    a. **A** is leading **B** by $\pi/3$ radians.
    b. **B** is leading **A** by $\pi/3$ radians.
    c. **A** is lagging **B** by $\pi/3$ radians.
    d. **B** is lagging **A** by $\pi/3$ radians.
    e. **B** has an instantaneous value of 10 V.

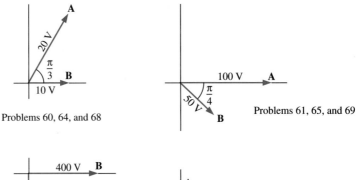

FIGURE 11-49
Chapter 11 Problems 60 through 71

61. In Problem 61 of Figure 11-49:
    a. **A** is leading **B** by $\pi/4$ radians.
    b. **B** is leading **A** by $\pi/4$ radians.
    c. **A** is lagging **B** by $\pi/4$ radians.
    d. **B** is lagging **A** by $\pi/4$ radians.
    e. **A** has a minimum voltage of -100 V.

62. In Problem 62 of Figure 11-49:
    a. **A** is leading **B** by $\pi/2$ radians.
    b. **B** is leading **A** by $\pi/2$ radians.
    c. **A** is lagging **B** by $\pi/2$ radians.
    d. **B** is lagging **A** by $\pi/2$ radians.
    e. **A** has a peak voltage of 200 V.

63. In Problem 63 of Figure 11-49:
    a. **A** is leading **B** by $\pi/2$ radians.
    b. **B** is leading **A** by $\pi/2$ radians.
    c. **A** is lagging **B** by $\pi/2$ radians.
    d. **B** is lagging **A** by $\pi/2$ radians.
    e. **A** will be at peak when **B** crosses zero.

64.–67. For each phasor diagram in Figure 11-49, sketch the corresponding sine waves.

68.–71. For each phasor diagram in Figure 11-49, write the equation of the sine wave.

In Problems 72 through 75, answer True or False to each statement.

72. In Problem 72 of Figure 11-50:

FIGURE 11-50
Chapter 11 Problems 72, 73, 76, 77, 80, and 81

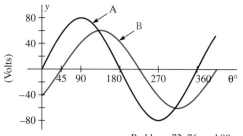

Problems 72, 76, and 80

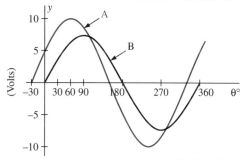

Problems 73, 77, and 81

    a. **A** is leading **B** by 45°.
    b. **B** is leading **A** by 45°.
    c. **A** is lagging **B** by 45°.
    d. **B** is lagging **A** by 45°.
    e. **B** has an minimum voltage of −60 V.

73. In Problem 73 of Figure 11-50:
    a. **A** is leading **B** by 30°.
    b. **B** is leading **A** by 30°.
    c. **A** is lagging **B** by 30°.
    d. **B** is lagging **A** by 30°.
    e. **A** has a minimum voltage of −100 V.

74. In Problem 74 of Figure 11-51:
    a. **A** is leading **B** by 90°.
    b. **B** is leading **A** by 90°.
    c. **A** is lagging **B** by 90°.
    d. **B** is lagging **A** by 90°.
    e. **A** has a peak voltage of 40 V.

75. In Problem 75 of Figure 11-51:
    a. **C** is leading **A** by 90°.
    b. **A** is leading **C** by 90°.
    c. **C** is lagging **A** by 90°.
    d. **A** is lagging **C** by 90°.
    e. **A** will be at peak when **C** crosses zero.

FIGURE 11-51
Chapter 11 Problems 74, 75, 78, 79, 82, and 83

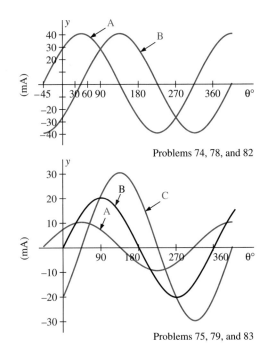

Problems 74, 78, and 82

Problems 75, 79, and 83

    f. **C** will be at peak when **A** crosses zero.
    g. **B** is leading **C** by 45°.
    h. **C** is leading **B** by 45°.
    i. **B** is lagging **C** by 45°.
    j. **C** is lagging **B** by 45°.
    k. **B** will be at peak when **C** crosses zero.
    l. **C** will be at peak when **B** crosses zero.
    m. **B** is leading **A** by 45°.
    n. **A** is leading **B** by 45°.
    o. **B** is lagging **A** by 45°.
    p. **A** is lagging **B** by 45°.
    q. **A** will be at peak when **B** crosses zero.
    r. **B** will be at peak when **A** crosses zero.

76., 77. For each set of sine waves in Figure 11-50, sketch the corresponding phasor diagram.

78., 79. For each set of sine waves in Figure 11-51, sketch the corresponding phasor diagram.

80., 81. For each set of sine waves in Figure 11-50, write the equation of each sine wave.

82., 83. For each set of sine waves in Figure 11-51, write the equation of each sine wave.

In Problems 84 through 86, answer True or False to each statement.

84. In Problem 84 of Figure 11-52:

FIGURE 11-52
Chapter 11 Problems 84 through 92

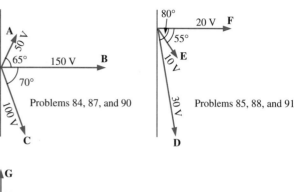

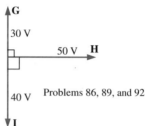

Problems 86, 89, and 92

a. **A** is leading **B** by 65°.
b. **A** is leading **C** by 135°.
c. **C** is lagging **B** by 65°.
d. **C** is lagging **A** by 135°.
e. **B** has a peak value of 100 V.

85. In Problem 85 of Figure 11-52:
    a. **E** is leading **D** by 25°.
    b. **D** is lagging **F** by 80°.
    c. **F** is lagging **D** by 80°.
    d. **E** is lagging **F** by 55°.
    e. **F** has a minimum voltage of −20 V.

86. In Problem 86 of Figure 11-52.
    a. **G** is leading **I** by 180°.
    b. **I** is leading **H** by 180°.
    c. **H** is lagging **G** by 90°.
    d. **H** is lagging **I** by 90°.
    e. **H** has a peak voltage of 50 V.

87.–89. For each phasor diagram in Figure 11-52, sketch the corresponding sine waves.

90.–92. For each phasor diagram in Figure 11-52, write the equation for each sine wave.

In Problems 93 through 95, answer True or False to each statement.

93. Problem 93 of Figure 11-53:
    a. **A** is leading **B** by 25°.
    b. **B** is leading **A** by 25°.

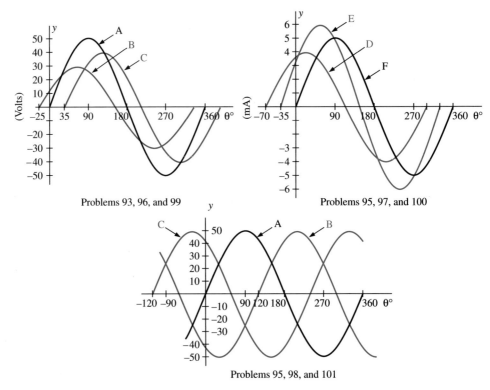

FIGURE 11-53
Chapter 11 Problems 93 through 101

    c. A is lagging B by 25°.
    d. B is lagging A by 25°.
    e. B has a minimum voltage of −50 V.
    f. B is leading C by 60°.
    g. C is leading B by 60°.
    h. B is lagging C by 60°.
    i. C is lagging B by 60°.
    j. B has a maximum voltage of 30 V.

94. In Problem 94 of Figure 11-53:
    a. D is leading E by 35°.
    b. E is leading D by 35°.
    c. D is lagging E by 35°.
    d. E is lagging D by 35°.
    e. D has a maximum current of 5 mA.
    f. F is leading D by 70°.
    g. D is leading F by 70°.
    h. F is lagging D by 70°.
    i. D is lagging F by 70°.
    j. F has a minimum current of −5 mA.

95. In Problem 95 of Figure 11-53:
   a. **A** is leading **B** by 120°.
   b. **B** is leading **C** by 120°.
   c. **A** is lagging **C** by 120°.
   d. **B** is lagging **A** by 120°.
   e. **A** has a peak voltage of 50 V.

96.–98. For each set of waveforms in Figure 11-53, sketch the corresponding phasor diagrams.

99.–101. For each set of waveforms in Figure 11-53, write the equation of each sine wave represented.

Calculate the frequency of sine waves generated by the phasors revolving at these angular frequencies.

102. 60 revolutions per second
103. 56 000 revolutions per minute
104. 3400 radians per second

12
Complex Numbers

CHAPTER OUTLINE

- **12.1** Complex Numbers
- **12.2** Graphing Complex Numbers
- **12.3** Rectangular Notation
- **12.4** Polar Notation
- **12.5** Converting From Polar Coordinates to Rectangular Coordinates
  - **12.5.1** Calculator P>R Key
- **12.6** Converting From Rectangular Coordinates to Polar Coordinates
  - **12.6.1** Calculator R>P Key
- **12.7** Adding Complex Numbers
  - **12.7.1** Graphical Interpretation
- **12.8** Subtracting Complex Numbers
  - **12.8.1** Graphical Interpretation
- **12.9** Multiplying Complex Numbers
  - **12.9.1** Polar Form
  - **12.9.2** Rectangular Form
- **12.10** Dividing Complex Numbers
  - **12.10.1** Polar Form
  - **12.10.2** Rectangular Form
- **12.11** Series AC Circuit Analysis

OBJECTIVES

After completing this chapter, you should be able to

1. describe complex numbers.
2. graph complex numbers.
3. convert from polar notation to rectangular notation using trigonometry.
4. convert from polar notation to rectangular notation using your calculator.
5. convert from rectangular notation to polar notation using trigonometry.
6. convert from rectangular notation to polar notation using your calculator.
7. add complex numbers.
8. subtract complex numbers.
9. multiply complex numbers.
10. divide complex numbers.
11. apply complex numbers to the solution of series alternating current circuits.

NEW TERMS TO WATCH FOR

complex numbers
real part
imaginary part
imaginary number
j operator, *i* operator
complex plane
rectangular coordinates, rectangular form, rectangular notation

polar coordinates, polar form, polar notation
complex conjugate
capacitor, C
inductor, L
inductive reactance, X_L
capacitive reactance, X_C
impedance, Z

In this chapter, we will apply a system called complex numbers to our work with vectors. Complex numbers provide a convenient way to represent vectors and to formalize the mathematics involved in adding, subtracting, multiplying, and dividing vectors. Complex numbers will be applied to series AC circuits.

12.1 ■ COMPLEX NUMBERS

Complex numbers can be written in the form $a+jb$, where a and b are any of the numbers we have worked with so far, any element of the set of real numbers that includes fractions, whole numbers, decimal numbers, and irrational numbers like π and $\sqrt{2}$. a is called the **real part** and *jb* is called the **imaginary part**. If a is equal to zero, then the complex number becomes *jb*, called an **imaginary number**. If b is equal to zero, then the complex number becomes a, a real number. For example, $3 + j6$, $2/9 - j5.1$, $5.7 - j56$, and $6/5 + j(7)^{1/2}$ are all complex numbers. Note that if the imaginary part is negative, the $-$ sign is written before the *j*. Mathematicians use *i* instead of *j* and write complex numbers in the form $a+bi$. Engineers and technicians use *j*, and *j* will be used in this book. *j* is defined as $(-1)^{1/2}$.

The square root of negative numbers can be rewritten in terms of *j*.

$$(-a)^{1/2} = [(-1)a]^{1/2} = (-1)^{1/2} \cdot a^{1/2} = j(a)^{1/2}$$

EXAMPLE 12.1 Express $(-25)^{1/2}$ in terms of *j*.

Solution
$$(-25)^{1/2} = [(-1)(25)]^{1/2} = (-1)^{1/2} \cdot 25^{1/2} = j5$$

EXAMPLE 12.2 Express $(-27)^{1/2}$ in terms of *j*.

Solution $(-27)^{1/2} = [(-1)(3)(9)]^{1/2} = (-1)^{1/2} \cdot 3^{1/2} \cdot 9^{1/2} = j3(3)^{1/2}$

If
$$j = (-1)^{1/2}$$
then
$$j^2 = (-1)^{1/2} \cdot (-1)^{1/2} = -1$$
$$j^3 = j^2 \cdot j = -1 \cdot j = -j$$
and
$$j^4 = j^2 \cdot j^2 = -1 \cdot -1 = 1$$

These facts lead to a graphical interpretation of j.

j is called an **operator**. j "operates" on a vector by rotating it 90° in the counterclockwise (CCW) direction. Think of a vector whose magnitude is 1 unit, lying on the positive x axis with its tail on the origin and its tip to the right. See Figure 12-1. $j1$ is that vector rotated 90° so that it points up. j^21 is rotated another 90° CCW, so that it now lies on the negative x-axis pointing left. That fits because $j^21 = -1$. j^31 is rotated another 90° CCW, so that it lies on the negative y-axis pointing down. $j^31 = -j1$. j^41 is rotated once more 90° CCW so that the vector lies once again on the x axis pointing to the right where it started. That fits because $j^41 = 1$.

FIGURE 12-1
j Operator

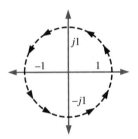

To summarize, positive real numbers lie on the positive x-axis: negative real numbers lie along the negative x-axis: positive imaginary numbers lie along the positive y-axis: and negative imaginary numbers lie along the negative y-axis. $+j$s go up: $-j$s go down.

12.2 ■ GRAPHING COMPLEX NUMBERS

Figure 12-2 shows a coordinate axis system with the horizontal axis labeled R for real and the vertical axis labeled j for imaginary. This system is called the **complex plane**. To plot a complex number in the complex plane, move left or right along the R-axis according to the real part of the complex number, and then up or down according to the imaginary part.

FIGURE 12-2
Complex Plane

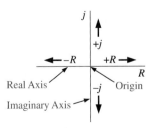

The complex number can represent a vector with its tail at the origin of the complex plane and its tip at the point plotted.

EXAMPLE 12.3 Graph the vectors represented by these complex numbers.

1. $A = 6 - j4$
2. $B = -5 + j2$
3. $C = -4.5 - j3.6$
4. $D = j4.5$ (real part = 0)
5. $E = -3.9$ (imaginary part = 0)

Solution See Figure 12-3.

FIGURE 12-3
Example 12.1

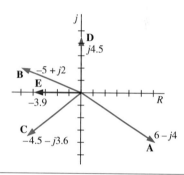

SECTIONS 12.1 and 12.2 EXERCISES

Express in terms of j.

1. $(-49)^{1/2}$
2. $(-375)^{1/2}$
3. $(-a^3b^4c^5)^{1/2}$
4. $(-27u^7v^5w^2)^{1/2}$

Plot these vectors.

5. $A = 3 + j5$
6. $B = -7 + j2$
7. $C = -2 + j6$
8. $D = -2 - j6$
9. $E = -6 - j1.5$
10. $F = 3.5 - j3.5$
11. $G = 7 - j2$
12. $H = 3 - j7$
13. $I = 6$
14. $J = j5$
15. $K = -6$
16. $L = -j5$

12.3 ■ RECTANGULAR NOTATION

You might have noticed that the real part of the complex number is the horizontal component of the vector and the imaginary part is the vertical component. Figure 12-4 shows the vector $45 - j37$ lb, where 45 is the horizontal component and −37 is the vertical component. Complex numbers can be thought of as "horizontal component + j vertical component." Complex numbers written in this form are expressed in **rectangular coordinates** or **rectangular form** or **rectangular notation**.

FIGURE 12-4
Rectangular Notation

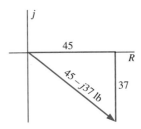

12.4 ■ POLAR NOTATION

Vectors can also be expressed as a magnitude acting at a specific angle. Figure 12-5 shows the vector 75 ∠110° lb. It is a force vector of magnitude 75 lb acting at an angle of 110° counterclockwise from the positive real axis. When written as a magnitude and angle, the vector is expressed in **polar coordinates** or **polar form** or **polar notation**. Rectangular notation states the horizontal and vertical components of a vector, and polar notation states the magnitude and direction.

FIGURE 12-5
Polar Coordinates

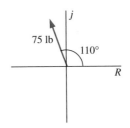

12.5 ■ CONVERTING FROM POLAR COORDINATES TO RECTANGULAR COORDINATES

Converting from polar notation to rectangular notation is equivalent to resolving a vector into its horizontal and vertical components. Figure 12-6 shows **A** expressed in polar notation, 72 ∠−52° lb. To convert to rectangular notation using trig, create a right triangle by drawing a vertical line from the tip of the vector to the real axis. This line, the vector, and a portion of the real axis create a right triangle. Use your knowledge of trig to perform the conversion. $A_x = A \cos \theta$ and $A_y = A \sin \theta$, just as in the last chapter.

FIGURE 12-6
Converting from Polar Coordinates to Rectangular Coordinates

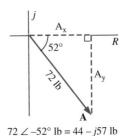

72 ∠−52° lb = 44 − j57 lb

12.5 CONVERTING FROM POLAR COORDINATES TO RECTANGULAR COORDINATES ▪ 397

EXAMPLE 12.4 Convert 72 ∠−52° lb to rectangular notation.

Solution **Given:** $\theta = -52°$, $A = 72$ lb. See Figure 12-6.
Find: A_x and A_y.

$$A_x = A \cos \theta \qquad A_y = A \sin \theta$$

$$A_x = 72 \text{ lb} \cdot \cos -52° \qquad A_y = 72 \text{ lb} \cdot \sin -52°$$

$$A_x = 44.327 \text{ lb} \qquad A_y = -56.736 \text{ lb}$$

Round A_x to 44 lb. Round A_y to −57 lb.

Significant digits: −52° is precise to the nearest degree; cos −52° and sin −52° have an accuracy of two. Round the results to two significant digits.

$$72 \angle -52° \text{ lb} = 44 - j57 \text{ lb}$$

12.5.1 Calculator P>R Key

Scientific calculators have polar-to-rectangular conversion keys that reduce the time and effort required for this conversion. Presented here are three of the procedures that are used on various calculators. Study your operator's guide to learn what your calculator is capable of and how to do this conversion. The time spent will be well invested.

Procedure 1 Two registers are loaded with data (magnitude and angle) using the *x*<>*y* register swap key, and then the calculator is told what to do with the data.

1. Enter the magnitude.
2. Press *x*<>*y*.
3. Enter the angle and its sign.
4. Press 2nd P>R (or INV P>R). The real (horizontal) component will appear in the display.
5. Press *x*<>*y*. The imaginary (vertical) component will appear.
6. Press *x*<>*y* again to return to the real component. The *x*<>*y* key toggles between the real and imaginary values.

EXAMPLE 12.5 Convert to rectangular coordinates:

$$72 \angle -52° \text{ lb. See Figure 12-6.}$$

Solution Enter this sequence: 72 *x*<>*y* 52 +/− 2nd P>R.

44.327 626 22 should appear in the display. That number rounded properly is the real coordinate. Press *x*<>*y*, and −56.736 774 26 should appear. That number rounded is the imaginary component. Use the rules of thumb for significant digits: angle to nearest degree, round to two significant digits.

$$72 \angle -52° \text{ lb} = 44 - j57 \text{ lb}$$

Procedure 2 In this scheme, the calculator is told what operation to perform after the magnitude is entered.

1. Enter the magnitude.
2. Press INV P>R (or 2nd P>R).
3. Enter the angle, including the sign.
4. Press =. The real component will appear.
5. Press x<>y swap register key. The vertical component will appear.
6. Press x<>y again to return to the real component. The x<>y key toggles between the real and imaginary values.

EXAMPLE 12.6 Convert from polar to rectangular coordinates:

$$72 \angle -52° \text{ lb. See Figure 12-6.}$$

Solution Enter this sequence: 72 INV P>R 52 +/− =. 44.327 626 should appear in the display. That number rounded to the proper number of significant digits is the real coordinate. Press x<>y, and the imaginary component will appear in the display, −56.736 774. Significant digits: nearest degree, two significant digits.

$$72 \angle -52° \text{ lb} = 44 - j57 \text{ lb}$$

Procedure 3 In some of the more sophisticated calculators, the function to be performed is selected from a list of operations, in this case P>R. The magnitude and angle are entered in parentheses, separated by a comma. For example, on the TI-81 graphing calculator:

1. Press the MATH key to get a list of operations. Press the down arrow to highlight P>R(. Note that an open parenthesis has been included.
2. Press ENTER.
3. P>R(appears on the display. Enter the magnitude, followed by ALPHA, the angle including the sign, and a closing parenthesis).
4. Press ENTER. The horizontal component will appear.
5. Press ALPHA Y to get the vertical component.

EXAMPLE 12.7 Convert from polar to rectangular coordinates: $72 \angle -52°$ lb. See Figure 12-6.

Solution Enter this sequence: MATH select P>R ENTER 72 ALPHA, (−) 52) ENTER. 44.327 626 should appear in the display. That number rounded to the proper number of significant digits is the real coordinate. Press ALPHA, and the imaginary component will appear in the display, −56.736 774. Significant digits: nearest degree, two significant digits.

$$72 \angle -52° \text{ lb} = 44 - j57 \text{ lb}$$

12.6 CONVERTING FROM RECTANGULAR COORDINATES TO POLAR COORDINATES

SECTION 12.5 EXERCISES

Use your calculator to convert from polar to rectangular coordinates. Graph each vector to see that the result is reasonable.

1. 124 ∠42° mi
2. 58 ∠106° kips
3. 83.4 ∠−200° in
4. 89.4 ∠−69° kΩ
5. 68.3 ∠83.5° ft
6. 200.4 ∠164° N
7. 34.7 ∠250° mm
8. 195 ∠310° ft/sec

12.6 ■ CONVERTING FROM RECTANGULAR COORDINATES TO POLAR COORDINATES

Converting from rectangular coordinates to polar coordinates is equivalent to combining a horizontal component and a vertical component into a resultant vector. For the vector $a + jb$, the resultant is

$$c = (a^2 + b^2)^{1/2} \text{ and } \theta = \arctan(b/a).$$

Look familiar? Remember that this procedure on many calculators yields an acute angle, the reference angle. Study the associated figure to determine the total angle.

EXAMPLE 12.8 Convert $82 - j46$ lb into polar coordinates. See Figure 12-7.

FIGURE 12-7
Converting from Rectangular Coordinates to Polar Coordinates

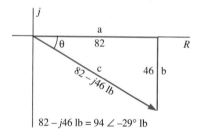

$82 - j46$ lb = 94 ∠−29° lb

Solution Given: $a = 82$ $b = -46$.
Find: c and θ.

$c = (a^2 + b^2)^{1/2}$ $\theta = \arctan(b/a)$

$c = [(82)^2 + (-46)^2]^{1/2}$ $\theta = \arctan(-46/82)$

$c = 94.021$ lb $\theta = -29.291°$

Round c to 94 lb.   Round θ to −29°.

Significant digits: a and b each have an accuracy of two, round θ to the nearest degree and the hypotenuse to two significant digits.

$$82 - j46 \text{ lb} = 94 \angle -29° \text{ lb}$$

12.6.1 Calculator R>P Key

Scientific calculators also have polar to rectangular conversion keys. The calculator sequences for converting polar to rectangular are quite similar to the rectangular to polar conversions. Presented here are three of the procedures that are used on various calculators. Study your operator's guide to learn what your calculator is capable of and how to do this conversion.

Procedure 1 Two registers are loaded with data (horizontal and vertical components) using the x<>y register swap key, and then the calculator is told what to do with the data.

1. Enter the horizontal component.
2. Press x<>y.
3. Enter the vertical component.
4. Press 2nd R>P (or INV R>P). The magnitude will appear in the display.
5. Press x<>y. The angle will appear.
6. Press x<>y again to return to the magnitude. The x<>y key toggles between the magnitude and angle.

EXAMPLE 12.9 Convert $82 - j46$ lb into polar coordinates. See Figure 12.7.

Solution Enter this sequence: 82 x<>y 46 +/− 2nd R>P = .94.021 should appear. Rounded off properly, it is the magnitude. Press x<>y to get the angle, −29.291°.

$$82 - j46 \text{ lb} = 94 \angle -29° \text{ lb}$$

Significant digits: The components given have a precision of two and so will the magnitude. The angle is determined by the ratio 46/82, which has an accuracy of two. Round the angle to the nearest degree.

Procedure 2 In this scheme, the calculator is told what operation to perform after the horizontal component is entered.

1. Enter the horizontal component.
2. Press INV R>P (or 2nd R>P).
3. Enter the vertical component.
4. Press =. The magnitude will appear.
5. Press the x<>y swap register key. The angle will appear.
6. Press x<>y again to return to the magnitude. The x<>y key toggles between the magnitude and angle.

EXAMPLE 12.10 Convert $82 - j46$ lb into polar coordinates. See Figure 12-7.

Solution Enter this sequence: 82 INV R>P 46 +/− = .94.021 274 should appear. Rounded off properly, it is the magnitude. Press x<>y to get the angle, −29.291 362°.

12.6 CONVERTING FROM RECTANGULAR COORDINATES TO POLAR COORDINATES ■ 401

$$82 - j46 \text{ lb} = 94 \angle -29° \text{ lb}$$

Significant digits: The components given have a precision of two and so will the magnitude. The angle is determined by the ratio 46/82, which has an accuracy of two. Round the angle to the nearest degree.

Procedure 3 In some of the more sophisticated calculators, the function to be performed is selected from a list of operations, in this case R>P. The rectangular coordinates are entered in parentheses, separated by a comma. For example, on the TI-81 graphing calculator:

1. Press the MATH key to get a list of operations. Press the down arrow to highlight R>P(. Note that an open parenthesis has been included.
2. Press ENTER.
3. R>P(appears on the display. Enter the real component, followed by ALPHA, followed by the imaginary component and a closing parenthesis,).
4. Press ENTER. The magnitude will appear.
5. Press ALPHA θ to get the angle.

EXAMPLE 12.11 Convert to polar: $82 - j46$ lb. See Figure 12-7.

Solution Enter this sequence: MATH select R>P ENTER 82 ALPHA, (−) 46) ENTER. 94.021 274 19 should appear in the display. That number rounded to the proper number of significant digits is the magnitude. Press ALPHA θ, and the angle will appear in the display, −29.291 362 17.

$$82 - j46 \text{ lb} = 94 \angle -29° \text{ lb}$$

Significant digits: The components given have a precision of two and so will the magnitude. The angle is determined by the ratio 46/82, which has an accuracy of two. Round the angle to the nearest degree.

EXAMPLE 12.12 Convert to polar: $-67.4 - j124$ lb

Solution 141.133 837 2 $\angle$118.526 220 3° lb rounds to 141 $\angle -118.5°$ lb

SECTION 12.6 EXERCISES

Use your calculator to convert from rectangular to polar coordinates. Sketch a figure to see that your result is reasonable.

1. $2.9 \times 10^{-3} - j4.8 \times 10^{-3}$ m
2. $-64.73 + j100.5$ mi
3. $-37 - j84$ N
4. $92.6 - j69.3$ lb
5. $83.7 + j65.3$ kips
6. $-500.5 + j788.4$ ft
7. $-933 - j655$ cm
8. $94 - j32$ km

12.7 ■ ADDING COMPLEX NUMBERS

Addition of complex numbers is performed using rectangular notation. Following the rules for signed numbers, add the real parts and add the imaginary parts.

$$(a + jb) + (c + jd) = (a + c) + j(b + d)$$

EXAMPLE 12.13 Add $4 + j9$ and $-6 - j7$.

Solution
$$(4 + j9) + (-6 - j7) = (4 + -6) + j(9 + -7)$$
$$= -2 + j2$$

EXAMPLE 12.14 Add $-8 - j4$ and $3 - j6$.

Solution
$$(-8 - j4) + (3 - j6) = (-8 + 3) + j(-4 + -6)$$
$$= -5 - j10$$

12.7.1 Graphical Interpretation

Adding complex numbers accomplishes the vector addition that was performed in the last chapter. In Figure 12-8, the two vectors used in the last example are graphed on the complex plane. A parallelogram is formed to add them together. The diagonal of the parallelogram is the vector sum. Its tip is $-5 - j10$ as predicted by the addition process.

FIGURE 12-8
Addition of Complex Numbers

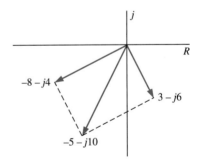

To add vectors, convert to rectangular notation, add the real parts, add the imaginary parts, and if desired, convert the result back to polar notation.

EXAMPLE 12.15 Add $89 \angle -35°$ lb to $73 \angle -155°$ lb.

Solution See Figure 12-9.

1. Convert to rectangular coordinates.

$$\mathbf{A} = 89 \angle -35° \text{ lb} = 72.9 - j51.0 \text{ lb}$$
$$\mathbf{B} = 73 \angle -155° \text{ lb} = -66.2 - j30.8 \text{ lb}$$

FIGURE 12-9
Example 12.13

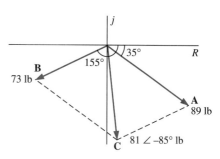

2. Add the real parts and add the imaginary parts.

$$\mathbf{C} = 6.7 - j81.8 \text{ lb}$$

3. Convert back to polar coordinates.

$$\mathbf{C} = 82 \angle -85° \text{ lb}$$

SECTION 12.7 EXERCISES

Add these vectors graphically and algebraically.

1. $(5 - j4) + (-3 + j7)$
2. $(-8 - j5) + (7 - j6)$
3. $34 \angle 109° \text{ lb} + 53 \angle 139° \text{ lb}$
4. $107 \angle -37° \text{ kips} + 291 \angle 48° \text{ kips}$

12.8 ■ SUBTRACTING COMPLEX NUMBERS

Subtraction of complex numbers is also performed using rectangular notation. Following the rules for subtraction of signed numbers, subtract the real parts and subtract the imaginary parts.

$$(a + jb) - (c + jd) = (a - c) + j(b - d)$$

EXAMPLE 12.16 Subtract $-6 - j12$ N from $4 - j6$ N.

Solution
$$(4 - j6) - (-6 - j12) = (4 - -6) + j(-6 - -12)$$
$$= 10 + j6 \text{ N}$$

EXAMPLE 12.17 $\mathbf{A} = -4.6 + j5.8 \text{ ft/sec}, \mathbf{B} = -6.2 - j3.6 \text{ ft/sec}.$

Subtract **A** from **B**.

Solution
$$\mathbf{C} = \mathbf{B} - \mathbf{A}$$
$$= (-6.2 - j3.6 \text{ ft/sec}) - (-4.6 + j5.8 \text{ ft/sec})$$
$$= (-6.2 - -4.6) + j(-3.6 - 5.8) \text{ ft/sec}$$
$$= -1.6 - j9.4 \text{ ft/sec}$$

12.8.1 Graphical Interpretation

Subtracting complex numbers accomplishes the vector subtraction performed in the last chapter. In Figure 12-10, the two vectors used in the last example are graphed on the complex plane. To subtract **A** from **B** graphically, the direction of **A** is reversed. A parallelogram is formed using **B** and **−A** as sides. The diagonal of the parallelogram is the vector **C** that will work along with **A** to produce the same effect as **B** acting alone. **A** + **C** = **B**. To subtract vectors mathematically, convert to rectangular notation, subtract the real parts, subtract the imaginary parts, and, if desired, convert the result back to polar notation.

FIGURE 12-10
Vector Subtraction

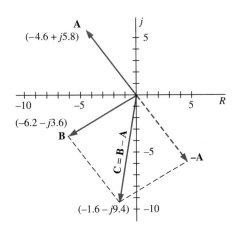

EXAMPLE 12.18 Subtract 54 ∠83° N from 37 ∠148° N.

Solution See Figure 12-11.

FIGURE 12-11
Example 12.18

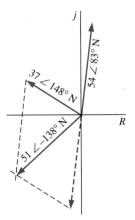

1. Convert to rectangular coordinates.

$$37 \angle 148° \text{ N} = -31.4 + j19.6 \text{ N}$$
$$54 \angle 83° \text{ N} = 6.58 + j53.6 \text{ N}$$

An extra significant digit was retained on these intermediate results.

2. Subtract the real parts and subtract the imaginary parts.

$$-37.98 - j34.0$$

3. Convert back to polar notation.
   Round to two significant digits and the nearest degree.

$$51 \angle -138° \text{ N}$$

SECTION 12.8 EXERCISES

Subtract these vectors graphically and algebraically.

1. $(13 + j12) - (21 + j26)$
2. $(-84 - j73) - (-34 + j52)$
3. $62 \angle -38° \text{ lb} - 54 \angle 36° \text{ lb}$
4. $99 \angle -160° \text{ kip} - 99 \angle -34° \text{ kips}$

12.9 ■ MULTIPLYING COMPLEX NUMBERS

Although multiplication of complex numbers can be performed in polar or rectangular notation, the process is easier with polar quantities.

12.9.1 Polar Form

To multiply quantities in polar form, multiply the magnitudes and add the angles, using the rules for signed numbers.

$$A \angle B° \cdot C \angle D° = A \cdot C \angle (B + D)°$$

EXAMPLE 12.19 Multiply $62 \angle 48° \cdot 12 \angle -81°$.

Solution $62 \cdot 12 \angle (48 - 81)° = 744 \angle -33°$

12.9.2 Rectangular Form

A complex number in rectangular form is a binomial. To multiply two complex numbers in rectangular form, follow the FOIL rule for multiplying two binomials: *first* term times first term, *outer* term times outer term, *inner* term times inner term, and *last* term times last term. Remember $j^2 = -1$.

$$(a + jb)(c + jd) = ac + jad + jbc - bd$$

The two middle terms are imaginary and will combine. The first and last terms are real and will combine.

$$= ac - bd + j(ad + bc)$$

EXAMPLE 12.20 Multiply $(4 - j7)(2 + j3)$.

Solution
$$(4 - j7)(2 + j3) = 8 + j12 - j14 - j^2 21$$
$$= 8 + 21 + j12 - j14$$
$$= 29 - j2$$

SECTION 12.9 EXERCISES

Multiply.

1. $592 \angle 35.2° \cdot 193 \angle -87.4°$
2. $3 \angle 16° \cdot 2 \angle 21° \cdot 4 \angle 112°$
3. $(7 - j13)(4 + j6)$
4. $(34.6 - j12.9)(45.6 + j88.3)$

12.10 ■ DIVIDING COMPLEX NUMBERS

Like multiplication of complex numbers, division of complex numbers can be accomplished in either polar form or rectangular form. The process is easier in polar form.

12.10.1 Polar Form

To divide complex numbers in polar form, divide the magnitudes and subtract the angle of the divisor from the angle of the dividend, following the rules of signed numbers.

$$\frac{A \angle B°}{C \angle D°} = \frac{A}{C} (B - D)°$$

In single line notation:

$$A \angle B° / C \angle D° = A/C \angle (B - D)°$$

EXAMPLE 12.21 Divide $17 \angle 43° / 31 \angle -16°$.

Solution
$$17 \angle 43° / 31 \angle -16° = 17/31 \angle (43 - -16°)$$
$$= 0.55 \angle 59°$$

12.10.2 Rectangular Form

Dividing complex numbers in rectangular form involves the concept of complex conjugate. To form a **complex conjugate** of a complex number, change the sign of the imaginary term. The complex conjugate of $3 - j6$ is $3 + j6$, and the complex conjugate of $12 + j17$ is $12 - j17$. When a complex number is multiplied by its complex conjugate, all imaginary terms disappear. For example,

$$(3 - j6)(3 + j6) = 9 + j18 - j18 - j^2 36 = 9 + j18 - j18 + 36 = 45.$$

To divide complex numbers in rectangular form, multiply both the numerator and the denominator by the complex conjugate of the denominator. This will eliminate all the imaginary terms in the denominator.

$$\frac{a + jb}{c + jd} \times \frac{c - jd}{c - jd} = \frac{ac - jad + jbc + bd}{c^2 - jcd + jcd + d^2} = \frac{ac + bd - j(ad + bc)}{c^2 + d^2}$$

EXAMPLE 12.22 $(4 + j5)/(3 - j4)$.

Solution

$$\frac{4 + j5}{3 - j4} \times \frac{3 + j4}{3 + j4} = \frac{12 + j16 + j15 - 20}{9 + j12 - j12 + 16}$$

$$= \frac{-8 + j31}{25}$$

$$= -0.32 + j1.24$$

SECTION 12.10 EXERCISES

Multiply each binomial by its complex conjugate.

1. $3 - j7$
2. $-5 + j7$
3. $-3 - j4$
4. $2.4 \times 10^{-3} - j1.6 \times 10^{-3}$

Divide.

5. $68.5 \angle 82.5° / 39.4 \angle 67.3°$
6. $394 \angle 112° / 34.5 \angle -26.7°$
7. $(15 - j38) / (18 + j37)$
8. $(-69 - j23) / (-46 - j42)$

12.11 ■ SERIES AC CIRCUIT ANALYSIS

Complex numbers are put to good use in the analysis of series alternating current circuits. If electronics is new to you, the terms in this section may seem a bit overwhelming at first. To develop a deep understanding of the new terms, you will need to take a course in DC and AC circuits. The intent here is to show how complex numbers are applied in a technical field. Work the exercises by following the examples, and you will develop skills in applying complex numbers.

Three main components of AC circuits are **resistors**, **capacitors (C)**, and **inductors (L)**. Each of these components impede the flow of current through the circuit in their own way. The opposition offered by each is measured in ohms. The symbol for ohms is the uppercase Greek letter omega, Ω. The opposition to current flow offered by the resistor is called **resistance R (R)**; the opposition offered by the inductor is called **inductive reactance** (X_L); and the opposition offered by the capacitor is called **capacitive reactance** (X_C). R, X_C, and X_L must be combined to know the total opposition of the circuit to current flow. Even though they are all measured in ohms, they cannot be added together directly because they are not in phase. R is in quadrature (90° out of phase) with X_L and X_C. They are combined using vector addition with complex numbers.

R is plotted on the positive real axis (right). X_L is plotted on the positive j-axis (up) and is written $+jX_L$. X_C is plotted on the negative j-axis (down and is written $-jX_C$. See Figure 12-12. Note the two R's in the figure. The R on the right is a label for the real axis and the R on the left is the name of the vector representing the resistance in the circuit. Since X_C and X_L are collinear (on the same straight line) they can be combined algebraically, $X_L - X_C$. The result will be positive, up, if X_L is greater in magnitude than X_C, and negative, down, if X_C is greater than X_L.

FIGURE 12-12
Graphical Representation of R, X_L, X_C

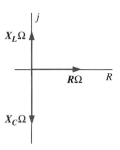

Figure 12-13 shows one resistor, R, one capacitor, C, and one inductor, L, connected in series across an AC voltage source, V_S. The order in which the three components are connected does not affect the operation of the circuit. In a series circuit, the vector sum of R, X_C, and X_L is called **impedance (Z)**, and it too is measured in ohms. In rectangular form,

$$Z = R + j(X_L - X_C)$$

FIGURE 12-13
Series RLC Circuit

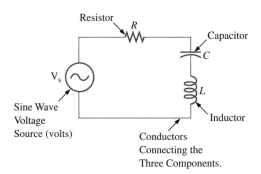

Using the techniques learned in Chapter 10, Z can be written in polar form. The magnitude of Z is:

$$Z = [R^2 + (X_L - X_C)^2]^{1/2}$$

and its phase angle is:

$$\theta = \arctan[(X_L - X_C)/R]$$

Using the techniques learned in Chapter 10, Z is written in rectangular form and converted into polar form using the calculator.

EXAMPLE 12.23 An inductor, a capacitor, and a resistor are connected in series as shown in Figure 12-14A. If X_L is 2300 Ω, X_C is 1300 Ω, and R is 4700 Ω, calculate Z and express the result in rectangular and polar coordinates.

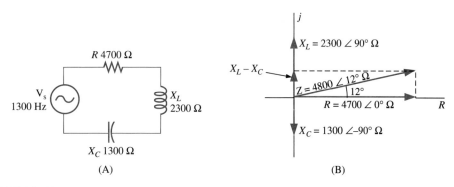

FIGURE 12-14
Example 12.23

Solution **Given:** $X_L = 2300\ \Omega$, $X_C = 1300\ \Omega$, R = 4700 Ω.
Find: Z. See Figure 12-14B.

$$Z = R + j(X_L - X_C)$$
$$Z = 4700 + j\,(2300 - 1300)\ \Omega$$
$$Z = 4700 + j1000\ \Omega\ \text{(rectangular form)}$$
$$Z = 4805.205\ \angle 12.011°\ \Omega$$

Use the R>P function on the calculator.

Round Z to 4800 ∠12° Ω. (polar form)

Significant digits: Assuming that the ohms given are precise to the nearest 100 ohms, $X_L - X_C = 1000\ \Omega$ with two significant digits. The results should be rounded to two significant digits and the angle to the nearest degree.

Some circuits do not have all three components. The circuit in the next example contains a resistor and a capacitor.

EXAMPLE 12.24 A 330 Ω resistor and a capacitor whose reactance is 570 Ω are connected in series. See Figure 12-15A. Find Z.

FIGURE 12-15
Example 12.24

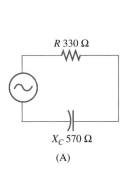

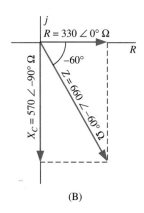

(A)   (B)

Solution **Given:** $R = 330\ \Omega$, $X_C = 570\ \Omega$.
Find: Z. See Figure 12-15B.

$$Z = R + j(X_L - X_C)$$
$$Z = 330 + j(0 - 570)\ \Omega$$
$$Z = 330 - j\,570\ \Omega$$
$$Z = 658.634\ \angle -59.931°\ \Omega$$
$$Z = 660\ \angle -60°\ \Omega$$

Significant digits: Assuming that R and X_C are precise to the nearest 10 ohms, they have an accuracy of two. Round the magnitude to two significant digits and the angle to the nearest degree.

Z represents the total opposition of the circuit to current flow. How much current will flow? Knowing Z and the source voltage, V_S, Ohm's law can be used to calculate current.

$$V_S = I \cdot Z$$

where V_S is the source voltage in volts,

I is the current in amps, and

Z is the impedance in ohms.

R, X_C, X_L, V_S, I, and Z have a magnitude and a phase shift. Each of these quantities can be represented by a phasor whose tail is at the origin of the complex plane. The phasor's angle of rotation from the positive real axis is the same as its phase shift. The current through each component in a series circuit is the same. All the current that leaves the signal generator, V_S, flows through each of the components, and all of the current flows back into the signal generator. It is advantageous to have the current as a reference with a phase shift of 0°. Then the angle that is calculated for the voltage across a component represents the phase shift of that voltage with respect to the current. In the following example, current is calculated by solving Ohm's law for I. $I = V_S/Z$. Since angles are subtracted during the process of division of vectors, V_S is assigned the same phase shift as Z. The resulting phase shift of I is 0° as desired. Why can an arbitrary phase shift be assigned to the applied voltage? We will see in a moment.

EXAMPLE 12.25 If 50 volts is applied to the circuit in the last example, how much current will flow?

Solution **Given:** $Z = 600 \angle -60° \; \Omega, V_S = 50$ volt.
Find: I.

$$V_S = IZ$$
$$I = V_S/Z$$
$$I = 50 \angle -60° \text{ V}/660 \angle -60° \; \Omega$$

V_S has been assigned a shift of −60° to match the phase shift of Z.

$$I = 0.07575 \angle 0° \text{ A}$$

Round I to 76 $\angle 0°$ mA

I has a shift of 0° as desired.

Knowing the current that flows around the circuit through each component, the voltage drop across each component can be found by using Ohm's law again. Voltage drops are found by multiplying the current through a component by the ohms provided by the component. They are represented by V's and are measured in volts.

Resistor: $V_R = IR$
Capacitor: $V_C = IX_C$
Inductor: $V_L = IX_L$

EXAMPLE 12.26 Using the circuit from the last two examples, calculate the voltage drop across the capacitor and the resistor.

Solution **Given:** $R = 330 \; \Omega, X_C = 570 \; \Omega, I = 76 \angle 0°$ mA.
Find: V_R and V_C. See Figure 12-16.

$$V_R = IR$$
$$V_R = 76 \times 10^{-3} \angle 0° \text{ A} \cdot 330 \angle 0° \; \Omega$$
$$V_R = 25.08 \angle 0° \text{ V}$$

Round V_R to 25 $\angle 0°$ V.

FIGURE 12-16
Example 12.26

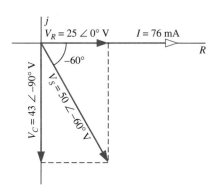

$V_C = IX_C$
$V_C = 76 \times 10^{-3} \angle 0°$ A $\cdot$ 570 $\angle -90°$ Ω
$V_C = 43.320 \angle -90°$ V
Round V_C to 43 $\angle -90°$ V.

The phase relationships of the quantities are summarized in Figures 12-15 and 12-16. Note the similarities between the two figures. X_C and V_C are both plotted down. R and V_R are both plotted to the right. Z and V_S are both at $-60°$. Z is the vector sum of R and X_C. V_S is the vector sum of V_R and V_C.

Imagine that the phasors in Figure 12-16 are rotating CCW. V_S crosses the real axis 60° after I crosses. Current is leading the voltage by 60°. In a series circuit with resistance and more capacitive reactance than inductive reactance, the current always leads the voltage. Also, V_R leads V_C by 90°, and I and V_R are in phase.

The phasors in Figure 12-16 represent the amplitude and phase shift of sinusoidal voltages. The voltages are graphed in Figure 12-17 as they would appear on an oscilloscope. Note that V_R starts up from zero first, looking from left to right. V_S lags behind by 60° and V_C lags behind by 90°.

FIGURE 12-17
Phase Shift

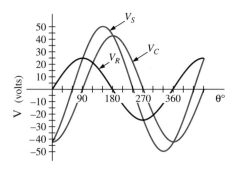

EXAMPLE 12.27 In a series circuit, R is 1200 Ω, X_L is 2800 Ω, and X_C is 1100 Ω. See Figure 12-18A. If 20 volts is applied, find the impedance, current, voltage drop across the resistor, voltage

drop across the capacitor, and voltage drop across the inductor. Draw the impedance diagram, the voltage/current phasor diagram, and the sine waves as they would appear on an oscilloscope.

Solution **Given:** $R = 1200\ \Omega$, $X_L = 2800\ \Omega$, $X_C = 1100\ \Omega$.
$V_S = 20$ volt

Find: Z, I, V_R, V_C, V_L. See Figure 12-18A, B, C, and D.

$$Z = R + j(X_L - X_C)$$
$$Z = 1200 + j(2800 - 1100)\ \Omega$$
$$Z = 1200 + j1700\ \Omega\ \text{(rectangular form)}$$
$$Z = 2080\ \angle 54.8°\ \Omega\ \text{(polar form)}$$
$$V_S = IZ$$
$$I = V_S/Z$$
$$I = 20\ \angle 54.8°\ \text{V} / 2080\ \angle 54.8\ \Omega$$

V_S is assigned an angle of 54.8° to make I the reference at 0°.

$$I = 9.615\ \angle 0°\ \text{mA}$$

Round I to 9.62 $\angle 0°$ mA (one extra significant digit)

$$V_R = IR$$
$$V_R = 9.62 \times 10^{-3}\ \angle 0°\ \text{A} \cdot 1200\ \angle 0°\ \Omega$$
$$V_R = 11.544\ \angle 0°\ \text{V}$$

Round V_R to 12 $\angle 0°$ V.

$$V_C = IX_C$$
$$V_C = 9.62 \times 10^{-3}\ \angle 0°\ \text{A} \cdot 1100\ \angle -90°\ \Omega$$
$$V_C = 10.58\ \angle -90°\ \text{V}$$

Round V_C to 11 $\angle -90°$ V.

$$V_L = IX_L$$
$$V_L = 9.62 \times 10^{-3}\ \angle 0°\ \text{A} \cdot 2800\ \angle 90°\ \Omega$$
$$V_L = 26.936\ \angle 90°\ \text{V}$$

Round V_L to 27 $\angle 90°$ V.

In Figure 12-18C and D, notice that there is no phase shift between V_R and I. They are always in phase. When V_R starts up from zero, V_L is already at positive peak, 90° ahead. V_C starts up from zero 90° after V_R. V_C lags V_R and the current by 90°. The applied voltage starts up from zero 55° ahead of V_R. V_S leads V_R and the current by 55°. In a series circuit, when X_L is greater in magnitude than X_C, the applied voltage will always lead the current.

So, what happens if the applied voltage is assigned an angle of 0°? If V_S is used as a reference instead of I, the entire voltage/current phasor diagram is rotated so that V_S lies

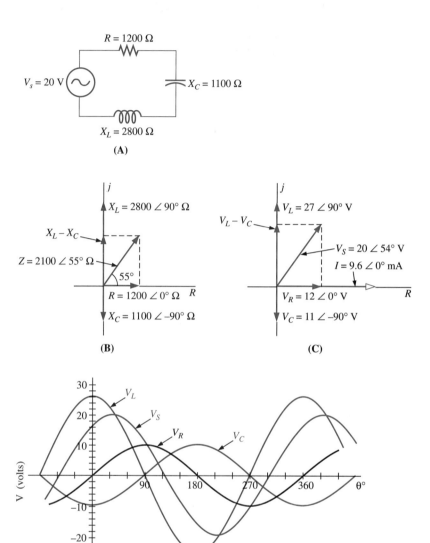

FIGURE 12-18
Example 12.27

on the positive real axis and I appears at an angle to the voltage. This alternate representation is shown in Figure 12-19. Note that V_C has been rotated into the third quadrant. As the phasors in Figure 12-19 rotate CCW, the same set of sine waves are generated as in Figure 12-18D.

SECTION 12.11 EXERCISES

1. In a series circuit, $R = 2200\ \Omega$, $X_C = 3700\ \Omega$, $X_L = 220\ \Omega$, and $V_S = 35$ volt. Find the impedance, the current, the voltage across the resistor, the voltage across the

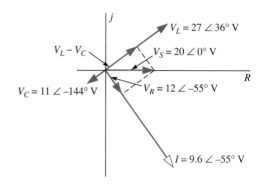

FIGURE 12-19
Voltage/Current Diagram with V_S as Reference

capacitor, and the voltage across the inductor. Draw a phasor diagram showing R, X_L, X_C, and Z. Draw another showing the current and the voltages. Sketch the sine waves as they would appear on an oscilloscope.

2. In a series circuit, $R = 560\ \Omega$, $X_C = 760\ \Omega$, $X_L = 2200\ \Omega$, and $V_S = 35$ volt. Find the impedance, the current, the voltage across the resistor, the voltage across the capacitor, and the voltage across the inductor. Draw a phasor diagram showing R, X_L, X_C, and Z. Draw another showing the current and the voltages. Sketch the sine waves as they would appear on an oscilloscope.

■ SUMMARY

1. Complex numbers can be written in the form $a+jb$, where a and b are real numbers. a is called the real part and jb is called the imaginary part.

2. j is called an operator. j "operates" on a vector by rotating it 90° in the counterclockwise (CCW) direction.

3. Positive real numbers lie on the positive horizontal-axis; negative real numbers lie along the negative horizontal-axis; positive imaginary numbers lie along the positive vertical-axis; and negative imaginary numbers lie along the negative vertical-axis. j's go up and $-j$'s go down.

4. The complex plane is a coordinate axis system with the horizontal axis labeled R for real and the vertical axis labeled j for imaginary. To plot a complex number in the complex plane, move left or right along the R-axis according to the real part of the complex number, and then move up or down according to the imaginary part. The complex number can represent a vector, with its tail at the origin of the complex plane and its tip at the point plotted. See Figure 12-2.

5. Complex numbers written in the form $a+jb$ are expressed in rectangular coordinates or rectangular form or rectangular notation.

6. Complex numbers written as a magnitude acting at a specific angle are expressed in polar coordinates or polar form or polar notation.

7. Converting from polar to rectangular notation is equivalent to resolving a vector into its horizontal and vertical components. Create a right triangle and use sine and cosine functions.

8. Converting from rectangular to polar coordinates is equivalent to combining a horizontal component and a vertical component into a resultant vector. For the vector $a+jb$, the resultant is

$$c = (a^2 + b^2)^{1/2} \text{ and } \theta = \arctan(b/a)$$

9. Addition of complex numbers is performed using rectangular notation. Following the rules for signed numbers, add the real parts and add the imaginary parts.

$$(a + jb) + (c + jd) = (a + c) + j(b + d)$$

10. Adding complex numbers accomplishes vector addition. See Figure 12-8.

11. Subtracting complex numbers is also performed using rectangular notation. Following the rules for subtraction of signed numbers, subtract the real parts and subtract the imaginary parts.

$$(a + jb) - (c + jd) = (a - c) + j(b - d)$$

12. Subtracting complex numbers accomplishes vector subtraction. See Figure 12-10.

13. To multiply quantities in polar form, multiply the magnitudes and add the angles, using the rules for signed numbers.

$$A \angle B° \cdot C \angle D° = A \cdot C \angle (B + D)°$$

14. A complex number in rectangular form is a binomial. To multiply two complex numbers in rectangular form, follow the rules for multiplying two binomials. Remember $j^2 = -1$.

$$(a + jb)(c + jd) = ac + jad + jbc - bd$$

The two middle terms are imaginary and will combine. The first and last terms are real and will combine.

$$= ac - bd + j(ad + bc)$$

15. To divide complex numbers in polar form, divide the magnitudes and subtract the angle of the divisor from the angle of the dividend, following the rules of signed numbers.

$$A \angle B° / C \angle D° = A/C \angle (B - D)°$$

16. To form a complex conjugate of a complex number, change the sign of the imaginary term.

17. When a complex number is multiplied by its complex conjugate, all imaginary terms disappear.

18. To divide complex numbers in rectangular form, multiply both the numerator and the denominator by the complex conjugate of the denominator. This will eliminate all the imaginary terms in the denominator.

19. The opposition of a resistor to current flow is called resistance (R), measured in ohms (Ω). R is graphed on the positive R-axis.

20. The opposition of a capacitor to current flow in an AC circuit is called capacitive reactance (X_C), measured in ohms (Ω). X_C is graphed on the negative j-axis (down).

21. The opposition of an inductor to current flow in an AC circuit is called inductive reactance (X_L), measured in ohms (Ω). X_L is graphed on the positive j-axis (up).

22. In a series circuit, the vector sum of R, X_C, and X_L is called impedance (Z), and it too is measured in ohms. In rectangular form,

$$Z = R + j(X_L - X_C)$$

In polar form, the magnitude of Z is:

$$Z = [R^2 + (X_L - X_C)^2]^{1/2}$$

and its phase angle is:

$$\theta = \arctan[(X_L - X_C)/R]$$

23. The current that flows in an AC circuit is calculated by Ohm's law.

$$I = V_S/Z$$

24. The voltage drop across each component in a series circuit can be found by using Ohm's law again.

Resistor: $V_R = IR$
Capacitor: $V_C = IX_C$
Inductor: $V_L = IX_L$

25. In a series circuit with more capacitive reactance than inductive reactance, the current always leads the voltage.

26. In a series circuit when X_L is greater in magnitude than X_C, the applied voltage will always lead the current.

SOLUTIONS TO CHAPTER 12 SECTION EXERCISES

Solutions to Sections 12.1 and 12.2 Exercises

1. $j7$
2. $j5(15)^{1/2}$
3. $jab^2c^2(ac)^{1/2}$
4. $j3u^3v^2w(3uv)^{1/2}$
5.–16. See Figure 12-20.

Solutions to Section 12.5 Exercises

1. $92 + j83$ mi
2. $-16 + j56$ kips
3. $-78 + j29$ in
4. $32 - j83$ kΩ
5. $7.73 + j67.9$ ft
6. $-190 + j55$ N
7. $-12 - j33$ mm
8. $130 - j150$ ft/sec

FIGURE 12-20
Solutions to Section 12.2 Exercises, Problems 5 through 16

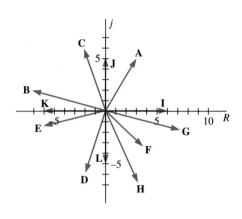

Solutions to Section 12.6 Exercises

1. 5.6 ∠−59° mm
2. 119.5 ∠122.78° mi
3. 92 ∠−114° N
4. 116 ∠−36.8° lb
5. 106 ∠38.0° kips
6. 933.8 ∠122.41° ft
7. 1140 ∠−144.9° cm
8. 99 ∠19° km

Solutions to Section 12.7 Exercises

1. $2 + j3$
2. $-1 - j11$
3. 34 ∠−148° lb
4. 319 ∠28° kips

Solutions to Section 12.8 Exercises

1. $-8 - j14$
2. $-50 - j125$
3. 70 ∠−86° lb
4. 180 ∠173° kips

Solutions to Section 12.9 Exercises

1. 114 000 ∠−52.1°
2. 24 ∠149°
3. $106 - j10$
4. $2720 + j2470$

Solutions to Section 12.10 Exercises

1. 58
2. 74
3. 25
4. 8.3×10^{-6}
5. 1.74 ∠15.2°
6. 11.4 ∠139°
7. $0.67 - j0.73$
8. $1.1 - j0.47$

Solutions to Section 12.11 Exercises

1. $Z = 2660$ ∠−34.3° Ω. See Figure 12-21.

   $I = 13.2$ ∠0° mA
   $V_R = 29.0$ ∠0° V
   $V_C = 48.8$ ∠−90° V
   $V_L = 29.0$ ∠90° V
   $V_S = 35$ ∠−34.3° V

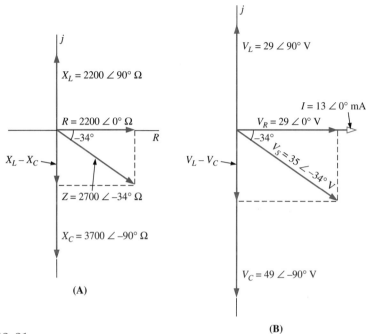

FIGURE 12-21
Solutions to Section 12.11 Exercises, Problem 1

2. $Z = 778 \angle -44.0° \, \Omega$. See Figure 12-22.

   $I = 32.1 \angle 0° \, \text{mA}$
   $V_R = 18.0 \angle 0° \, \text{V}$
   $V_C = 24.4 \angle -90° \, \text{V}$
   $V_L = 7.1 \angle 90° \, \text{V}$
   $V_S = 25 \angle -44° \, \text{V}$

CHAPTER 12 PROBLEMS

Graph these vectors.

1. $A = -7 - j4$
2. $B = 5 + j7$
3. $C = -4 + j6$
4. $D = 8 - j6$

Use your calculator to convert to polar notation.

5. $72 - j82$
6. $-18 - j51$
7. $-18 + j18$

Use your calculator to convert to rectangular coordinates.

8. $912 \angle -81° \, \text{lb}$
9. $24.6 \angle 17.5° \, \text{kips}$
10. $2.6 \angle -165° \, \text{mi}$

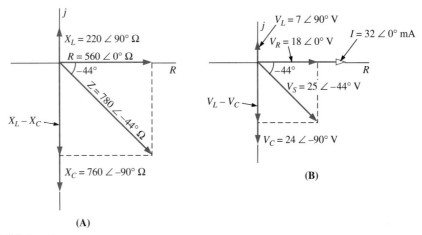

FIGURE 12-22
Solution to Section 12.11 Exercises, Problem 2

Add these vectors.

11. $18 + j45 + -25 - j36$
12. $-74.7 - j23.8 + 34.6 - j91.5$
13. $81 \angle 67° + 36 \angle -38°$
14. $73.5 \angle 69.3° + 69.4 \angle 54.7°$

Subtract these vectors.

15. $(55.5 + j72.6) - (-34.9 - j72.6)$
16. $(27 - j23) - (39 - j72)$
17. $(44 \angle -71°) - (55 \angle -34°)$
18. $(92 \angle -11.5) - (83 \angle 112°)$

Multiply these vectors.

19. $(3 - j6)(2 + j4)$
20. $(12 + j7)(12 + j7)$
21. $34.2 \angle 78° \cdot 23.6 \angle -124°$
22. $78.3 \angle 23° \cdot 99.1 \angle 45°$

Multiply by the complex conjugate.

23. $(4 - j8)$
24. $(-5 - j7)$
25. $(7 + j2)$

Divide these vectors.

26. $(3 - j6)/(-6 + j3)$
27. $(-4 - j5)/(9 + j3)$
28. $26.7 \angle -36° / 25.4 \angle 89°$
29. $246 \angle 77.4° / 567 \angle 38.5°$

For these series circuits, calculate Z.

30. $R = 586\ \Omega$, $X_C = 923\ \Omega$, and $X_L = 375\ \Omega$.
31. $R = 44\ k\Omega$, $X_C = 33\ k\Omega$, and $X_L = 55\ k\Omega$.
32. In a series circuit, $R = 270\ \Omega$, $X_C = 370\ \Omega$, $X_L = 220\ \Omega$, and $V_S = 5.0$ volt. Find the impedance, the current, the voltage across the resistor, the voltage across the capacitor, and the voltage across the inductor. Draw a phasor diagram showing

R, X_L, X_C, and Z. Draw another showing the current and the voltages. Sketch the sine waves as they would appear on an oscilloscope.

33. In a series circuit, $R = 56$ kΩ, $X_C = 76$ kΩ, $X_L = 22$ kΩ, and $V_S = 25$ volt. Find the impedance, the current, the voltage across the resistor, the voltage across the capacitor, and the voltage across the inductor. Draw a phasor diagram showing R, X_L, X_C, and Z. Draw another showing the current and the voltages. Sketch the sine waves as they would appear on an oscilloscope.

34. In a series AC circuit with $R = 7700 \; \Omega$ and $Z = 9800 \angle -37° \; \Omega$, calculate the amount of reactance. Is the reactance capacitive or inductive?

35. In a series AC circuit with $R = 47 \; \Omega$ and $Z = 86 \angle 67° \; \Omega$, calculate the amount of reactance. Is the reactance capacitive or inductive?

Unscramble these letters:

ARLE
OLRPA
EORPAORT
YMNIAIGRA
AONECTUGJ
ARTNUGRLAEC

Now unscramble the circled letters to solve the puzzle

_ _ _ _ _ _ X

13
Vector Applications

CHAPTER OUTLINE

- **13.1** Surveying
  - **13.1.1** Bearing of Perpendicular Vectors
  - **13.1.2** Bearing of a Vector in the Opposite Direction
  - **13.1.3** Traverse
  - **13.1.4** Closure (Closed Traverse)
- **13.2** Mechanics
  - **13.2.1** Free Body Diagram
  - **13.2.2** Moment
- **13.3** Navigation

OBJECTIVES

After completing this chapter, you should be able to

1. write the bearing of a line perpendicular to a given line.
2. write the bearing of a line heading in the opposite direction from a given line.
3. calculate closure (closed traverse).
4. calculate an unknown force in a static situation.
5. state three conditions for equilibrium.
6. draw and label a wind triangle.
7. calculate unknown vectors in wind triangle problems.

NEW TERMS TO WATCH FOR

complementary angles
traverse
closed traverse
open traverse
mechanics
statics
dynamics
free body diagram
moment arm

moment
drift correction angle
wind correction angle (WCA)
planned true heading (TH)
true air speed (TAS)
ground speed (GS)
true course (TC)
wind triangle

In this chapter, the skills you have developed working with vectors will be applied to the fields of surveying, mechanics, and airplane navigation.

13.1 ■ SURVEYING

Bearing, departure, and latitude were introduced in Chapter 10. We will use those concepts here in surveying applications.

13.1.1 Bearing of Perpendicular Vectors

First, let's develop a method to write the bearing of a vector perpendicular to a given vector. Figure 13-1 shows vector **A** with a bearing of Nθ°E. The angle formed by the north axis and vector **A** measures θ°. Both **B** and **C** are perpendicular to **A**. The angle formed by the north axis and **B** measures $\theta + 90°$. The north axis and south axis form a straight angle of 180°. The bearing of **B** is the angle formed by the south axis and **B**. It measures $180° - (\theta + 90°)$.

$$180° - (\theta + 90°) = 180° - \theta - 90° = 90° - \theta$$

FIGURE 13-1
Perpendicular Vectors

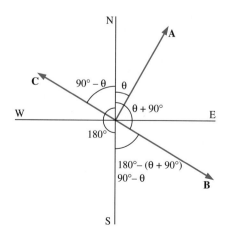

Two angles that add to 90° are called **complementary angles**. To find the bearing of **B**, take the complement of θ (subtract it from 90°) and reference it to south, S(90−θ)°E.

Since **A** and **C** are perpendicular, they form an angle whose measure is 90°. If **A** forms an angle of θ° with the north axis, then **C** must form an angle of 90° − θ with the north axis. To find the bearing of vector **C**, take the complement of θ, and change the departure from east to west. The bearing of **C** is N(90−θ)°W.

So, to find the bearing of a perpendicular vector, take the complement of the angle and change the direction of either the latitude or departure.

EXAMPLE 13.1 Find the bearing of two vectors perpendicular to a vector whose bearing is N84°E.

Solution Find the complement of the given angle: 90° − 84° = 6°. Change the latitude: S6°E. Change the departure: N6°W.

EXAMPLE 13.2 Find the bearing of two vectors perpendicular to a vector whose bearing is S24°E.

Solution Find the complement of the given angle: 90° − 24° = 66°. Change the latitude: N66°E. Change the departure: S66°W.

13.1.2 Bearing of a Vector in the Opposite Direction

Figure 13-2 shows vector **B** running in the opposite direction from a given vector **A**. Vectors **A** and **B** form a straight angle of 180°. That line intersects the N–S axis at the origin and forms vertical angles α and β. Vertical angles are always equal.

$$\angle \alpha = \angle \beta$$

β is the bearing of **B**. Therefore, to find the bearing of a vector that heads in the opposite direction from a given vector, use the same angle, but change both the latitude and the departure.

FIGURE 13-2
Vector Running in the Opposite Direction

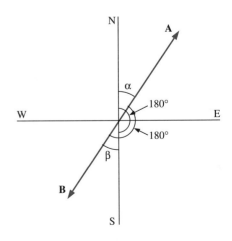

EXAMPLE 13.3 Find the bearing of a vector running opposite to a vector whose bearing is S34°W.

Solution Leave the angle alone, change both the latitude and the departure.

N34°E.

SECTION 13.1.1 and 13.1.2 EXERCISES

Find the bearing of two vectors perpendicular to the given vector. Sketch the three vectors.

1. N28°W 2. S16°E 3. S56°W 4. N79°E

Find the bearing of the vector running in the opposite direction from the given vector. Sketch the two vectors.

5. N28°W 6. S16°E 7. S56°W 8. N79°E

13.1.3 Traverse

A **traverse** is a succession of vectors representing distance and bearing connected together tip to tail. If the traverse forms a closed area, it is called a **closed traverse**. Otherwise, it is called an **open traverse**.

SECTION 13.1.3 EXERCISES

Figure 13-3 shows area ABCDEFA. The bearing of vector **AB** is N76°E, and the bearing of **CB** is S36°W. **CD** is perpendicular to **CB**; **DE** is parallel to **CB**; **EF** is parallel to **AB**; and **FA** is perpendicular to **EF**. Calculate the bearing of each vector to complete a closed traverse ABCDEFA. (CB must be reversed.)

FIGURE 13-3
Section 13.1.3 Exercise

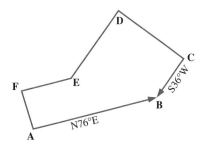

13.1.4 Closure (Closed Traverse)

Calculating closure means finding the distance and bearing from the tip of the last known vector of a traverse back to the starting point. This task involves finding the latitude and departure of each vector given. Latitudes toward the south and departures toward the west are considered negative. The unknown vector brings the total latitude and departure to zero.

EXAMPLE 13.4 The distance between home plate and first base on a baseball diamond is 90.0 feet. If the bearing is N37°E, find the bearing and distance from first base to second base.

Solution See Figure 13-4A. A baseball diamond is a square; the angles are right angles and the sides are equal length.

426 ■ CHAPTER 13 VECTOR APPLICATIONS

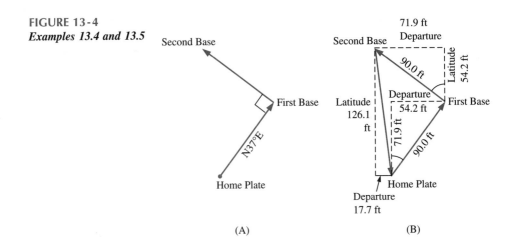

FIGURE 13-4
Examples 13.4 and 13.5

The vector from first base to second base is perpendicular to the vector from home plate to first base. To find its bearing, take the complement of the given angle and change the departure from east to west.

$$90° - 37° = 53°$$

The bearing of the first base to second base vector is N53°W, and the distance is again 90.0 ft.

EXAMPLE 13.5 Use the distance and bearing from home plate to first base and first base to second base to find the distance and bearing from second base to home plate.

Solution See Figure 13-4B.

Given: Home plate to first base: 90.0 ft at N37°E.

First base to second base: 90.0 ft at N53°W.

Find: Bearing and distance from second base to home plate.

To resolve a vector into latitude and departure, start at the tail and draw the latitude and then the departure. A right triangle is formed that has the bearing of the vector included as one of the angles.

Vector	Latitude	Departure
Home plate to first base	90 cos 37° = 71.9 ft	90 sin 37° = 54.2 ft
First base to second base	90 cos 53° = 54.2 ft	90 sin 53° = −71.9 ft
	126.1 ft	−17.7 ft
Second base to home plate	−126.1 ft	+17.7 ft
Total	0	0

Use the Pythagorean theorem to find the distance from second base to home plate.

$$d = (\text{latitude}^2 + \text{departure}^2)^{1/2}$$
$$d = (-126.1^2 + 17.7^2)^{1/2}$$
$$d = 127.336 \text{ ft}$$

Round d to 127.3 ft.

Use arctan to find the angle.

$$\theta = \arctan \text{departure/latitude}$$
$$\theta = \arctan 17.7/126.1$$
$$\theta = 7.990°$$

Round θ to 8.0°.

127.3 ft at S8.0°E

EXAMPLE 13.6

This same baseball field sits on a lot as shown in Figure 13-5. How much clearance is there between first base and the side of the lot (perpendicular distance from the edge of the lot to first base)?

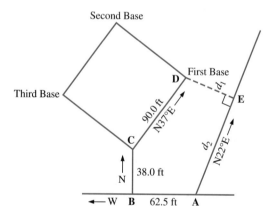

FIGURE 13-5
Example 13.6

Solution   **Given:**

Vector	Bearing	Distance
AB	W	62.5 ft
BC	N	38.0 ft
CD	N37°E	90.0 ft
AE	N22°E	d_2

Find: d_1.

First calculate the bearings as needed to form a continuous path (traverse) around the area ABCDEA. The bearing of DE must be calculated and the heading of line AE must be reversed to form a clockwise loop around the area. See Figure 13-6.

FIGURE 13-6
Example 13.6

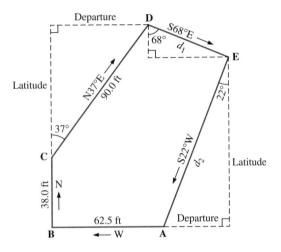

The bearing of **EA** is the reverse of bearing **AE**, S22°W. **DE** is perpendicular to **EA**; find its bearing.

$$90° - 22° = 68°. \quad S68°E.$$

Resolve each vector into latitude and departure. The distance from D to E is our unknown d_1. The latitude and departure of **DE** will be written in terms of d_1. The distance from E to A is also unknown. It has been assigned the variable d_2. The latitude and departure of **EA** will be written in terms of d_2.

Vector	Distance (ft)	Bearing	Latitude (ft)	Departure (ft)
AB	62.5	due west	0	−62.5
BC	38	N	38	0
CD	90	N37°E	$90 \cos 37° = 71.9$	$90 \sin 37 = 54.2$
DE	d_1	S68°E	$d_1 \cos 68° = -0.375 \, d_1$	$d_1 \sin 68 = 0.927 \, d_1$
EA	d_2	S22°W	$d_2 \cos 22° = -0.927 \, d_2$	$d_2 \sin 22 = -0.375 \, d_2$

The sum of the latitudes equals zero.

$$(1) \quad 38 + 71.9 - 0.375 \, d_1 - 0.927 \, d_2 = 0$$

The sum of the departures equals zero.

$$(2) \quad -62.5 + 54.2 + 0.927 \, d_1 - 0.375 \, d_2 = 0$$

We have a system of equations—two equations, two unknowns.

(1) $0.375 d_1 + 0.927 d_2 = 109.9$
(2) $0.927 d_1 - 0.375 d_2 = 8.3$

Divide equation (1) by 0.927 and equation (2) by 0.375 to get $+ d_2 - d_2$. These two terms will add to 0 when equations (1) and (2) are added.

(1) $0.405 d_1 + d_2 = 118.6$
(2) $2.47 d_1 - d_2 = 22.1$
(1) + (2) $2.875 d_1 = 140.7$
$d_1 = 48.939$ 13 ft

Round d_1 to 48.9 ft.

d_1 can be substituted into equation (1) or (2) to find d_2, but d_2 is not required in this problem.

SECTION 13.1.4 EXERCISES

Calculate closure graphically and algebraically in each situation.

1. **AB** = 254 ft at N80°E, **BC** = 304 ft at S25°E (Find **CA**).
2. **AB** = 38 m at S56°W, **BC** = 42 m at S41°E, **CD** = 56 m at N49°E (Find **DA**).
3. A 50-ft by 40-ft rectangular building is to be positioned on a lot as shown in Figure 13-7. One side of the lot runs due north, another side N60°W. The building is to be positioned parallel to and 24 ft from the lot line that runs north. The building must be 15 ft or more from the lot line to satisfy the local building code. How far is the southwest corner of the building from the N60°W lot line?

FIGURE 13-7
Section 13.1.4 Exercises, Problem 3

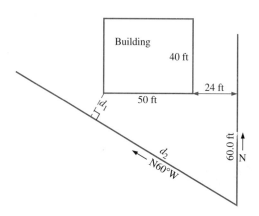

13.2 ■ MECHANICS

Mechanics is a branch of applied science that deals with the energy and forces acting on bodies. **Statics** is the branch of mechanics that deals with the energy and forces acting on a body at rest; and **dynamics** is the branch that deals with bodies in motion.

13.2.1 Free Body Diagram

In statics and dynamics, the system or component of the system under study is represented by a **free body diagram**. In the diagram, all of the forces acting on the body under study are represented by vectors. Using the vectors and principles of mechanics, forces of compression and tension throughout the system can be calculated. For example, Figure 13-8A shows a 315-lb block being supported by two cables. Figure 13-8B shows a free body diagram of the three forces acting on the block. The weight of the block is shown concentrated at the center of mass, acting down. As shown in Figure 13-8C, Force 1, F_1, can be resolved into a vertical component acting up and a horizontal component acting to the left. Force 2, F_2, can be resolved into a vertical component acting up and a horizontal component acting to the right.

For the block to remain at rest, the sum of the vertical forces provided by the two cables must equal the weight of the block acting down. The horizontal components of F_1 and F_2 are the only two horizontal forces. They must be equal in magnitude and opposite in direction for the block to remain at rest.

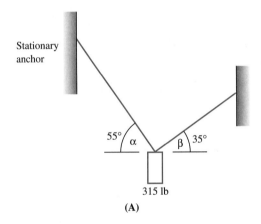

(A)

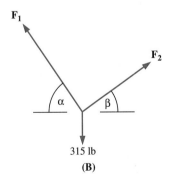

(B)

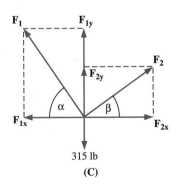

(C)

FIGURE 13-8
Example 13.7

EXAMPLE 13.7   Find the force of tension in each cable in Figure 13-8.

Solution   **Given:** $W = 315$ lb, $\alpha = 55°$, and $\beta = 35°$.
Find: F_1 and F_2.

Resolve F_1 into its horizontal and vertical components. See Figure 13-8C.

$$\cos \alpha = F_{1X}/F_1 \qquad \sin \alpha = F_{1Y}/F_1$$
$$F_{1X} = F_1 \cos A \qquad F_{1Y} = F_1 \sin A$$
$$F_{1X} = F_1 \cos 55° \qquad F_{1Y} = F_1 \sin 55°$$
$$F_{1X} = F_1 \cdot 0.574 \qquad F_{1Y} = F_1 \cdot 0.819$$
$$F_{1X} = 0.574\, F_1 \qquad F_{1Y} = 0.819\, F_1$$

F_1 is one of our unknowns, so F_{1X} and F_{1Y} can only be found in terms of F_1. Resolve F_2 into its horizontal and vertical components.

$$\cos \beta = F_{2X}/F_2 \qquad \sin \beta = F_{2Y}/F_2$$
$$F_{2X} = F_2 \cos \beta \qquad F_{2Y} = F_2 \sin \beta$$
$$F_{2X} = F_2 \cos 35° \qquad F_{2Y} = F_2 \sin 35°$$
$$F_{2X} = F_2 \cdot 0.819 \qquad F_{2Y} = F_2 \cdot 0.574$$
$$F_{2X} = 0.819\, F_2 \qquad F_{2Y} = 0.574\, F_2$$

If the sum of the vertical components of F_1 and F_2 is greater than the weight of the block, the block will rise; if smaller, the block will fall. The sum must equal the weight of the block for the block to be static.

$$F_{1Y} + F_{2Y} = W$$
$$0.819\, F_1 + 0.574\, F_2 = 315 \text{ lb}$$

The horizontal components of F_1 and F_2 must be equal. If not, the block will be pulled to one side or the other.

$$F_{1X} = F_{2X}$$
$$0.574\, F_1 = 0.819\, F_2$$

We have developed a system of equations—two equations each with two unknowns.

(1) $0.819\, F_1 + 0.574\, F_2 = 315$ lb
(2) $0.574\, F_1 = 0.819\, F_2$

Solve equation (2) for F_1.

(2) $F_1 = 0.819\, F_2/0.574 = 1.43\, F_2$

Substitute 1.43 F_2 for F_1 in equation (1).

(1) $0.819 (1.43\ F_2) + 0.574\ F_2 = 315$

$1.17\ F_2 + 0.574\ F_2 = 315$

$1.74\ F_2 = 315$

$F_2 = 315/1.74$

$F_2 = 181\text{ lb}$

Round F_2 to 180 lb.

Significant digits. The angles are precise to the nearest degree. Their sines and cosines are accurate to two digits. The results should be limited to two significant digits. Substitute the value of F_2 into (2) to find F_1.

(2) $F_1 = 1.43\ F_2$

$F_1 = 1.43 \cdot 181\text{ lb}$

$F_1 = 259\text{ lb}$

Round F_1 to 260 lb.

Significant digits: The angles are precise to the nearest degree. Their sines and cosines are accurate to two digits. The result should be limited to two significant digits.

13.2.2 Moment

In the last example, the block was put into equilibrium by setting the sum of the vertical forces to zero and the sum of the horizontal forces to zero. Actually, there is one more condition that must be fulfilled to ensure equilibrium. It involves the concept of moment of a force.

In Figure 13-9, force **F** will cause a rotation or moment about the point P. The perpendicular distance d from the force vector **F** to point P is called the **moment arm**. The **moment** of a force about a point is the product of the force and the moment arm. The moment of force **F** about point P, M_P is

$$M_P = d \cdot \mathbf{F}$$

where M_P is the moment of force **F** about point P, and d is the perpendicular distance from F to point P.

FIGURE 13-9
Moment and Moment Arm

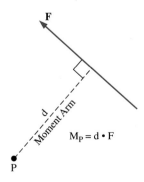

13.2 MECHANICS ▪ 433

The units for M are distance times force. The standard US unit is foot-pounds, and the standard SI unit is newton-meter.

EXAMPLE 13.8 In Figure 13-10, calculate the moment of force **F** about points A, B, and C.

FIGURE 13-10
Example 13.8

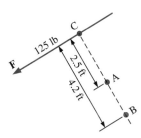

Solution **Given:** F = 125 lb, d_1 = 2.5 ft, d_2 = 4.2 ft, and d_3 = 0 ft.
Find: M_A, M_B, and M_C.

$M_A = d_1 \cdot F$ $M_B = d_2 \cdot F$ $M_C = d_3 \cdot F$
$M_A = 2.5 \text{ ft} \cdot 125 \text{ lb}$ $M_B = 4.2 \text{ ft} \cdot 125 \text{ lb}$ $M_C = 0 \cdot 125 \text{ lb}$
$M_A = 62.5 \text{ ft-lb}$ $M_B = 525 \text{ ft-lb}$ $M_C = 0 \text{ ft-lb}$

Round to two significant digits.

$M_A = 63 \text{ ft-lb}$ $M_B = 530 \text{ ft-lb}$ $M_C = 0 \text{ ft-lb}$

Significant digits: d_1 and d_2 limit the result to an accuracy of two. **F** produces a counterclockwise rotation or moment about point A and point B.

The third condition of equilibrium is that the sum of the moments of all forces acting about any point must be zero. Clockwise moments are considered negative, and counterclockwise moments are considered positive. In the last example, all three forces acting on the block had a moment arm of 0, so moments did not enter into the solution.

In summary, here are the three conditions that must be satisfied for a body to be at rest. The uppercase Greek letter sigma, Σ, is used to denote the summation process.

1. The sum of all vertical forces = 0. $\Sigma F_V = 0$
2. The sum of all horizontal forces = 0. $\Sigma F_H = 0$
3. The sum of moments about any point = 0. $\Sigma M_A = 0$

EXAMPLE 13.9 A 175-lb person is standing on the end of a 10.0-ft diving board. The board weighs 200 lb (nearest 10 lb). It is pinned at one end and rests on a support 4.0 ft from the pinned end. Calculate the forces acting on the board. See Figure 13-11A.

FIGURE 13-11
Example 13.9

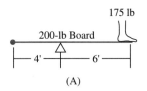

(A)

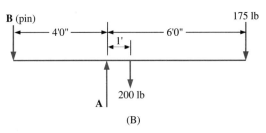

(B)

Solution Draw a free body diagram as shown in Figure 13-11B. The center of the board is one foot to the right of the support. The 200-lb weight of the board is shown concentrated at the center. The weight of the board and the weight of the diver are both acting down to the right of support **A**, producing clockwise moments. The pin at the left end of the board must push down with force **B** to provide a counterclockwise moment to counteract the clockwise moments. Find forces **A** and **B**.

Summation of moments about A = 0:

$$\Sigma\, M_A = 0$$

$$-1\text{ ft} \cdot 200\text{ lb} - 6\text{ ft} \cdot 175\text{ lb} + 4\text{ ft} \cdot \mathbf{B} = 0$$

$$\mathbf{B} = 1250\,/\,4$$

$$\mathbf{B} = 312.5\text{ lb downward}$$

Round **B** to 310 lb downward. (two significant digits)

The weight of the board, the 175-lb weight of the person, and force **B** are all acting downward. Force **A** must push up with a force equal to the sum of the three.

Summation of forces in the vertical direction = 0:

$$\Sigma\, F_V = 0$$

$$\mathbf{A} - 200 - 175 - 312.5 = 0$$

$$\mathbf{A} = 687.5\text{ lb upward}$$

Round **A** to 690 lb upward. (two significant digits)

There are no forces acting in the horizontal direction.

EXAMPLE 13.10 The sign for Joe's Diner weighs 200 lb (nearest ten lb), is 6.0 ft wide, and is hung on a 50-lb (nearest lb) bar that is pinned into the wall. The bar is also supported by a cable that is attached 5.0 ft from the wall. The end of the sign is 2.0 ft from the wall. See Figure 13-12A. Calculate the forces on the rod supplied by the cable and the wall.

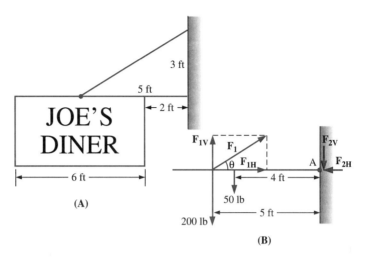

FIGURE 13-12
Example 13.10

Solution Draw a free body diagram as shown in Figure 13-12B. The weight of the sign is shown as 200 lb acting down at the center of the sign, 5 ft from the wall. The weight of the rod is shown as a 50-lb force acting down at the center of the part of the rod that extends from the building, 4 ft from the wall. The force supplied by the cable is shown as F_1 pulling up and to the right on the sign. F_1 is shown resolved into horizontal, F_{1H}, and vertical, F_{1V}, components. The right end of the rod is pinned to the wall, and the pin must provide horizontal, F_{2H}, and vertical, F_{2V}, forces to produce equilibrium. F_{2V} is shown acting down. That is just a guess at this point. If its value comes out negative, it is actually acting up.

1. Sum moments about the rod at the wall, point A.

$$\Sigma M_A = 0$$
$$200 \text{ lb} \cdot 5 \text{ ft} + 50 \text{ lb} \cdot 4 \text{ ft} - F_{1V} \cdot 5 \text{ ft} = 0$$

The 200-lb weight and the 50-lb weight both produce counterclockwise (positive) moments. F_{1V} produces a clockwise (negative) moment. The rest of the forces have a moment arm of 0 and do not need to be taken into account. This equation has only one unknown, F_{1V}. Solve it.

$$1000 + 200 = 5 F_{1V}$$
$$1200 = 5 F_{1V}$$
$$F_{1V} = 240 \text{ lb up}$$

2. Sum forces in the vertical direction—up is positive, down is negative.

$$\Sigma F_V = 0$$
$$240 \text{ lb} - 200 \text{ lb} - 50 \text{ lb} - F_{2V} = 0$$
$$-10 - F_{2V} = 0$$
$$F_{2V} = -10 \text{ lb}$$
$$F_{2V} = 10 \text{ lb up}$$

F_{2V} was assumed downward originally. Its value of -10 lb indicates that it is actually acting upward.

3. F_{1H} can be found by trig.

$$\tan \theta = 3/5$$

also

$$\tan \theta = F_{1V}/F_{1H}$$
$$3/5 = F_{1V}/F_{1H}$$
$$3/5\, F_{1H} = F_{1V}$$
$$F_{1H} = 5/3 \cdot F_{1V}$$
$$F_{1H} = 5/3 \cdot 240 \text{ lb}$$
$$F_{1H} = 400 \text{ lb}$$

4. Sum forces in the horizontal direction—positive to the right, negative to the left.

$$\Sigma F_H = 0$$
$$F_{1H} - F_{2H} = 0$$
$$400 - F_{2H} = 0$$
$$F_{2H} = 400 \text{ lb}$$

The wall must push up on the rod with 10 lb of force and out on the rod with 400 lb of force.

5. Combine F_{1H} and F_{1V} to find the resultant tension in the cable.

$$F_1 = (F_{1H}^2 + F_{1V}^2)^{1/2}$$
$$F_1 = (400^2 + 240^2)^{1/2}$$
$$F_1 = 466 \text{ lb}$$
$$F_1 = 470 \text{ lb (two significant digits)}$$

SECTION 13.2.2 EXERCISES

Draw a free body diagram and solve.

1. An 11.0-ft diving board weighs 100 lb (nearest ten lb). It is pinned at one end and rests on a support 5.0 ft from the pinned end. If a 150-lb person is standing on the end of the board, find the forces acting on the board.

2. A sign for Acme Tool Inc. weighs 158 lb and is 4.0 ft wide. It is hung 1 foot from the wall on a 58-lb, 5.0-ft bar that is pinned at the wall. The bar is supported by a cable that is attached to the bar 4.0 feet from the wall. The other end of the cable is attached to the wall 2.0 ft above the bar. Calculate all forces acting on the bar and the resultant force in the cable. See Figure 13-13.

3. An 1800-lb (nearest 100 lb) block is suspended by two cables as shown in Figure 13-14. Calculate the resultant forces in each cable.

4. 95-lb Boney Maroney is sitting on a swing as shown in Figure 13-15. Calculate the forces in the chains.

FIGURE 13-13
Section 13.2.2 Exercises, Problem 2

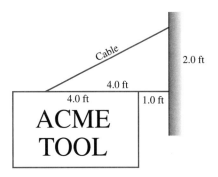

FIGURE 13-14
Section 13.2.2 Exercises, Problem 3

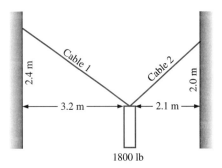

FIGURE 13-15
Section 13.2.2 Exercises, Problem 4

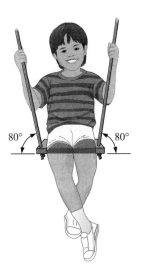

13.3 ■ NAVIGATION

Pilots use north as 0° and measure direction clockwise, 0°–360°. See Figure 13-16. Northeast is 045°; east is 090°; southeast is 135°; south is 180°; southwest is 225°; west is 270°; and northwest is 315°. Wind is reported as the angle from which wind is blowing, followed by a slash and the wind speed in knots. 090°/35kn indicates a 35-knot wind blowing east to west. In planning a flight, the pilot takes into account the reported wind, which is used as a vector.

FIGURE 13-16
Navigation Coordinates

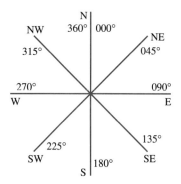

The wind causes the plane to drift as it flies. The angle of drift is called **drift correction angle** or **wind correction angle (WCA)**. The direction that the pilot actually points the plane is called the **planned true heading (TH)**. The speed at which the plane cruises is called **true air speed (TAS)**. TAS is used as the magnitude of a vector, with TH as its angle. The magnitude of the vector sum of the wind vector and the TAS vector describes the speed of the plane with respect to the ground, called **ground speed (GS)**. The angle of the vector sum describes the **true course (TC)** of the plane over land. This "wind triangle" is shown in Figure 13-17. The TAS vector plus the wind vector yields the GS vector. Likewise, the GS vector minus the TAS vector yields the wind vector. You will probably need to refer back to these definitions as you work through the following examples.

FIGURE 13-17
Wind Triangle

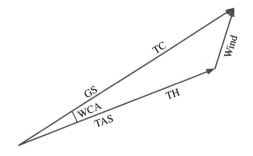

The examples and exercises present several scenarios. The first example supposes a pilot is in flight without information on the prevailing wind.

EXAMPLE 13.11 A pilot is flying at 081° with an airspeed of 145 knots in unknown wind conditions. The "naviguesser" determines that the plane is actually traveling at 090° at 163 kn. Find the direction and speed of the wind.

13.3 NAVIGATION ■ 439

Solution 1—Graphical

Given: TH = 081°, TAS = 145 kn. (vector **A**).
TC = 090°, GS = 163kn. (vector **C**).
Find: Wind (vector **B**). See Figure 13-18A.

Since **A** + **B** = **C**, **B** is the vector that runs from the tip of **A** to the tip of **C**. Draw **B**. See Figure 13-18B. Its magnitude measures about 30 knots and its angle measures about 50° with respect to the positive x-axis. See Figure 13-18C. Now calculate the direction from which the wind is blowing. Starting from north and working clockwise, the wind is blowing from 90° + 50° + 180° = 320°. The wind is approximately 30 knots at 320°. **B** = 320°/30 kn.

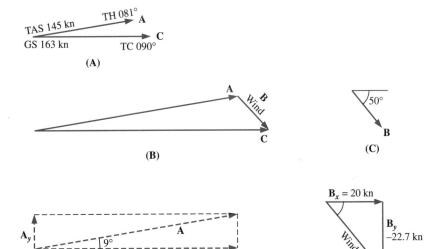

FIGURE 13-18
Example 13.11

Solution 2—Algebraic

Since **A** + **B** = **C**, **B** = **C** − **A**. In this solution, the horizontal and vertical components of **A** will be subtracted from the corresponding components of **C**.

Given: TH = 081°, TAS = 145 kn, TC = 090°, and GS = 163 kn.
Find: Wind speed and wind direction (vector **B**).

Resolve **A** into its horizontal and vertical components. See Figure 13-18D.

$$\cos\theta = A_x/A \qquad \sin\theta = A_y/A$$
$$A_x = A\cos\theta \qquad A_y = A\sin\theta$$
$$A_x = 145\cos 9° \qquad A_y = 145\sin 9°$$
$$A_x = 143 \text{ kn} \qquad A_y = 22.7 \text{ kn}$$

C has only a horizontal component. $C_x = 163$ kn. Subtract the horizontal components.

$$B_x = C_x - A_x$$
$$B_x = 163 \text{ kn} - 143 \text{ kn}$$
$$B_x = 20 \text{ kn}$$
$$B_y = C_y - A_y$$
$$B_y = 0 - 22.7 \text{ kn}$$
$$B_y = -22.7 \text{ kn}$$

Combine B_x and B_y into a resultant vector, the wind. See Figure 13-18E.

$$B = (B_x^2 + B_y^2)^{1/2}$$
$$B = [20^2 + (-22.72)^2]^{1/2}$$
$$B = 30.253\ 76 \text{ kn}$$

Round B to 30 kn.

$$\theta = \arctan B_y/B_x$$
$$\theta = \arctan -22.7/20$$
$$\theta = -48.6°$$

θ is measured with respect to the horizontal axis. Reference the direction from which the wind is blowing to north.

$$\beta = 90.0° + 48.6° + 270.0° = 318.6°$$

Wind is 318.6°/30 kn.

In the next example, the pilot knows the reported wind, planned true air speed, and planned true course. The task is to find the planned true heading and the ground speed. This scenario is more difficult to handle because we are given the angle of one vector but not the magnitude, and the magnitude of another vector but not the angle.

EXAMPLE 13.12 A pilot wishes to fly to a city 200 miles to the northwest, maintaining an airspeed of 120 kn. The reported wind is 35 kn at 223°. Find the true heading to fly and the resulting ground speed. See Figure 13-19A.

Solution 1—Graphical

Given: TC = 360° − 45° = 315°, wind = 223°/35kn, and TAS = 120 kn.
Find: TH and GS.

GS determines the length of vector **C**, and GS is not known. Choose a point to be the tip of vector **C**. See Figure 13-19B. Draw a line at 315° to represent vector **C**. Its length will be determined. Graph the reported wind, with the tip of the wind vector at the tip of **C**. Set the point of a compass on the tail of vector **B** and swing an arc with a radius equal to TAS, 120. The intersection with vector **C** determines the tail of **C** and the heading of **A**. If you do not have a compass handy, swing a scale to find the point of

13.3 NAVIGATION ▪ 441

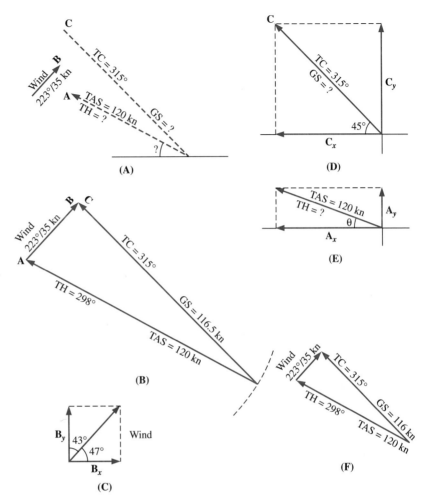

FIGURE 13-19
Example 13.12

intersection. Measure the GS (length of **C**) and TH (heading of **A**). GS = 116.5 kn, and TH = 298°.

Solution 2—Algebraic

Given: TC = 360° − 45° = 315°, wind = 223°/35kn, TAS = 120 kn.
Find: TH and GS. See Figure 13-19A.

Translate the wind vector over to the origin and find its reference angle. See Figure 13-19C.

$$223° − 180° = 43°$$
$$90° − 43° = 47°$$

Resolve **B** into its horizontal and vertical components.

$$B_x = B \cos 47° \qquad B_y = B \sin 47°$$
$$B_x = 35 \text{ kn} \cdot \cos 47° \qquad B_y = 35 \text{ kn} \cdot \sin 47°$$
$$B_x = 23.9 \text{ kn} \qquad B_y = 25.6 \text{ kn}$$

Find the reference angle of **C**. See Figure 13-19D.

$$315° - 270° = 45°$$

Resolve C into its components.

$$C_x = C \cos 45° \qquad C_y = C \sin 45°$$
$$C_x = -.707\,C \qquad C_y = .707\,C$$

Let θ be the reference angle of **A**. See Figure 13-19E. Resolve A into its components.

$$A_x = A \cos \theta \qquad A_y = A \sin \theta$$
$$A_x = -120 \cos \theta \qquad A_y = 120 \sin \theta$$

Add the horizontal components of **A** and **B** to get the horizontal component of **C**.

$$A_x + B_x = C_x$$
$$(1)\ -120 \cos \theta + 23.9 \text{ kn} = -0.707\,C$$

Add the vertical components of **A** and **B** to get the vertical component of **C**.

$$A_y + B_y = C_y$$
$$(2)\ 120 \sin \theta + 25.6 = 0.707\,C$$
$$(1)\ -120 \cos \theta + 23.9 = -0.707\,C$$
$$(2)\ 120 \sin \theta + 25.6 = 0.707\,C$$

We have once again generated a system of equations. These two equations have three unknowns: $\sin \theta$, $\cos \theta$, and C. This system is not solvable unless a third equation with the same unknowns can be written. One of the trig identities studied in the graphing chapter, $\sin^2 \theta + \cos^2 \theta = 1$, is needed here as our third equation.

$$(1)\ -120 \cos \theta + 23.9 = -0.707\,C$$
$$(2)\ 120 \sin \theta + 25.6 = 0.707\,C$$
$$(3)\ \sin^2 \theta + \cos^2 \theta = 1$$

The right sides of equations (1) and (2) are the same except for signs. Add (2) to (1) to eliminate C. (Left side plus left side equals right side plus right side.)

$$-120 \cos \theta + 23.9 + 120 \sin \theta + 25.6 = 0$$
$$-120 \cos \theta + 120 \sin \theta + 49.5 = 0$$

Solve for $\cos \theta$.

$$120 \cos \theta = 120 \sin \theta + 49.5$$

Divide both sides by 120.

$$\cos \theta = \sin \theta + 0.4125$$

Square both sides to get an expression for $\cos^2\theta$.

$$(4)\ \cos^2\theta = (\sin \theta + 0.4125)^2$$
$$(4)\ \cos^2\theta = \sin^2\theta + 0.825 \sin \theta + 0.170$$

Solve equation (3) for $\cos^2\theta$.

$$(3)\ \cos^2\theta = 1 - \sin^2\theta$$

Subtract (3) from (4) to eliminate $\cos^2\theta$.

$$\cos^2\theta - \cos^2\theta = \sin^2\theta + 0.825 \sin \theta + 0.170 - (1 - \sin^2\theta)$$
$$0 = \sin^2\theta + 0.825 \sin \theta + 0.170 - (1 - \sin^2\theta)$$
$$0 = \sin^2\theta + 0.825 \sin \theta + 0.170 - 1 + \sin^2\theta$$

$2 \sin^2\theta + 0.825 \sin \theta - 0.83 = 0$

Divide both sides by 2.

$$\sin^2\theta + 0.4125 \sin \theta - 0.415 = 0$$

By making a substitution, this equation becomes a quadratic equation, which will be studied in a later chapter.

$$\text{Let } u = \sin \theta, \text{ then } u^2 = \sin^2\theta.$$

The equation becomes $u^2 + 0.4125u - 0.415 = 0$.

An equation of this form is called a quadratic equation. Four methods of solving an equation will be presented in Chapter 15. We will solve it here by using the quadratic formula.

Our equation is of the form $au^2 + bu + c = 0$, where $a = 1, b = 0.4125$, and $c = -0.415$. This equation can be solved by substituting a, b, and c into the quadratic formula:

$$u = [-b +/- (b^2 - 4ac)^{1/2}]/(2a)$$

The +/− sign means $(b^2 - 4ac)^{1/2}$ will be added to −b for one solution and subtracted from −b for a second solution. We will have to determine which solution is valid for our problem.

$$u = [-0.4125 +/- (0.4125^2 - 4 \cdot 1 \cdot -0.415)^{1/2}]/(2 \cdot 1)$$
$$u = [-0.4125 +/- 1.353]/2$$
$$u = 0.940/2, -1.7655/2$$
$$u = 0.470 \text{ and } u = -0.882\ 75$$

Replace u with $\sin \theta$. (We made a substitution $u = \sin \theta$.)

$$\sin \theta = 0.470 \qquad \sin \theta = -0.882\ 75$$
$$\theta = \arcsin 0.470, \qquad \theta = \arcsin -0.882\ 75$$
$$\theta = 28° \qquad \theta = -62°$$

28° is the reference angle for vector **A** (TH). With respect to north, TH = 28° + 270° = 298°.

Now, to find **C**, which represents ground speed, substitute 28° for θ in equation (1) or (2).

(2) $\quad 120 \sin\theta + 25.6 \text{ kn} = 0.707 C$

$$0.707 C = 120 \sin\theta + 25.6 \text{ kn}$$
$$C = (120 \sin\theta + 25.6 \text{ kn}) / 0.707$$
$$C = (120 \sin 28° + 25.6 \text{ kn}) / 0.707$$
$$C = 115.89 \text{ kn}$$

Round C to 116 kn.

TH = 298° and GS = 116 kn. These results agree nicely with the graphical solution. See Figure 13-19F.

EXAMPLE 13.13 At 116 kn, how long will it take the plane in the last example to travel the 200 miles?

Solution **Given:** rate, $r = 116$ kn; distance, $d = 200$ miles. Caution: Unit mismatch!
Find: time, t.

$$d = r \cdot t$$

where d is distance, r is rate of travel, and t is time.

$$t = d/r$$
$$t = 200 \text{ mi}/116 \text{ nautical mph}$$

Convert mi to nautical mi.

$$t = 200 \text{ mi} \cdot \frac{5280 \text{ ft}}{1 \text{ mi}} \cdot \frac{1 \text{ nau mi}}{6067 \text{ ft}} \cdot \frac{\text{hr}}{116 \text{ nau mi}}$$

Note that in the last fraction, dividing by nautical miles per hour is the same as multiplying by hours over nautical hours.

$$t = 1.50 \text{ hr}$$

SECTION 13.3 EXERCISES

Solve these problems graphically and algebraically.

1. A pilot is flying at 254° with an airspeed of 217 knots, in unknown wind conditions. The naviguesser determines that the plane is actually traveling at 245° at 210 knots. Find the direction and speed of the wind.

2. A pilot wishes to fly to an airport on a course of 125° and maintain a ground speed of 182 knots. The reported wind is due west at 25 knots. Calculate the heading that should be flown and the resulting airspeed.

3. A pilot is flying a heading of 337° with an airspeed of 150 knots. The wind is 27°/25kn. Find the ground speed and true course of the plane.

4. A pilot wishes to fly to a city 683 miles to the southeast, maintaining an airspeed of 120 kn. The reported wind is 35 kn at 223°. Find the true heading to fly and the resulting ground speed. Note: The algebraic solution to this problem is difficult and may be hazardous to your health. It involves algebra to be covered later in the book. Use Example 13.12 as a guide.

5. If the average ground speed of a plane is 180 knots, how long will it take to fly 683 miles?

■ SUMMARY

1. Two angles that add to 90° are called complementary angles.
2. To find the bearing of a vector perpendicular to a given vector, take the complement of the angle and change the direction of either the latitude or the departure.
3. To find the bearing of a vector that heads in the opposite direction from a given vector, use the same angle, but change both the latitude and the departure.
4. A traverse is a succession of vectors representing distance and bearing connected together tip to tail. If the traverse forms a closed area, it is called a closed traverse. Otherwise, it is called an open traverse.
5. Calculating closure means finding the distance and bearing from the tip of the last known vector of a traverse back to the starting point. This task involves finding the latitude and departure of each vector given. Latitudes toward the south and departures toward the west are considered negative. The unknown vector brings the total latitude and departure to zero.
6. Mechanics is a branch of applied science that deals with the energy and forces acting on bodies. Statics is the branch of mechanics that deals with the energy and forces acting on a body at rest; dynamics is the branch that deals with bodies in motion.
7. In statics and dynamics, the system or component of the system under study is represented by a free body diagram. In the diagram, all of the forces acting on the body under study are represented by vectors. Using the vectors and principles of mechanics, forces of compression and tension throughout the system can be calculated.
8. A force acting through a distance will cause a rotation or moment about a point of rotation. The moment of force F about point P, M_P, is

$$M_P = d \cdot F$$

9. There are three conditions that must be satisfied for a body to be at rest.
   1. The sum of all vertical forces = 0. $\Sigma F_V = 0$
   2. The sum of all horizontal forces = 0. $\Sigma F_H = 0$
   3. The sum of moments about any point = 0. $\Sigma M_A = 0$

10. Pilots use north as 0° and measure direction clockwise, 0°–360°.
11. Prevailing wind is reported as the angle from which wind is blowing, followed by a slash and the wind speed in knots. 090°/35kn indicates a 35-knot wind blowing east to west.

12. The wind causes the plane to drift as it flies. The angle of drift is called drift correction or wind correction angle (WCA).
13. The direction that the pilot actually points the plane is called the planned true heading (TH). The speed at which the plane cruises is called true air speed (TAS).
14. The magnitude of the vector sum of the wind vector and the TAS vector describes the speed of the plane with respect to the ground, called ground speed (GS). The angle of the vector sum describes the true course (TC) of the plane over land. The TAS vector plus the wind vector yields the GS vector.

SOLUTIONS TO CHAPTER 13 SECTION EXERCISES

Solutions to Section 13.1.1 and 13.1.2 Exercises

1. N62°E and S62°W
2. S74°W and N74°E
3. S34°E and N34°W
4. N11°W and S11°E
5. S28°E
6. N16°W
7. N56°E
8. S79°W

Solutions to Section 13.1.3 Exercises

AB N76°E
BC N36°E
CD N54°W
DE S36°W
EF S76°W
FA S14°E

Solutions to Section 13.1.4 Exercises

1. 444 ft at N59°W
2. 41.6 m at N67°W
3. 15 ft

Solution to Section 13.2.2 Exercises

1. See Figure 13-20. **A** = 190 lb down, **B** = 440 lb up.
2. See Figure 13-21. F_2 = 346 lb, F_{1H} = 310 lb right, F_{1V} = 61 lb up.
3. See Figure 13-22. F_1 = 1300 lb, F_2 = 1500 lb.
4. See Figure 13-23. $F_1 = F_2$ = 48 lb.

FIGURE 13-20
Solution to Section 13.2.2 Exercises, Problem 1

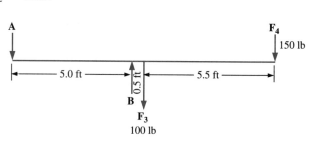

FIGURE 13-21
Solution to Section 13.2.2 Exercises, Problem 2

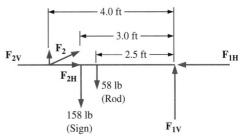

FIGURE 13-22
Solution to Section 13.2.2 Exercises, Problem 3

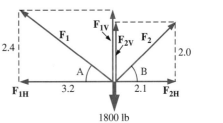

FIGURE 13-23
Solution to Section 13.2.2 Exercises, Problem 4

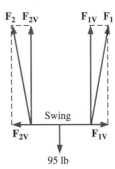

Solutions to Section 13.3 Exercises:

1. 228°/34 kn
2. 130°/162 kn
3. 329°/135 kn
4. GS = 116 kn, TH = 118°
5. 3.30 hr

CHAPTER 13 PROBLEMS

Surveying

Find the bearing of two lines perpendicular to the given line.

1. N75°W
2. S33°E

Find the bearing of the line running in the opposite direction.

3. N62°W
4. S28°E

Calculate closure graphically and algebraically in Problems 5 and 6.

5. **AB** = 78 m at N59°W, **BC** = 89 m at S37°W (Find **CA**)
6. **AB** = 47 m at N42°E, **BC** = 36 m at N62°E, **CD** = 63 m at N55°E (Find **DA**)
7. A 40-ft by 80-ft rectangular building is to be positioned on a lot as shown in Figure 13-24. The West lot line runs N12°E and the East lot line runs N25°W. The North and South lot lines run due east. The building is to be positioned parallel to the lot line that runs N25°W. The northeast corner of the building is to be 20 ft south of the North lot line and 20 ft west of the northeast corner of the lot.

   a. How far is the northwest corner of the building from the lot line on the west?
   b. How far is the southwest corner of the building from the lot line on the south?

FIGURE 13-24

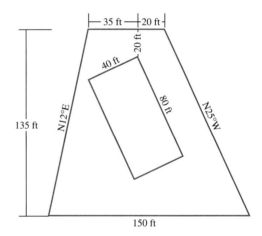

Mechanics

Draw a free body diagram and solve.

1. A 10.0-foot diving board weighs 175 lb. It is pinned at one end and rests on a support 4.5 ft from the pinned end. If a 163-lb person is standing on the end of the board, find the forces acting on the board.
2. A sign for "Someplace Else Bar and Grill" weighs 233 lb and is 6.5 ft wide. It hangs from and is centered along a 78 lb, 8.5-ft bar that is pinned into the wall at one end. The bar is supported by a cable that is attached to the bar 6.5 feet from the wall. The other end of the cable is attached to the wall 4.5 ft above the bar. Calculate all forces acting on the bar and the resultant force in the cable.
3. A 3.22-kip block is suspended by two cables. One cable forms an angle of 29.7° with the horizontal and the other forms an angle of 33.5°. Calculate the resultant forces in each cable.
4. A 65-pound child is sitting on a swing. If the swing chains form an angle of 32° with the vertical, calculate the forces in the chains.

Navigation

Sketch the wind triangle and solve graphically and algebraically.

1. A pilot is flying at 195° with an airspeed of 256 knots, in unknown wind conditions. The navigator determines that the plane is actually traveling at 180° at 250 kn. Find the direction and speed of the wind.

2. A pilot wishes to fly to an airport on a course of 244° and maintain a ground speed of 152 knots. The reported wind is 119°/32 kn. Calculate the heading that should be flown and the resulting airspeed.

3. A pilot is flying a heading of 070° with an airspeed of 215 knots. The wind is 212°/34 kn. Find the ground speed and true course of the plane.

4. A pilot wishes to fly to a city at a heading of 127° while maintaining an airspeed of 147 kn. The reported wind is 41 kn at 222°. Find the true heading to fly and the resulting ground speed. Note: The algebraic solution to this problem is difficult. Use Example 13.12 as a guide.

5. If a plane is to cover 683 mi in 4 hr, calculate the average ground speed in knots.

14
Logarithms

CHAPTER OUTLINE

- **14.1** Logarithmic Notation
- **14.2** Common Logs
- **14.3** Significant Digits
- **14.4** Sound Level
- **14.5** Power Gain (G_P)
- **14.6** Properties of Logarithms
- **14.7** Voltage Gain (G_V)
- **14.8** Log Graph Paper
- **14.9** Natural Logarithms
- **14.10** Universal Time Constant Chart

OBJECTIVES

After completing this chapter, you should be able to

1. find the common logs of numbers.
2. find antilogs.
3. convert an expression in exponential notation to logarithmic notation.
4. convert an expression in logarithmic notation to exponential notation.
5. perform sound intensity and sound level calculations.
6. perform power gain calculations.
7. apply the three principles of logarithms.
8. perform voltage gain calculations.
9. create a graph on semilog and log-log graph paper.
10. discuss the difference between natural logs and common logs.
11. calculate time constants.
12. use the time constant chart to solve capacitor charge problems.
13. use the time constant equation to solve capacitor charge problems.

NEW TERMS TO WATCH FOR

- exponential notation
- logarithmic notation
- logarithm
- base number
- common logs
- characteristic
- mantissa
- antilog
- sound intensity (I)
- threshold of hearing (I_0)
- threshold of pain
- sound level (β)
- bels (B)
- decibels (dB)
- power gain (G_P)
- voltage gain (G_V)
- linear graph paper
- cycle
- semi-log graph paper
- log-log graph paper
- natural log (ln)
- charge (Q)
- time constant (τ)
- universal time constant chart

14.1 ■ LOGARITHMIC NOTATION

Here are examples of statements written in **exponential notation**. We have used this notation throughout the book, and you should be comfortable with it.

$$10^3 = 1000 \quad 10^{-2} = 0.01$$

The same information can be stated in another form called **logarithmic notation**.

$$\log_{10} 1000 = 3 \quad \log_{10} 0.01 = -2$$

These are read "the log of 1000, base 10, equals 3" and "the log of 0.01, base 10, equals −2". So, a **logarithm** is an exponent. In exponential notation, there is a **base number** and an exponent. In logarithmic notation, there is a base number and a logarithm. See Figure 14-1.

FIGURE 14-1
Logarithmic Notation versus Exponential Notation

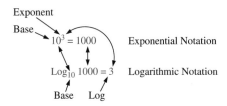

SECTION 14.1 EXERCISES

Write in logarithmic form.

1. $10^4 = 10\,000$
2. $10^{-4} = 0.0001$
3. $2^8 = 256$
4. $10^x = 12$
5. $10^7 = x$

Write in exponential form.

6. $\log_{10} 0.001 = -3$ **7.** $\log_{10} 3162 = 3.5000$ **8.** $\log_2 65\,536 = 16$
9. $\log_{10} x = 5.36$ **10.** $\log_{10} 492 = x$

14.2 ■ COMMON LOGS

Any positive number other than 1 can be used as a base for a system of logarithms.

$$\text{If } m^n = N, \text{ then } \log_m N = n.$$

When 10 is used as the base of a logarithm system, the logs are called **common logs**. For example, $\log_{10} 10\,000 = 4$. If a base number is not written after log, then the base is understood to be 10 and the log is a common log: $\log 10\,000 = 4$.

EXAMPLE 14.1 Find the common log of 0.000 01.

Solution
$$10^{-5} = 0.000\,01$$
$$\log 0.000\,01 = -5$$

Not all logs are integers.

EXAMPLE 14.2 Find the log of 4682.

Solution On an algebraic calculator, enter the number 4682 and press the LOG key. 3.6704 should appear in the display.

$$\log 4682 = 3.6704$$

The decimal point divides a logarithm into two parts. The part to the left of the decimal point is called the **characteristic**. The part to the right of the decimal point is called the **mantissa**. In the last example, 3 is the characteristic and 0.6704 is the mantissa. What about the inverse process: given a logarithm, find the corresponding number? This process is called finding the **anti-log**.

EXAMPLE 14.3 Find the number whose log is 5.

Solution On an algebraic calculator, enter the log, 5, and press the 10^x key. 10^x is often a second function above the LOG key. 100 000 should appear.

$$\log 100\,000 = 5$$

EXAMPLE 14.4   Find the anti-log of 2.4500.

Solution   Use the 10^x key.
$$10^{2.4500} = 281.8$$

EXAMPLE 14.5 $\log x = -2.350$ Solve for x.

Solution   Rewrite the expression in exponential form.
$$10^{-2.350} = x$$
$$x = 10^{-2.350} = 0.004\ 47$$

Logs of numbers that lie between 0 and 1 are negative.

SECTION 14.2 EXERCISES   Complete these statements.

1. $\log 100 =$
2. $\log 0.000\ 001 =$
3. $\log \underline{\quad} = 6$
4. $\log \underline{\quad} = -4$
5. $\log 456.2 =$
6. $\log \underline{\quad} 0.445 =$
7. $\log \underline{\quad} = 2.8876$
8. $\log \underline{\quad} = -2.8876$
9. $\log 10^3 =$
10. $\log 10^{4.5} =$

14.3 ■ SIGNIFICANT DIGITS

Let's find the logarithms of a variety of numbers and use the results to write a rule of thumb for determining the proper number of significant digits.

Evaluate log 36. If 36 is a measurement, it ranges from 35.5 to 36.5.

$$\log 35.5 = 1.550\ 228\ 35$$
$$\log 36 = 1.556\ 302\ 5$$
$$\log 36.5 = 1.562\ 298\ 6$$

These extremes begin to differ in the hundredth's place. Round to the hundredth's place.

$$\log 36 = 1.56$$

Find the common log of 1245. If 1245 is a measurement, it is precise to the unit's place and varies from 1244.5 to 1245.5.

$$\log 1244.5 = 3.094\ 994\ 901$$
$$\log 1245 = 3.095\ 169\ 35$$
$$\log 1245.5 = 3.095\ 343\ 732$$

These extremes are close to the ten-thousandth's place. Round the result to the ten-thousandth's place. log 1245 = 3.0952.

Are you ready for a shortcut? In these examples, logs were found of numbers whose accuracies are, 2 and 4. In each case, the number of significant digits in the number was the same as the number of decimal places in the mantissa. So, as a rule of thumb, **round the mantissa of the logarithm to the same number of places as significant digits in the original number.**

EXAMPLE 14.6 Evaluate log 1.000 06.

Solution 1.000 06 has an accuracy of 6, so include six places in the mantissa.

$$\log 1.000\ 06 = 0.000\ 026$$

Check: log 1.000 055 = 0.000 023 8 and log 1.000 065 = 0.000 028 28. These differ first in the millionth's place.

Let's use the same process for anti-logs.

Find the number whose log is 4.344. 4.344 ranges from 4.3435 to 3.3445.

$$10^{4.3435} = 22\ 054.64136$$
$$10^{4.344} = 22\ 080.0473$$
$$10^{4.3445} = 22\ 105.4825$$

The extremes begin to differ in the hundred's place. Round the result to the hundred's place.

$$\log N = 4.344 \quad N = 22\ 100$$

Find the number whose log is 2.8356. 2.8356 ranges from 2.835 55 to 2.835 65.

$$10^{2.835\ 55} = 684.778\ 31$$
$$10^{2.8356} = 684.857\ 159$$
$$10^{2.835\ 65} = 684.936\ 01$$

The extremes begin to differ in the tenth's place. Round the result to that place.

$$\log N = 2.8356 \quad N = 684.9$$

These examples use logs whose mantissas are precise to the thousandth's and ten-thousandth's place. In each case, the accuracy of the corresponding number is the number of decimal places in the mantissa of the given logarithm. So, **given a logarithm, the accuracy of the corresponding number is equal to the number of decimal places in the mantissa of the logarithm.**

EXAMPLE 14.7 Solve for N. log N=3.455 (Find the antilog of 3.455.)

Solution

$$N = 10^{3.455} = 2851.018$$

The mantissa of the log has three decimal places. The result has an accuracy of three.

Round N to 2850.

Check: 3.455 ranges from 3.4545 to 3.4555.

$$10^{3.4545} = 2847.737\ 800$$
$$10^{3.4555} = 2854.302\ 513$$

These extremes vary in the ten's place. Round to 2850. It checks!

SECTIONS 14.3 EXERCISES

Round each result to the proper number of significant digits.

1. log 28 476 =
2. log 0.289 =
3. log N = 2.1111   Find N.
4. log N = −2.111   Find N.

14.4 ■ SOUND LEVEL

The human ear is sensitive to sound waves over a tremendous range of intensities. Logarithms are used to compare one intensity to another. The **intensity** (I) of a sound wave is the rate at which sound energy flows through a unit area A that is perpendicular to the direction of travel of the wave.

$$\text{intensity} = \text{power/area} \qquad I = P/A$$

Sound intensity is measured in watts per meter^2. The most faint sound audible to human beings has an intensity of about 1×10^{-12} W/m^2. This intensity is called the **threshold of hearing**. Its symbol is I_0. It is used as a reference for the measurements of other intensities. On the other end of the scale is the loudest sound the human ear can stand, about 1 W/m^2. This intensity is called the **threshold of pain**. Logarithms are used to measure sound levels within this tremendous range of intensities. To calculate the **sound level** (β) of a sound intensity, divide the intensity by the threshold of hearing and take the log (base 10) of the quotient.

$$\beta = \log(I/I_0)$$

β is measured in **bels**, in honor of Alexander Graham Bell. The bel is a large unit and is usually broken into units 1/10 as large, the **decibel** (**dB**). Unitwise, the units in the numerator and denominator cancel, and the log of a number is another number, no units. A decibel is dimensionless.

$$\beta = 10 \log(I/I_0)$$

where β is measured in dB, I is the intensity of the sound wave being measured, and I_0 is the intensity of the threshold of hearing.

EXAMPLE 14.8 A sound intensity is measured at 1.57 mW/m². Calculate its sound level.

Solution **Given:** $I = 1.57$ mW/m².
Find: β.

$$\beta = 10 \log(I/I_0)$$
$$\beta = 10 \log(1.57 \times 10^{-3}/1 \times 10^{-12})$$
$$\beta = 10 \log(1.57 \times 10^{9})$$
$$\beta = 91.958 \text{ dB}$$

Round β to 91.96 dB.

Significant digits: 1.57 has an accuracy of three. Its log has three decimal places in its mantissa. After multiplying by 10, there are two decimal places to the right of the decimal.

Extended exposure to sound levels at 90 dB and higher are considered harmful to the ear. Ear plugs are recommended.

EXAMPLE 14.9 If the sound level at a rock concert is 115 dB, calculate the sound intensity.

Solution **Given:** $\beta = 115$ dB.
Find: I.

$$\beta = 10 \log(I/I_0)$$
$$\beta/10 = \log(I/I_0)$$
$$\log(I/I_0) = \beta/10$$

Rewrite this logarithmic expression in exponential form.

$$10^{\beta/10} = I/I_0$$

Solve for I.

$$I_0 \cdot 10^{\beta/10} = I$$
$$I = I_0 \cdot 10^{\beta/10}$$
$$I = 1 \times 10^{-12} \text{ W/m}^2 \cdot 10^{115/10}$$
$$I = 1 \times 10^{-12} \text{ W/m}^2 \cdot 10^{11.5}$$
$$I = 10^{-0.5} \text{ W/m}^2$$
$$I = 0.3162 \text{ W/m}^2$$

Round I to 0.3 W/m².

Significant digits: Our log is 0.5; one decimal place in the mantissa. Round the result to one significant digit.

SECTION 14.4 EXERCISES

For each of the following sound intensities, calculate the corresponding sound level in decibels.

1. 3.16 nW/m² whisper
2. 0.316 mW/m² heavy traffic
3. Threshold of hearing
4. Threshold of pain

For each of the following sound levels, calculate the corresponding sound intensity in W/m².

5. 95 dB subway
6. 135 dB machine gun

14.5 ■ POWER GAIN (G_P)

In electronics, the power delivered to a load is sometimes compared to a standard, just as sound levels are compared to the threshold of hearing. In power measurements, one of the standards used is 1 milliwatt delivered to a 600 Ω load. The comparison is called **power gain (G_P)**, measured in decibels.

$$G_P = 10 \log (P_2/P_1)$$

where P_2 is the power being measured in watts,
P_1 is the standard 1 mW, and G_P is the power gain in dB.

Since this measurement is in reference to a standard 1 mW, the units are given as dBm.

EXAMPLE 14.10 4.50 mW is delivered to a 600 Ω load. Calculate the power gain in dBm.

Solution **Given:** $P_2 = 4.50$ mW and $P_1 = 1$ mW.
Find: G_P in dBm.

$$G_P = 10 \log (P_2/P_1)$$
$$G_P = 10 \log (4.50 \text{ mW}/1 \text{ mW})$$
$$G_P = 10 \log (4.50)$$
$$G_P = 6.5321 \text{ dBm}$$

Round G_P to 6.53 dBm.

EXAMPLE 14.11 Calculate the power level that corresponds to −3.0 dBm.

Solution **Given:** $G_P = -3.0$ dBm and $P_1 = 1$ mW.
Find: P_2.

$$G_P = 10 \log(P_2/P_1)$$
$$G_P/10 = \log(P_2/P_1)$$
$$\log(P_2/P_1) = G_P/10$$

Rewrite this logarithmic expression in exponential form.

$$10^{G_P/10} = P_2/P_1$$

Solve for P_2.

$$P_1 \cdot 10^{G_P/10} = P_2$$
$$P_2 = P_1 \cdot 10^{G_P/10}$$
$$P_2 = 1 \text{ mW} \cdot 10^{-3.0/10}$$
$$P_2 = 1 \text{ mW} \cdot 10^{-0.30}$$
$$P_2 = 0.5011 \text{ mW}$$

Round P_2 to 0.50 mW.

In the equation

$$G_P = 10 \log(P_2/P_1)$$

P_1 does not have to be a standard like 1 mW. It can be any power level that we want to compare to another power, P_2. For example, the power level of an input signal into an amplifier can be compared to the power level of the output signal of the amplifier. Or, in an amplifier, the power delivered to the load varies with the frequency of the applied signal. The output power at one frequency can be compared to the output power at another frequency. Both of these measurements are made in decibels.

EXAMPLE 14.12 Calculate the power gain of an amplifier if the input power is 15.0 mW and the output power is 195.0 mW.

Solution **Given:** $P_1 = 15.0$ mW and $P_2 = 195.0$ mW.

Find: G_P.

$$G_P = 10 \log (P_2/P_1)$$
$$G_P = 10 \cdot \log (195.0 \times 10^{-3}/15.0 \times 10^{-3})$$
$$G_P = 10 \cdot \log (195.0/15.0)$$

Calculator sequence: 13 log × 10 =

$$G_P = 11.1394 \text{ dB}$$

Round G_P to 11.14 dB.

Significant digits: (195.0/15.0) has an accuracy of three; the log will have three decimal places in the mantissa, 1.114. 10 is exact. The final product is 11.14.

An increase in power level yields a positive G_P.

14.5 POWER GAIN (G_p)

EXAMPLE 14.13 The output power of an amplifier at the mid-range frequencies is 1.0 watt, and the output power at 100 kHz is 0.55 watts. Calculate the power gain at 100 kHz with respect to the mid-range frequencies.

Solution **Given:** $P_1 = 1.0$ W and $P_2 = 0.55$ W.
Find: G_P.

$$G_P = 10 \cdot \log (P_2 / P_1)$$

Use the mid-range gain as P_1.

$$G_P = 10 \cdot \log (0.55 / 1.0)$$
$$G_P = 10 \cdot \log (0.55)$$
$$G_P = -2.5963 \text{ dB}$$
Round G_P to -2.6 dB.

0.55 has an accuracy of two; round its log so that the mantissa has two decimal places. −0.26. 10 is exact, so the final product is −2.6.

At 100 kHz, the output power has decreased by 2.6 dB as compared to the mid-range gain. A decrease in power level results in a negative power gain.

EXAMPLE 14.14 If an amplifier provides a 20 dB gain, and the output power is 2.50 watts, calculate the required input power.

Solution **Given:** $G_P = 20$ dB and $P_2 = 2.50$ watts.
Find: P_1.

$$G_P = 10 \cdot \log (P_2/P_1)$$
$$G_P/10 = \log (P_2/P_1)$$

Express in exponential form.

$$10^{G_P/10} = P_2/P_1$$

Solve for P_1.

$$P_1 = P_2/10^{G_P/10}$$
$$P_1 = 2.50 \text{ W}/10^{20/10}$$
$$P_1 = 0.025 \text{ W} = 25 \text{ mW}$$

SECTION 14.5 EXERCISES

1. Calculate the power gain of an amplifier if the input power is 45.6 m watts and the output power is 1.00 watt.

2. Calculate the power gain of an amplifier if the input power is 56.2 m watts and the output is 20.5 m watts.

3. If power doubles, what is the power gain?

4. If power gets cut in half, what is the power gain?
5. If a filter attenuates an input signal by −33.0 dB, and the output power is 47 mW, calculate the required input power.
6. If an amplifier provides a 48.0 dB power gain, and the input power is 2.16 W, calculate the output power.

14.6 ■ PROPERTIES OF LOGARITHMS

Property 1: $\log N^n = n \cdot \log N$
The log of a number raised to a power is equal to the exponent times the log of the number. In other words, the exponent comes out front as a multiplier. Logs reduce the process of raising a quantity to a power to multiplication. For example,

$$\log 10^2 = 2 \cdot \log 10 = 2 \cdot 1 = 2$$

Check: $\log 100 = 2$

$$\log 10^{-3} = -3 \cdot \log 10 = -3 \cdot 1 = -3$$

Check: $\log 0.001 = -3$

$$\log 45.6^{3.4} = 3.4 \cdot \log 45.6 = 5.640$$

Check: $\log 436\,999 = 5.64$

Property 2: $\log (A \cdot B) = \log A + \log B$
The log of a product is the sum of the logs of the multiplicands. Logs reduce multiplication to addition. For example,

$$\log (76.3 \cdot 21.7) = \log 76.3 + \log 21.7 = 3.22$$

Check: $\log (76.3)(21.7) = \log 1656 = 3.22$

Property 3: $\log (A/B) = \log A - \log B$
The log of a quotient is equal to the difference of the logs. Logs reduce division to subtraction. For example,

$$\log (456\,000/123\,000) = \log 456\,000 - \log 123\,000 = 0.569$$

Check: $\log(456\,000/123\,000) = \log 3.707 = 0.569$

$$\log (34.6/78.9) = \log 34.6 - \log 78.9 = -0.358$$

Check: $\log (34.6/78.9) = \log 0.438\,53 = -0.358$

SECTION 14.6 EXERCISES

Use these facts to evaluate the following expressions.

$$\log 2 = 0.301 \quad \log 3 = 0.477 \quad \log 5 = 0.699$$

For example,

$$\log 6 = \log (2 \cdot 3) = \log 2 + \log 3 = 0.301 + 0.477 = 0.778$$

1. log 10
2. log 3/2
3. log 2/3
4. log 25
5. log 100
6. log 60
7. log 4/9
8. log 27
9. log 125
10. log 1024

14.7 ■ VOLTAGE GAIN (G_V)

By using the properties of logarithms, the power gain equation can be modified for use with voltage or current levels. The following sequence derives an equation for **voltage gain** (G_V).

Watt's law can be written in three forms:

$$P = IV, \, P = I^2R, \text{ and } P = V^2/R$$

We will use $P = V^2/R$ in this derivation.

Let P_2 be the output power, and V_2 be the output voltage across resistor R_2. Let P_1 be the input power, and V_1 be the input voltage across resistor R_1.

$$G_P = 10 \cdot \log(P_2/P_1)$$

Substitute V_2^2/R_2 for P_2 and V_1^2/R_1 for P_1.

$$G_P = 10 \cdot \log[(V_2^2/R_2)/(V_1^2/R_1)]$$

We are now dealing with a voltage ratio. Call the expression G_V, the voltage gain.

$$G_V = 10 \cdot \log[(V_2^2/R_2)/(V_1^2/R_1)]$$

To divide by a fraction, invert and multiply.

$$G_V = 10 \cdot \log[(V_2^2/R_2) \cdot (R_1/V_1^2)]$$
$$G_V = 10 \cdot \log[(V_2^2/V_1^2) \cdot (R_1/R_2)]$$

Apply Property 2.

$$G_V = 10 \cdot [\log(V_2^2/V_1^2) + \log(R_1/R_2)]$$
$$G_V = 10 \cdot \log(V_2^2/V_1^2) + 10\log(R_1/R_2)$$

If $R_1 = R_2$, the second term becomes 10 log 1. log 1 = 0 ($10^0 = 1$). The second term disappears.

$$G_V = 10 \log(V_2^2/V_1^2)$$

The numerator and denominator are both squared. Express as a fraction squared.

$$G_V = 10 \log(V_2/V_1)^2$$

Apply Property 1.

$$G_V = 2 \cdot 10 \log(V_2/V_1)$$
$$G_V = 20 \log(V_2/V_1)$$

where G_V is the voltage gain in decibels, V_1 is the initial or input voltage in volts, and V_2 is the final or output voltage in volts.

Voltage gain is 20 times the log of the voltage ratio; power gain is 10 times the log of the power ratio. Both are measured in decibels.

EXAMPLE 14.15 An amplifier amplifies an input voltage of 126 mV to produce an output of 1.48 volts. Calculate the voltage gain.

Solution Given: $V_1 = 126 \times 10^{-3}$ V and $V_2 = 1.48$ V.
Find: G_V.

$$G_V = 20 \log(V_2/V_1)$$
$$G_V = 20 \log(1.48 \text{ V}/126 \times 10^{-3} \text{ V})$$
$$G_V = 21.3978 \text{ dB}$$

Calculator sequence: 1.48/126 EXP 3 +/− = LOG × 20 =

Round G_V to 21.40 dB.

Significant digits: 1.48 and 126 m each have an accuracy of three, as does their quotient 11.7. Log 11.7 has three decimal places in its mantissa. That number times 20 has two digits in its mantissa.

EXAMPLE 14.16 The desired output from an amplifier is 34.5 volts, and the voltage gain of the circuit is 15 dB. How large must the input voltage be?

Solution Given: $V_2 = 34.5$ volts and $G_V = 15$ dB.
Find: V_1.

$$G_V = 20 \log (V_2/V_1)$$

The unknown V_1 is not only in the denominator of a fraction, it is "locked up" inside of a log. First divide both sides by 20 to isolate the log.

$$G_V/20 = \log (V_2/V_1)$$

Convert from log form into exponential form.

$$10^{G_V/20} = V_2/V_1$$

Solve for V_1.

$$V_1 \cdot 10^{G_V/20} = V_2$$
$$V_1 = V_2/10^{G_V/20}$$
$$V_1 = 34.5 \text{ V}/10^{15/20}$$
$$V_1 = 34.5 \text{ V}/10^{0.75}$$
$$V_1 = 6.1350 \text{ V}$$
$$V_1 = 6.1 \text{ V}$$

Significant digits: 0.75 has two decimal places in the mantissa. The corresponding number has an accuracy of two. Round the result to two significant digits.

SECTION 14.7 EXERCISES

1. A three-stage amplifier amplifies an input voltage of 550 µV to produce an output of 3.5 volts. Calculate the voltage gain.

2. The desired output from an amplifier is 8.6 volts, and the voltage gain of the circuit is 22 dB. How large must the input voltage be?

3. If the input voltage to an amplifier measures 0.48 V, and the gain of the amplifier is 25 dB, calculate the output voltage.

14.8 ■ LOG GRAPH PAPER

Techniques for graphing experimental data were studied in an earlier chapter. In those graphs, equally spaced divisions were used to represent a unit of measure. For example, each division on the horizontal axis represented 1 second. The graph paper was **linear**. Suppose one, or both, of the variables being measured extends over a wide range. For example, the following chart shows voltage amplification of an amplifier measured over a wide range of frequencies. Voltage amplification is the ratio of output voltage to input voltage. It is dimensionless.

Frequency in Hertz	Voltage Amplification
100	2
200	6
400	10
1000	14
2000	17
4000	19
10000	21
20000	21
40000	21
100 k	21
200 k	19
400 k	17
1 M	14
2 M	11
4 M	7
10 M	3

This data would be a challenge to graph on linear graph paper because the frequency ranges from 100 to 10 MHz. If your intervals are wide enough to accommodate the high end, then the low-end numbers will all be smashed together. The solution is to graph the log, base 10, of the frequency. The data now becomes:

Log Frequency	Voltage Amplification
2	2
2.3	6
2.6	10
3	14
3.3	17
3.6	19
4	21
4.3	21
4.6	21
5	21
5.3	19
5.6	17
6	14
6.3	11
6.6	7
7	3

Now our horizontal data ranges from 2 to 7. Divide the horizontal axis into six equal pieces and temporarily label them 2, 3, 4, 5, 6, and 7. See Figure 14-2A.

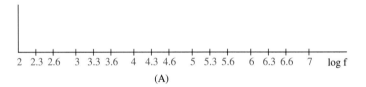
(A)

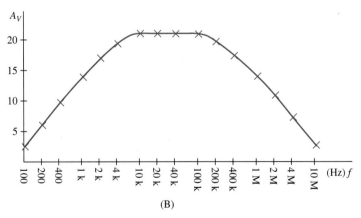
(B)

FIGURE 14-2
Homemade Semilog Graph Paper

Place tic marks at each of the log frequencies in the table and graph the data in the table. Now change the temporary labels to the actual frequencies they represent: 100, 1000, 10 000, and so on. Plot the points and sketch the curve. See Figure 14-2B. Congratulations. You have just created a sheet of 5-cycle, semilog graph paper. Five-**cycle** means five decades or powers of ten are represented. **Semilog** refers to the fact that only one axis is graphed logarithmically. The other is linear. If both axes are logarithmic, the graph paper is called **log-log**. Commercially prepared log graph paper can be purchased in both semilog and log-log form in a variety of cycles.

FIGURE 14-3
3-cycle Semilog Graph Paper

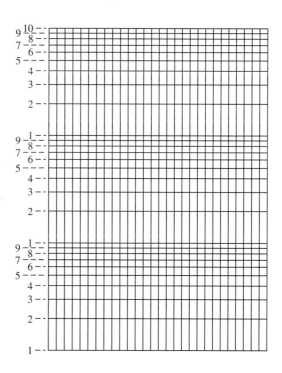

Figure 14-3 shows a simplified version of a sheet of commercially prepared, 3-cycle, semilog graph paper. Each cycle begins with 1 and ends with 10. The numbers printed along one edge are actually the mantissas of the logs represented. The user chooses the characteristics for each cycle to suit the data being graphed. For example, the three cycles could begin with 10, 100, 1000 or 1 M, 10 M, 100 M. The other axis is linear and is divided into 20 divisions per inch.

EXAMPLE 14.17 Label the major divisions of the log axis of a sheet of 3-cycle graph paper to represent frequencies that range from 100 Hz to 100 kHz.

Solution See Figure 14-4.

FIGURE 14-4
Example 14.17

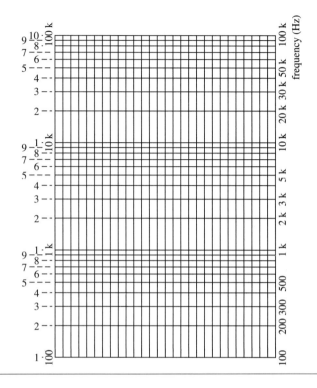

SECTION 14.8 EXERCISE Graph voltage amplification versus frequency on homemade semilog graph paper and on commercially prepared, semilog graph paper.

Frequency	Input Voltage	Output Voltage	Amplification
(hertz)	(volts)	(volts)	(V_{in} / V_{out})
1 k	100 μ	150 μ	
2 k	100 μ	300 μ	
5 k	100 μ	400 μ	
11 k	100 μ	1.2 m	
14 k	50 μ	1.0 m	
20 k	50 μ	1.1 m	
49 k	50 μ	1.0 m	
70 k	50 μ	1.0 m	
100 k	50 μ	1.0 m	
120 k	50 μ	0.9 m	
140 k	50 μ	820 μ	
200 k	100 μ	300 μ	
400 k	100 μ	100 μ	

14.9 ■ NATURAL LOGARITHMS

The **natural log (ln)** system uses the number ϵ (lower case Greek letter epsilon) as a base. To see the value of ϵ, find the e^x key on your calculator. Enter 1 and press the e^x key. The number 2.718 281 828 46 should appear. This number has its origin and definition in calculus, so we won't worry about where it comes from. However, we will use the number as a base in logs and investigate its applications. The general approach is the same as before. Here is an equation written in exponential form.

$$\epsilon^3 = 20.09$$

To express the same information in log form, use ϵ as the base. 3 becomes the log.

$$\log_\epsilon 20.09 = 3$$

This equation is read "log of 20.09, base ϵ, equals 3." ϵ is the base of the natural log system, so the last equation can also be read "natural log of 20.09 equals 3." In natural log expressions, $\log_\epsilon$ is usually written "ln." The last equation becomes

$$\ln 20.09 = 3$$

To confirm that $\epsilon^3 = 20.09$, enter 3 and press the e^x key. 20.09 should appear.

EXAMPLE 14.18 Evaluate $\epsilon^{5.3} =$.

Solution Enter 5.3 and press the e^x key. 200.336 80 should appear.

$$\epsilon^{5.3} = 200.336\ 80$$

What about significant digits? The rule of thumb developed for common logs will be satisfactory for natural logs also.

Round $\epsilon^{5.3}$ to 200.

EXAMPLE 14.19 Evaluate ln 789.3 =.

Solution Enter 789.3 and press the ln key. 6.671 14 should appear.

$$\ln 789.3 = 6.671\ 14$$

Round ln 789.3 to 6.6711.

EXAMPLE 14.20 Evaluate ln $\epsilon^3 =$.

Solution 1

$$\text{Let } \ln \epsilon^3 = n$$

Write in exponential form.

$$\epsilon^n = \epsilon^3$$
$$n = 3$$

Solution 2 "ln ϵ^3" means "what exponent do we have to raise ϵ to to get ϵ^3?" The answer is 3.

SECTION 14.19 EXERCISES

Write in log form.

1. $\epsilon^{4.9999} = 148.4$
2. $\epsilon^{2.7801} = 16.12$

Write in exponential form.

3. ln $1097 = 7.0000$
4. ln $561 = 6.330$

Evaluate.

5. $\epsilon^{5.300} =$
6. $\epsilon^{1.0000} =$
7. $\epsilon^{-1.0000} =$
8. ln $98.7 =$
9. ln $322.2 =$
10. ln $\epsilon^4 =$
11. ln $\epsilon^{3.33} =$

14.10 ■ UNIVERSAL TIME CONSTANT CHART

Graph the exponential function $y = 1 - \epsilon^{-x}$.

x	$y = 1 - \epsilon^{-x}$
0	0
1	0.632
2	0.865
3	0.950
4	0.982
5	0.993
6	0.998
7	0.999

See Figure 14-5.

This curve is analyzed in the following chart. Column A shows x incrementing from 0 to 6. Column B shows the corresponding value of y from the equation $y = 1 - \epsilon^{-x}$. Figure 14-5 shows that the maximum value of y is 1. Column C shows how far y has to go to get to 1, $1 - y$. Column D is the vertical distance covered by y during the next increment of x. It is calculated by subtracting the next entry in column C from the current entry in column C. This value also results from column B, subtracting the current value of y from the next entry for y. Column E calculates the percent of vertical distance covered in the present increment compared to the vertical distance remaining to get to 1.

FIGURE 14-5
$y = 1 - \epsilon^{-x}$

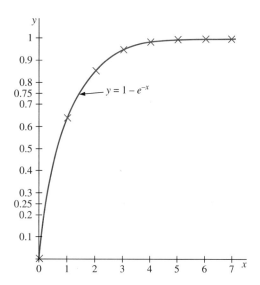

A very remarkable result appears in column E. In every case, 63.2% of the remaining vertical distance is covered in the next increment of x. From any point on the graph, during the next increment of x the y value increases 63.2% of the remaining distance to the maximum value.

A	B	C (Distance Remaining to Top)	D (Vertical Distance Covered during Next Increment of x)	E (% of Remaining Distance Covered)
(x)	(y)	(1 − B)		D/C·100%
0	0	1	1.00000 − 0.36788 = 0.63212	0.63212/1.00000 = 63.2%
1	0.63212	0.36788	0.36788 − 0.13534 = 0.23254	0.23254/0.36788 = 63.2%
2	0.86466	0.13534	0.13534 − 0.04979 = 0.08555	0.08555/0.13534 = 63.2%
3	0.95021	0.04979	0.04979 − 0.01832 = 0.03147	0.03147/0.04979 = 63.2%
4	0.98168	0.01832	0.01832 − 0.00674 = 0.01158	0.01158/0.01832 = 63.2%
5	0.99326	0.00674	0.00674 − 0.00248 = 0.00426	0.00426/0.00674 = 63.2%
6	0.99752	0.00248		

This curve is the basis of the next application. Figure 14-6A shows a series circuit containing a 10 kΩ resistor and a 33 µF capacitor in series with a switch and a 100 V battery. When the switch is closed, current flows. See Figure 14-6B. Electrons are repelled by the negative terminal of the battery and stored on the side of the capacitor connected to ground. The electrons are represented by the − signs on the capacitor.

FIGURE 14-6
Charging a Capacitor

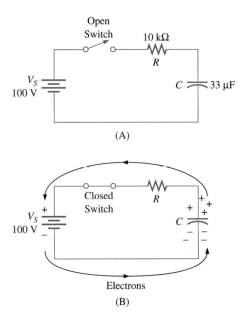

For each electron stored on one side of the capacitor, an electron is stripped from the other side of the capacitor to flow into the positive terminal of the battery. The absence of electrons is represented by + signs. This difference in electrons is the **charge (Q)** stored on the capacitor. As the charge builds, so does the voltage across the capacitor. The product of the resistance that the current flows through to charge the capacitor times the capacitance being charged is called the **time constant (τ)**. τ (lowercase Greek letter tau) is measured in seconds.

$$\tau = RC$$

In this case, $R = 10 \text{ k}\Omega$ and $C = 33 \text{ }\mu\text{F}$

$$\tau = 10 \text{ k}\Omega \cdot 33 \text{ }\mu\text{F}$$
$$\tau = 330 \text{ msec} = 0.33 \text{ sec}$$

During the first time constant, 0.33 sec, the capacitor charges from 0 V to 63.2% of the battery voltage. During the next time constant, the capacitor charges 63.2% of the voltage that is left. This continues for 5 time constants, at which time the capacitor is charged almost to the supply voltage.

Let's modify the graph in Figure 14-5 so that the horizontal axis represents time, marked off in time constants, and the vertical axis represents the voltage across the capacitor, marked off in percent of the applied voltage. See Figure 14-7. The graph in this form is called the **universal time constant chart**.

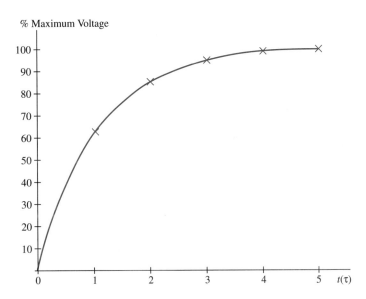

FIGURE 14-7
Universal Time Constant Chart

EXAMPLE 14.21 For the circuit in Figure 14-6, determine the voltage across the capacitor one time constant after the switch is closed.

Solution From the 1 time constant mark on the horizontal axis, go vertical to intersect the graph and then horizontal to intersect the y-axis at 63%. After 1 time constant, 330 ms is this case, the voltage across the capacitor will be 63% of the applied voltage.

$$V_C = 0.63 \cdot 100 \text{ V} = 63 \text{ V}$$

At the end of the first time constant, the capacitor has 37 volts to go before it has charged to the 100 V applied voltage. During the second time constant, the capacitor charges to 63% of the remaining voltage.

$$63\% \text{ of } 37 \text{ V} = 23.3 \text{ V}$$

After the second time constant, the voltage across the capacitor is 86.3 V (63V + 23.3V). To verify this on the graph, find 2τ on the horizontal axis. Go vertical to intersect the graph and then horizontal to read the percentage, 87%. During each time constant the capacitor will charge 63% of the remaining voltage. Theoretically, it will never fully charge, but practically speaking, it is fully charged in 5 time constants.

EXAMPLE 14.22 For the circuit in Figure 14-6, determine the voltage across the capacitor one second after the switch is closed. The capacitor begins charging from 0V.

Solution **Given:** $R = 10\ \text{k}\Omega$, $C = 33\ \mu\text{F}$, $V_S = 100\ \text{V}$, and $t = 1\ \text{s}$.
Find: V_C.

To use the universal time constant chart, we need to know the number of time constants in 1 second.

$$\tau = RC$$
$$\tau = 10\ \text{k}\Omega \cdot 33\ \mu\text{F}$$
$$\tau = 330\ \text{ms}$$
$$t = 1\ \text{sec}/(330\ \text{msec}/\tau)$$
$$t = 3.0303\ \tau$$
Round t to $3.0\ \tau$.

Consult the chart.

$3.0\ \tau$ corresponds to 95%.
95% of 100 V = $0.95 \cdot 100\ \text{V} = 95\ \text{V}$.

EXAMPLE 14.23 For the circuit in Figure 14-6, determine how long it takes for the voltage across the capacitor to charge from 0V to 85 V.

Solution **Given:** $R = 10\ \text{k}\Omega$, $C = 33\ \mu\text{F}$, $V_S = 100\ \text{V}$, and $V_C = 85\ \text{V}$.
Find: t.

To use the universal time constant chart, we need to know the percent of applied voltage across the capacitor.

$$85\ \text{V}/100\ \text{V} = 0.85 = 85\%$$

Consult the chart.

85% corresponds to $1.8\ \tau$.

Calculate the time.

$$t = 1.8\ \tau$$
$$t = 1.8\ RC$$
$$t = 1.8 \cdot 10\ \text{k}\Omega \cdot 33\ \mu\text{F}$$
$$t = 0.59\ \text{sec}$$

The equation for this curve is

$$y = 1 - \epsilon^{-t/RC}$$

where t is the time elapsed since the switch was closed,
R is the resistance that the charging current is flowing through,
and C is the capacitance being charged from 0V.

14.10 UNIVERSAL TIME CONSTANT CHART

Note that the exponent is the time elapsed divided by the time constant RC. It is the number of time constants that have passed since the switch was closed.

EXAMPLE 14.24 Calculate the charge on the capacitor in Figure 14-6 after 1 constant.

Solution Given: $R = 10\ \text{k}\Omega$, $C = 33\ \mu\text{F}$, $V_S = 100\ \text{V}$, and $t = 1\ RC$.
Find: $y = 1 - \epsilon^{-t/RC}$.

$$y = 1 - \epsilon^{-1\ RC/RC}$$
$$y = 1 - \epsilon^{-1}$$
$$y = 0.632\ 120$$

From the graph, we estimated that the capacitor would be charged to 63% of the applied voltage. With the equation, we get a more accurate result, 63.2%.

The equation gives us the percentage of the applied voltage; let's modify the equation one more time to include the applied voltage.

$$V_C = V_S(1 - \epsilon^{-t/RC})$$

where V_C is the voltage across capacitor C t seconds after it begins to charge from 0 volts through resistor R, and V_S is the applied voltage.

EXAMPLE 14.25 Calculate the voltage across the capacitor in Figure 14-6 100 ms after the switch is closed. The capacitor is initially discharged.

Solution Given: $R = 10\ \text{k}\Omega$, $C = 33\ \mu\text{F}$, $V_S = 100\ \text{V}$, and $t = 100\ \text{msec}$.
Find: V_C.

$$V_C = V_S(1 - \epsilon^{-t/RC})$$
$$V_C = 100\ \text{V}[1 - \epsilon^{-100\ \text{ms}/(10\ \text{k}\Omega\ \cdot\ 33\ \mu\text{F})}]$$
$$V_C = 26.1423\ \text{V}$$

Calculator sequence: Start with the exponent.

$$0.1\ +/-\ x \div 10\ 000 \div 33\ \text{EXP}\ 6\ +/- = \epsilon^x\ +/- + 1 = \times 100 =$$
Round V_C to 26 V.

EXAMPLE 14.26 A 75 μF capacitor is being charged through a 4700 Ω resistor by a 25.0 V DC supply. Calculate the voltage across the capacitor 0.11 seconds after charging begins.

Solution Given: $C = 75\ \mu\text{F}$, $R = 4700\ \Omega$, $V_S = 25.0\ \text{V}$, and $t = .11\ \text{sec}$.
Find: V_C.

$$V_C = V_S(1 - \epsilon^{-t/RC})$$
$$V_C = 25.0 \text{ V}[1 - \epsilon^{-0.11 \text{ s}/(4700 \text{ }\Omega \cdot 75 \text{ }\mu\text{F})}]$$
$$V_C = 6.7015 \text{ V}$$

Calculator sequence:

$$.11 +/- \div 4700 \div 75 \text{ EE } 6 +/- = \epsilon^x +/- + 1 = \times 25 =$$

Round V_C to 6.7 V.

Significant digits: The exponent of ϵ has two digits in its mantissa. ϵ to that exponent has two significant digits, .73. 1 − .73 has a precision of hundredths, an accuracy of two. The product has an accuracy of two.

EXAMPLE 14.27 How long does it take the capacitor in the last example to charge from 0V to 25.0 V?

Solution **Given:** $C = 75 \text{ }\mu\text{F}, R = 4700 \text{ }\Omega, V_S = 25.0 \text{ V}$, and $V_C = 25.0 \text{ V}$.
Find: t.

We cannot use the equation to solve this problem, because theoretically the capacitor doesn't ever charge completely. Practically speaking, it charges in 5 time constants.

$$t = 5\tau$$
$$t = 5 \cdot R \cdot C$$
$$t = 5 \cdot 4700 \text{ }\Omega \cdot 75 \text{ }\mu\text{F}$$
$$t = 1.7625 \text{ sec}$$

Round t to 1.8 sec.

EXAMPLE 14.28 How long does it take to charge a 2.2 µF cap from 0V to 12.0 V through a 7500 Ω resistor? The supply voltage is 15.0 V.

Solution **Given:** $C = 2.2 \text{ }\mu\text{F}, R = 7500 \text{ }\Omega, V_S = 15.0 \text{ V}$, and $V_C = 12.0 \text{ V}$.
Find: t.

$$V_C = V_S(1 - \epsilon^{-t/RC})$$

Our unknown t is part of the exponent of ϵ. First isolate $\epsilon^{-t/RC}$.

$$V_C/V_S = 1 - \epsilon^{-t/RC}$$
$$\epsilon^{-t/RC} = 1 - V_C/V_S$$

Express in logarithmic form.

$$\ln(1 - V_C/V_S) = -t/RC$$

Solve for t.

$$RC \cdot \ln(1 - V_C/V_S) = -t$$
$$t = -RC \cdot \ln(1 - V_C/V_S)$$
$$t = -7500\ \Omega \cdot 2.2\ \mu F \cdot \ln(1 - 12.0\ V/15.0\ V)$$
$$t = 2.655\ 57 \times 10^{-2} \text{sec.}$$

Round t to 27 msec.

SECTION 14.10 EXERCISES

Calculate the RC time constant.

1. $R = 3.3\ k\Omega$, $C = 1.0\ \mu F$
2. $R = 100\ \Omega$, $C = 550\ pF$ (pico = 10^{-12})

Calculate the time required to charge the capacitor completely.

3. $R = 3.3\ k\Omega$, $C = 1.0\ \mu F$
4. $R = 100\ \Omega$, $C = 550\ pF$

Use the universal time constant chart to solve these problems.

5. Estimate the voltage across a 47 µF capacitor 5.5 sec after it begins charging through a 75 kΩ resistor. The applied voltage is 12.0 V.
6. How long will it take to charge a 10.0 µF capacitor through a 390 Ω resistor to 27.0 V from a 30.0 V DC supply?
7. and 8. Use the capacitor charge equation to solve Problems 5 and 6.
9. A capacitor of unknown value charges to 35.5 V through a 10 kΩ resistor in 2.3 sec. The applied voltage is 50.0 V. Calculate the capacitance of the capacitor.

■ SUMMARY

1. $10^{-2} = 0.01$ exponential notation
   $\log_{10} 0.01 = -2$ logarithmic notation
2. A log is an exponent.
3. When 10 is used as the base of a logarithm system, the logs are called common logs. If no base number is indicated, it is understood to be base 10. log 1000 = 3.
4. Round the mantissa of the logarithm to the same number of places as significant digits in the original number.
5. Given a logarithm, finding the corresponding number is called finding the antilog.
6. Given a logarithm, the accuracy of the corresponding number is equal to the number of decimal places in the mantissa of the logarithm.
7. The intensity (I) of a sound wave is the rate at which sound energy flows through a unit area A that is perpendicular to the direction of travel of the wave.

$$\text{intensity} = \text{power/area} \quad I = P/A$$

Sound intensity is measured in watts per meter^2.

8. The most faint sound that is audible to human beings has an intensity of about 1×10^{-12} W/m². This intensity is called the threshold of hearing. Its symbol is I_0. It is used as a reference for the measurements of other intensities.

9. The sound level, β, of a sound intensity, is calculated by

$$\beta = 10 \log(I/I_0) \text{ dB}$$

10. The comparison between two power levels is called power gain (G_P), measured in decibels.

$$G_P = 10 \log (P_2/P_1) \text{ dB}$$

11. Here are three properties of logarithms:

$$\text{Property 1: } \log N^n = n \cdot \log N$$
$$\text{Property 2: } \log (A \cdot B) = \log A + \log B$$
$$\text{Property 3: } \log (A/B) = \log A - \log B$$

12. The comparison between two voltage levels is called voltage gain (G_V). G_V is measured in decibels.

$$G_V = 20 \log (V_2/V_1)$$

13. Log graph paper makes it possible to graph a wide range of values on the same axis.

14. The natural log system uses the number ϵ as a base. To see the value of ϵ, enter 1 and press the ϵ key on your calculator. The number 2.718 281 828 46 should appear.

15. The graph of the exponential function $y = 1 - \epsilon^{-x}$ is called the universal time constant chart.

16. A time constant τ is measured in seconds.

$$\tau = RC$$

17. A capacitor completely charges or discharges in 5 time constants.

SOLUTIONS TO CHAPTER 14 SECTION EXERCISES

Solutions to Section 14.1 Exercises

1. $\log_{10} 10\,000 = 4$
2. $\log_{10} 0.0001 = -4$
3. $\log_2 256 = 8$
4. $\log_{10} 12 = x$
5. $\log_{10} x = 7$
6. $10^{-3} = 0.001$
7. $10^{3.5000} = 3162$
8. $2^{16} = 65\,536$

9. $10^{5.36} = x$
10. $10^x = 492$

Solutions to Section 14.2 Exercises

1. 2
2. −6
3. 1 000 000
4. 0.000 1
5. 2.6592
6. −0.352
7. 772.0
8. 0.001 295
9. 3
10. 4.5

Solutions to Section 14.3 Exercises

1. 4.454 48
2. −0.539
3. 129.2
4. 0.007 743

Solutions to Section 14.4 Exercises

1. 35.00 dB
2. 85.00 dB
3. 0 dB
4. 120 dB
5. 3 mW/m²
6. 30 W/m²

Solutions to Section 14.5 Exercises

1. 13.41 dB
2. −4.38 dB
3. 3.010 299 9 dB
4. −3.010 299 9 dB
5. 94 mW
6. 63 W

Solutions to Section 14.6 Exercises

1. log 10 = 1.000
2. 0.176
3. −0.176
4. 1.398
5. 2.000
6. 1.778
7. −.352
8. 1.431
9. 2.097
10. 3.01

Solutions to Section 14.7 Exercises

1. 76 dB
2. 0.68 V
3. 8.5 V

Solutions to Section 14.8 Exercises

1. See Figure 14-8.

Solutions to Section 14.10 Exercises

1. ln 148.4 = 4.9999
2. ln 16.12 = 2.7801
3. $\epsilon^{7.0000} = 1097$
4. $\epsilon^{6.330} = 561$
5. 200
6. 2.718
7. 0.3679
8. 4.592
9. 5.7752
10. 4
11. 3.33

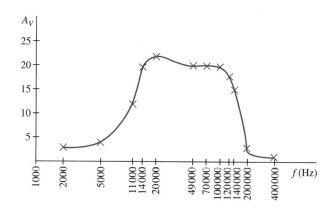

FIGURE 14-8
Solution to Section 14.8
Exercises, Problem 1

Solutions to Section 14.11 Exercises

1. 3.3 msec
2. 55 nsec
3. 16.5 msec
4. 275 nsec
5 and 7. 9.5 V
6 and 8. 9.0 msec
13. $C = 186 \ \mu F$

CHAPTER 14 PROBLEMS

Find the common log of each number.

1. 1000
2. 100
3. 10
4. 1
5. 0.1
6. 0.01

Find the common log of each number. Use the rule of thumb to round each answer to the proper number of significant digits.

7. 5896
8. 589.6
9. 58.96
10. 5.896
11. 0.5896
12. 0.05896

Solve for N. Use the rule of thumb to round each number to the proper number of significant digits.

13. $\log N = 2.7773$
14. $\log N = 1.0033$
15. $\log N = 0.5882$
16. $\log N = -1.4886$
17. $\log N = -2.8532$
18. $\log N = -3.6444$

Write in logarithmic form.

19. $10^4 = 10\ 000$
20. $10^{2.45} = 282$
21. $5^3 = 125$
22. $6^4 = 1296$
23. $2^{10} = 1024$
24. $3^{2.550} = 16.5$

Write in exponential form.

25. $\log_{10} 0.01 = -2$
26. $\log_{10} 603 = 2.780$
27. $\log 1.778\,279 \times 10^{-3} = -2.750$
28. $\log_4 128 = 3.500$
29. $\log_2 104\,857\,6 = 20$
30. $\log_{100} 100\,000\,0 = 3$
31. A sound intensity is measured at 6.87 mW/m². Calculate its sound level.
32. A sound intensity is measured at 15.9 nW/m². Calculate its sound level.
33. If the sound level at a stock car race is 103 dB, calculate the sound intensity.
34. If the sound level created by an insect is measured at 45 dB, calculate the sound intensity.
35. If 350 mW of power is delivered to a 600 Ω load, calculate the power gain in dBm.
36. If 8.46 W of power is delivered to a 600 Ω load, calculate the power gain in dBm.
37. Calculate the power level that corresponds to −8.00 dBm.
38. Calculate the power level that corresponds to 6.87 dBm.
39. Calculate the power gain of an amplifier if the input power is 150 mW and the output power is 20.0 W.
40. Calculate the power gain of an amplifier if the input power is 3.4 W and the output power is 86 W.
41. If the output power of an amplifier at the mid-range frequencies is 3.5 watt, and the output power at 20 kHz is 1.5 W, calculate the power gain at 20 kHz with respect to the mid-range frequencies.
42. If the output power of an amplifier at the mid-range frequencies is 680 mW and the output power at 200 Hz is 330 mW, calculate the power gain at 200 Hz with respect to the mid-range frequencies.
43. If an amplifier provides a 28.0 dB power gain and the input power is 2.16 W, calculate the output power.
44. If an amplifier provides a 12.2 dB power gain and the input power is 330 mW, calculate the output power.
45. If an amplifier provides 38.0 dB gain and the output power is 680 mW, calculate the required input power.
46. If an amplifier provides 29.0 dB gain and the output power is 8.5 W, calculate the required input power.
47. If a filter attenuates an input signal by −33.0 dB and the input power is 55 mW, calculate the output power.
48. If a filter attenuates an input signal by −18.0 dB and the input power is 780 mW, calculate the output power.

Use these facts to evaluate the expressions in Problems 67 through 76.

$$\log 2 = 0.301 \quad \log 3 = 0.477 \quad \log 5 = 0.699 \quad \log 7 = 0.845$$

49. $\log 21$

50. $\log 49$

51. $\log (5/7)$

52. $\log (14/15)$

53. $\log 63$

54. $\log 343$

55. $\log 21000$

56. $\log 945$

57. $\log 448$

58. $\log 44\,100$

59. An amplifier amplifies an input voltage of 550 mV to produce an output of 7.50 V. Calculate the voltage gain in dB.

60. An amplifier amplifies an input voltage of 2.50 V to produce an output of 15.6 V. Calculate the voltage gain in dB.

61. If the input voltage to an amplifier measures 1.86 V and the gain of the amplifier is 24.0 dB, calculate the output voltage.

FIGURE 14-9
Chapter 14 Problems, Problems 65 and 66

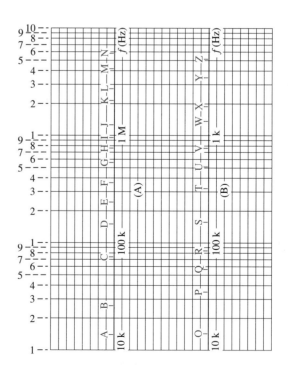

62. If the input voltage to an amplifier measures 447 mV and the gain of the amplifier is 29.0 dB, calculate the output voltage.
63. The desired output from an amplifier is 14.5 V and the voltage gain of the circuit is 49 dB. How large must the input voltage be?
64. The desired output from an amplifier is 6.8 V and the voltage gain of the circuit is 33.0 dB. How large must the input voltage be?
65. Determine what frequency is represented by each letter in Figure 14-9A.
66. Determine what frequency is represented by each letter in Figure 14-9B.
67. Graph the following voltage amplifications versus frequency on semilog graph paper.

Frequency (hertz)	Input Voltage (volts)	Output Voltage (volts)	Amplification (V_{out}/V_{in})
10	0.1	0.08	
30	0.1	0.8	
60	0.1	1.0	
100	0.1	2.4	
300	0.1	7.6	
600	0.1	12.0	
1 k	0.1	12.0	
3 k	0.1	12.0	
6 k	0.1	12.0	
10 k	0.1	12.0	
30 k	0.1	5.3	
60 k	0.1	3.0	
100 k	0.1	1.9	
300 k	0.1	0.6	
600 k	0.1	0.2	

68. Use your graph from the last problem to estimate these quantities.

f(Hz)	A_v
80	_____
500	_____
2 k	_____
7 k	_____
17 k	_____
53 k	_____
450 k	_____
_____ _____	20
_____ _____	35
_____ _____	50
_____ _____	90
_____ _____	110

69. The following data was measured from a band-reject active filter. Graph the amplifications versus frequency on semilog graph paper.

Frequency (hertz)	Input Voltage (volts)	Output Voltage (volts)	Amplification (V_{out}/V_{in})
10 k	0.1	8.6	
16 k	0.1	5.4	
30 k	0.1	1.4	
50 k	0.1	0.5	
100 k	0.1	0.14	
126 k	0.1	0.1	
300 k	0.1	0.1	
500 k	0.1	0.22	
1 M	0.1	0.86	
2.5 M	0.1	5.4	
5.0 M	0.1	8.6	

70. Use your graph from the last problem to estimate these quantities.

f(Hz)	A_v
14 k	____
33 k	____
330 k	____
750 k	____
1.4 M	____
2.7 M	____
3.8 M	____
____	74
____	62
____	33
____	11.5
____	6

71. Evaluate $\epsilon^{4.7} =$

72. Evalutate $\epsilon^{-7.9} =$

73. Evaluate $\ln 348.6 =$

74. Evaluate $\ln 0.0556 =$

Write in log form.

75. $\epsilon^{9.1049} = 9000$

76. $\epsilon^{-2.4165} = 0.089\,23$

Write in exponential form.

77. ln 162 755 = 12.000 000
78. ln 5100 = 8.54

Calculate the RC time constant.

79. $R = 420\ \Omega, C = 4.7\ \mu F$
80. $R = 100\ \Omega, C = 550\ pF\ (pico = 10^{-12})$

Calculate the time required to charge the capacitor completely.

81. $R = 420\ \Omega, C = 4.7\ \mu F$
82. $R = 100\ \Omega, C = 550\ pF$

Solve the following problems using the universal time constant chart.

83. For the circuit in Figure 14-10, determine the voltage across the capacitor one time constant after the switch is closed.

FIGURE 14-10
Chapter 14 Problems, Problems 83 through 90

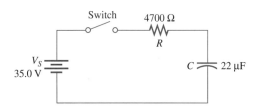

84. For the circuit in Figure 14-10, determine the voltage across the capacitor two time constants after the switch is closed.
85. For the circuit in Figure 14-10, determine the voltage across the capacitor 3.5 time constants after the switch is closed.
86. For the circuit in Figure 14-10, determine the voltage across the capacitor 4.5 time constants after the switch is closed.
87. For the circuit in Figure 14-10, determine the voltage across the capacitor 0.15 seconds after the switch is closed.
88. For the circuit in Figure 14-10, determine the voltage across the capacitor 0.30 seconds after the switch is closed.
89. For the circuit in Figure 14-10, determine how long it takes for the voltage across the capacitor to charge to 17 V.
90. For the circuit in Figure 14-10, determine how long it takes for the voltage across the capacitor to charge to 25 V.

Use the capacitor charge equation to solve the following problems.

91. For the circuit in Figure 14-11, calculate the voltage across the capacitor 0.30 seconds after the switch is closed.

FIGURE 14-11
Chapter 14 Problems, Problems 91 through 94

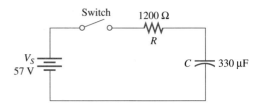

92. For the circuit in Figure 14-11, calculate the voltage across the capacitor 1.00 seconds after the switch is closed.

93. For the circuit in Figure 14-11, calculate how long it takes for the voltage across the capacitor to charge to 55 V.

94. For the circuit in Figure 14-11, calculate how long it takes for the voltage across the capacitor to build to 25 V.

15
Quadratic Equations

CHAPTER OUTLINE

- **15.1** Quadratic Equations
- **15.2** Finding the Roots by Factoring
- **15.3** Finding the Roots by Completing the Square
- **15.4** Finding the Roots by Using the Quadratic Formula
- **15.5** Graphing Parabolas
- **15.6** Applications of Quadratic Equations

OBJECTIVES

After completing this chapter, you should be able to

1. define quadratic equation.
2. find the roots of a quadratic by factoring.
3. find the roots of a quadratic by completing the square.
4. find the roots of a quadratic by using the quadratic formula.
5. use the discriminant to predict the nature and number of roots of a quadratic.
6. graph a quadratic equation by finding the axis of symmetry, the vertex, the roots, and the y-intercept.
7. find the focus and directrix of a parabola whose vertex is located at the origin.
8. draw a tangent and normal to a parabola.
9. describe the use of a paraboloid as a light reflector or dish antenna.

NEW TERMS TO WATCH FOR

quadratic equation
parabola
zeros
roots
finding the roots
perfect square
completing the square
quadratic formula
discriminant

axis of symmetry
vertex
focus
directrix
standard form
tangent
normal
paraboloid

15.1 ■ QUADRATIC EQUATIONS

Equations of the form $y = ax^2 + bx + c$ are called **quadratic equations**. A quadratic equation contains a second-degree term (the variable is squared), a first-degree term (the variable is raised to the first power), and a constant term. The letter a is used to represent the coefficient of the x^2 term, b is used to represent the coefficient of the x term, and c is used to represent the constant term.

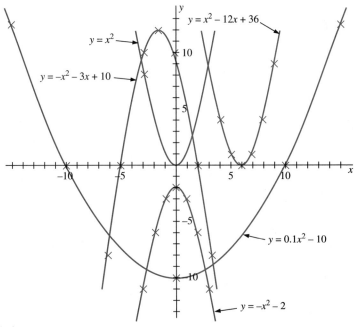

FIGURE 15-1
Parabolas

$y = 4x^2 - 3x - 7$ is a quadratic equation with $a = 4$, $b = -3$, and $c = -7$.
$y = -8x^2 + 3$ is a quadratic equation with $a = -8$, $b = 0$, and $c = 3$.
$y = 16x^2$ is a quadratic equation with $a = 16$ and $b = c = 0$.

b and c can be zero, but a cannot. A quadratic must be of the second degree.

Since this equation is of the second degree, its graph is not a straight line. The graph of a quadratic is called a **parabola**. Several parabolas and their corresponding equations are shown in Figure 15-1.

Two of the parabolas in Figure 15-1 intersect the x-axis in two places; two touch the x-axis in one place; and one parabola lies totally below the x-axis. These x-intercepts are called the **zeros** or **roots** of the quadratic equation. The process of determining the x-intercepts is called **finding the roots** of the equation. We will study three ways to find the roots of quadratic equations.

15.2 ■ FINDING THE ROOTS BY FACTORING

In this approach, the roots of the quadratics are found by setting y equal to 0 and factoring the resulting equation.

Step 1. Set $y = 0$. $ax^2 + bx + c = 0$.

Step 2. Factor the equation. $(x + d)(x + e) = 0$. Not all quadratics will factor easily. If you cannot find the factors, use one of the other methods presented in this chapter.

Step 3. Set each factor equal to zero. If the product of two factors = 0, then one of the two factors = 0. If $(x + d)(x + e) = 0$, then either $(x + d) = 0$, or $(x + e) = 0$.

$$x + d = 0 \qquad x + e = 0$$

Step 4. Solve each equation for x.

$$x = -d \qquad x = -e$$

Step 5. Check each solution by plugging back into the original equation. y should equal zero for each of the roots.

Step 6. In many applications, only one of the roots will make sense. Only one will solve the problem at hand. Test the two roots to determine which it is.

EXAMPLE 15.1 Find the roots: $y = x^2 + x - 6$.

Solution **Step 1.** Set $y = 0$.

$$x^2 + x - 6 = 0$$

Step 2. Factor the equation.

$$(x + 3)(x - 2) = 0$$

Step 3. Set each factor equal to zero.
$$x + 3 = 0 \qquad x - 2 = 0$$

Step 4. Solve each equation for x.
$$x = -3 \qquad x = 2$$

Step 5. Check each solution by substituting back into the original equation.
$$y = x^2 + x - 6 \qquad\qquad y = x^2 + x - 6$$
$$y = (-3)^2 - 3 - 6 \qquad\qquad y = 2^2 + 2 - 6$$
$$y = 0 \qquad\qquad y = 0$$

EXAMPLE 15.2 Find the roots: $y = 6x^2 + 7x - 20$.

Solution **Step 1.** $6x^2 + 7x - 20 = 0$
Step 2. $(3x - 4)(2x + 5) = 0$
Step 3. $3x - 4 = 0 \qquad 2x + 5 = 0$
Step 4. $3x = 4 \qquad 2x = -5$
$$x = 4/3 \qquad x = -5/2$$
Step 5.
$$y = 6x^2 + 7x - 20 \qquad\qquad y = 6x^2 + 7x - 20$$
$$y = 6(4/3)^2 + 7(4/3) - 20 \qquad y = 6(-5/2)^2 + 7(-5/2) - 20$$
$$y = 6(16/9) + 28/3 - 20 \qquad y = 150/4 - 35/2 - 20$$
$$y = 32/3 + 28/3 - 60/3 \qquad y = 150/4 - 70/4 - 80/4$$
$$y = 0 \qquad\qquad y = 0$$

SECTION 15.2 EXERCISES

Find the roots of each of these quadratics by factoring.

1. $y = x^2 + 2x - 8$
2. $y = x^2 + 16x + 60$
3. $y = x^2 - 13x + 40$
4. $y = 2x^2 + 9x - 18$
5. $y = 12x^2 - 19x + 5$

If a quadratic equation cannot be factored easily, an alternate approach called completing the square can be used.

15.3 ■ FINDING THE ROOTS BY COMPLETING THE SQUARE

In some quadratic equations, the x term is not present; its coefficient is zero. $y = ax^2 + c$. These equations are solved by isolating x^2 and then taking the square root of each side.

EXAMPLE 15.3 Find the roots: $y = x^2 - 49$.

Solution **Step 1.** Set $y = 0$.
$$x^2 - 49 = 0$$

Step 2. Isolate x^2.
$$x^2 = 49$$

Step 3. Take the square root of each side. $7^2 = 49$ and $(-7)^2 = 49$. So, the square root of 49 is ± 7.
$$x = \pm 7$$

Step 4. Check the solution.

$$y = x^2 - 49 \qquad y = x^2 - 49$$
$$y = 7^2 - 49 \qquad y = (-7)^2 - 49$$
$$y = 0 \qquad y = 0$$

If the x term is present, preliminary steps are taken to modify the quadratic equation so that one side is a **perfect square**. This process is called **completing the square**. For example, $x^2 - 10x + 25$ is a perfect square and can be factored into $(x - 5)^2$. Then the square root of each side can be taken. To complete the square, follow this procedure.

Step 1. Set $y = 0$.
Step 2. Force the coefficient of the x^2 term to be 1. If the coefficient of the x^2 term is other than 1, divide each side of the equation by that coefficient.
Step 3. Transpose the constant to the other side of the equation.
Step 4. Take 1/2 the coefficient of the x term, square it, and add it to each side of the equation.
Step 5. The left side of the equation is now a perfect square; factor it.
Step 6. Take the square root of each side.
Step 7. Solve for x.
Step 8. Check each root.

EXAMPLE 15.4 Find the roots: $y = x^2 + 12x - 13$.

Step 1. Set $y = 0$.
$$x^2 + 12x - 13 = 0$$

Step 2. Coefficient of the x^2 term is already 1.

Step 3. Transpose the constant to the other side of the equation.
$$x^2 + 12x = 13$$

Step 4. Take 1/2 the coefficient of the x term, square it, and add it to each side of the equation. $12/2 = 6 \quad 6^2 = 36 \quad$ Add 36 to each side.

$$x^2 + 12x + 36 = 13 + 36$$
$$x^2 + 12x + 36 = 49$$

Step 5. The left side of the equation is now a perfect square; factor it.

$$(x + 6)^2 = 49$$

Step 6. Take the square root of each side.

$$x + 6 = \pm 7$$

Step 7. Solve for x.

$$x = -6 + 7 \qquad x = -6 - 7$$
$$x = 1 \qquad x = -13$$

Step 8. Check each root.

$y = x^2 + 12x - 13 \qquad y = x^2 + 12x - 13$
$y = 1 + 12 - 13 \qquad y = (-13)^2 + 12(-13) - 13$
$y = 0 \qquad y = 169 - 156 - 13$
$\qquad\qquad\qquad\qquad y = 0$

EXAMPLE 15.5 Find the roots: $y = 5x^2 - x - 10$.

Solution
Step 1. $5x^2 - x - 10 = 0$
Step 2. $x^2 - x/5 - 10/5 = 0$
Step 3. $x^2 - x/5 = 2$
Step 4. $x^2 - x/5 + 1/100 = 2 + 1/100$
Step 5. $(x - 1/10)^2 = 200/100 + 1/100$
Step 6. $x - 1/10 = (201/100)^{1/2}$
Step 7. $x = 1/10 \pm 201^{1/2}/10$

$x = 1/10 + 201^{1/2}/10 \qquad x = 1/10 - 201^{1/2}/10$
$x = [1 + 201^{1/2}]/10 \qquad x = [1 - 201^{1/2}]/10$

Step 8.

$y = 5x^2 - x - 10$
$y = 5[(1 + 201^{1/2})/10]^2 - (1 + 201^{1/2})/10 - 10$
$y = 5[1 + 2(201)^{1/2} + 201]/100 - (1 + 201^{1/2})/10 - 10$
$y = [202 + 2(201)^{1/2}]/20 - 1/10 - 201^{1/2}/10 - 10$
$y = 101/10 + 201^{1/2}/10 - 1/10 - 201^{1/2}/10 - 10$
$y = 0$

$$y = 5x^2 - x - 10$$
$$y = 5[(1 - 201^{1/2})/10]^2 - (1 - 201^{1/2})/10 - 10$$
$$y = 5[1 - 2(201)^{1/2} + 201]/100 - (1 - 201^{1/2})/10 - 10$$
$$y = [202 - 2(201)^{1/2}]/20 - 1/10 + 201^{1/2}/10 - 10$$
$$y = 101/10 - 201^{1/2}/10 - 1/10 + 201^{1/2}/10 - 10$$
$$y = 0$$

SECTION 15.3 EXERCISES

Find the roots of each of these quadratics by completing the square.

1. $y = x^2 + 2x - 8$
2. $y = x^2 + 16x + 60$
3. $y = x^2 - 13x + 40$
4. $y = 2x^2 + 9x - 18$
5. $y = 12x^2 - 19x + 5$

15.4 ■ FINDING THE ROOTS BY USING THE QUADRATIC FORMULA

Let's begin with the general quadratic equation and follow the procedure for completing the square. It's a good exercise in algebra, and the resulting equation is quite useful. It's called the **quadratic formula**.

$$y = ax^2 + bx + c$$

Step 1. $ax^2 + bx + c = 0$
Step 2. $x^2 + bx/a + c/a = 0$
Step 3. $x^2 + bx/a = -c/a$
Step 4. $x^2 + bx/a + [b/(2a)]^2 = [b/(2a)]^2 - c/a$
Step 5. $[x + b/(2a)]^2 = b^2/(4a^2) - c/a$
Step 6. $x + b/(2a) = \pm [b^2/(4a^2) - c/a]^{1/2}$
Step 7. $x = -b/(2a) \pm [b^2/(4a^2) - c/a]^{1/2}$
$x = -b/(2a) \pm [b^2/(4a^2) - c(4a)/a(4a)]^{1/2}$
$x = -b/(2a) \pm [b^2/(4a^2) - 4ac/(4a^2)]^{1/2}$
$x = -b/(2a) \pm [(b^2 - 4ac)/(4a^2)]^{1/2}$
$x = -b/(2a) \pm (b^2 - 4ac)^{1/2}/(2a)$
$x = [-b \pm (b^2 - 4ac)^{1/2}]/(2a)$ quadratic formula

There you have it, the quadratic formula. The roots can always be found using the formula. Here is the quadratic formula written in fractional form.

$$x = \frac{-b \pm \sqrt{b^2 - 4ac}}{2a}$$

where a is the coefficient of the x^2 term, b is the coefficient of the x term, and c is the constant term of a quadratic equation.

EXAMPLE 15.6 Use the quadratic formula to find the roots of $y = x^2 + x - 6$. The roots of this quadratic were found in Example 15.1 by factoring.

Solution $a = 1, b = 1, c = -6$.

$$x = \frac{-b \pm \sqrt{b^2 - 4ac}}{2a}$$

$$x = \frac{-1 \pm \sqrt{1^2 - 4 \cdot 1 \cdot -6}}{2 \cdot 1}$$

$$x = \frac{-1 \pm \sqrt{25}}{2}$$

$$x = \frac{-1 \pm 5}{2}$$

$$x = \frac{-1 + 5}{2} \qquad x = \frac{-1 - 5}{2}$$

$$x = 2 \qquad x = -3$$

EXAMPLE 15.7 Use the quadratic formula to find the roots of $y = 6x^2 + 7x - 20$. The roots of this quadratic were found in Example 15.2 by factoring.

Solution $a = 6, b = 7, c = -20$.

$$x = \frac{-b \pm \sqrt{b^2 - 4ac}}{2a}$$

$$x = \frac{-7 \pm \sqrt{7^2 - 4 \cdot 6 \cdot -20}}{2 \cdot 6}$$

$$x = \frac{-7 \pm \sqrt{49 + 480}}{12}$$

$$x = \frac{-7 \pm 23}{12}$$

$$x = \frac{-7 + 23}{12} \qquad x = \frac{-7 - 23}{12}$$

$$x = \frac{16}{12} = \frac{4}{3} \qquad x = -\frac{30}{12} = -\frac{5}{2}$$

EXAMPLE 15.8 Use the quadratic formula to find the roots of $y = 5x^2 - x - 10$. The roots of this quadratic were found in Example 15.5 by completing the square.

Solution $a = 5, b = -1, c = -10$

$$x = \frac{-b \pm \sqrt{b^2 - 4ac}}{2a}$$

$$x = \frac{--1 \pm \sqrt{(-1)^2 - 4 \cdot 5 \cdot -10}}{2 \cdot 5}$$

15.4 FINDING THE ROOTS BY USING THE QUADRATIC FORMULA

$$x = \frac{1 \pm \sqrt{1 + 200}}{10}$$

$$x = \frac{1 + \sqrt{201}}{10} \quad x = \frac{1 - \sqrt{201}}{10}$$

In the quadratic formula, the quantity under the radical, $b^2 - 4ac$, is called the **discriminant**. The discriminant in the last three examples is a positive number. In the quadratic formula, the square root of the discriminant is taken. When the discriminant is positive, the square root can be + or −. This leads to two distinct real roots.

If the quadratic equation is a perfect square, the discriminant is equal to zero. The quadratic formula reduces to $x = -b/(2a)$, and only one real root exists. See Example 15.9.

EXAMPLE 15.9 Use the quadratic formula to find the roots of $y = x^2 - 12x + 36$.

Solution $a = 1, b = -12, c = 36$.

$$x = \frac{-b \pm \sqrt{b^2 - 4ac}}{2a}$$

$$x = \frac{--12 \pm \sqrt{(-12)^2 - 4 \cdot 1 \cdot 36}}{2 \cdot 1}$$

$$x = \frac{12 \pm \sqrt{144 - 144}}{2}$$

$$x = \frac{12 \pm 0}{2}$$

$$x = 6$$

So, if the discriminant is positive, the quadratic equation has two real roots. Its graph intersects the x-axis in two places. In Figure 15-1, $y = -x^2 - 3x + 10$ has two roots, -5 and $+2$, and $y = 0.1x^2 - 10$ has two roots -10 and $+10$. If the discriminant is 0, the quadratic has one real root. Its graph touches the x-axis in one place. In Figure 15-1, $y = x^2$ has one root, 0, and $y = x^2 - 12x + 36$ has one root, 6. What happens if the discriminant is negative? The quadratic formula yields two imaginary numbers. The graph of the corresponding quadratic does not cross the x-axis. It is either completely above the x-axis or completely below it. In Figure 15-1, $y = -x^2 - 2$ is completely below the x-axis and has no real roots.

EXAMPLE 15.10 Use the quadratic formula to find the roots of $y = x^2 + 6x + 11$.

Solution $a = 1, b = 6, c = 11$.

$$x = \frac{-b \pm \sqrt{b^2 - 4ac}}{2a}$$

$$x = \frac{-6 \pm \sqrt{6^2 - 4 \cdot 1 \cdot 11}}{2 \cdot 1}$$

$$x = \frac{-6 \pm \sqrt{36-44}}{2}$$

$$x = \frac{-6 \pm \sqrt{-8}}{2}$$

$$x = \frac{-6 \pm j2\sqrt{2}}{2}$$

$$x = -3 \pm j\sqrt{2}$$

$$x = -3 + j\sqrt{2} \quad x = -3 - j\sqrt{2}$$

The discriminant is negative and there are two imaginary roots (no x-intercepts).

SECTION 15.4 EXERCISES

Use the quadratic formula to find the roots of each of these quadratic equations.

1. $y = x^2 + 2x - 8$
2. $y = x^2 + 16x + 60$
3. $y = x^2 - 13x + 40$
4. $y = 2x^2 + 9x - 18$
5. $y = 12x^2 - 19x + 5$

In Problems 6 through 11, calculate the discriminant of each quadratic equation. Use the result to predict the nature of the roots (two real, one real, two imaginary) and number of x-intercepts.

6. $y = x^2 + 6x + 2$
7. $y = -4x^2 + 6x + 2$
8. $y = x^2 - 6x + 9$
9. $y = -x^2 + 5x - 7$
10. $y = x^2 - 6x + 11$
11. $y = -5x^2 - 6x + 4$

Now that we can find the roots of a quadratic, let's look at graphing.

15.5 ■ GRAPHING PARABOLAS

The graph of a quadratic is called a parabola. Since a parabola is not linear, more than three points are needed to sketch the graph. Here are some facts that help streamline the task.

1. The roots are the x-intercepts.
2. The parabola opens either upward or downward, which is determined by the sign of "a" (coefficient of the squared term). If a is positive, the curve opens upward. If a is negative, the curve opens downward.
3. The parabola is symmetrical about a vertical line called the **axis of symmetry**. The axis of symmetry lies halfway between the x-intercepts. The x-intercepts are predicted by the quadratic formula.

$$x_1 = \frac{-b + \sqrt{b^2 - 4ac}}{2a} \qquad x_2 = \frac{-b - \sqrt{b^2 - 4ac}}{2a}$$

To develop an expression for the line of symmetry, average the values of the x-intercepts.

$$x = \frac{\dfrac{-b + \sqrt{b^2 - 4ac}}{2a} + \dfrac{-b - \sqrt{b^2 - 4ac}}{2a}}{2}$$

$$x = \frac{\dfrac{-2b}{2a}}{2} = \frac{-b}{2a}$$

$x = -b/(2a)$ is a vertical line that passes through the point $(-b/(2a), 0)$.

4. The maximum or minimum point of the parabola will occur on the axis of symmetry. To find that point, plug $x = -b/(2a)$ into the original equation and find the corresponding value of y.
5. The y-intercept is found by setting $x = 0$.

$$y = x^2 + bx + c$$
$$y = 0^2 + b \cdot 0 + c$$
$$y = c$$

c is always the y intercept.

So, to graph a parabola, follow this procedure.

Step 1. Calculate the roots by factoring, completing the square, or by applying the quadratic formula. Plot the roots.

Step 2. Calculate the axis of symmetry by averaging the roots or by substituting values for a and b into the equation $x = -b/(2a)$.

Step 3. Calculate the maximum or minimum point by substituting the value $x = -b/(2a)$ into the original equation. Plot the maximum or minimum point.

Step 4. Plot the y-intercept, c. Use symmetry to locate the corresponding point on the opposite side of the axis of symmetry.

Step 5. Choose more values of x as needed to provide a good spread of values and to help determine how wide or narrow the parabola is. Locate the corresponding points on the opposite side of the axis of symmetry.

EXAMPLE 15.11 Graph $y = x^2 - x - 6 = 0$.

Solution $a = 1$, $b = -1$, $c = -6$.

Step 1. Calculate the roots by factoring.

$$x^2 - x - 6 = 0$$
$$(x + 2)(x - 3) = 0$$
$$x + 2 = 0 \quad x - 3 = 0$$
$$x = -2 \quad x = 3$$

Step 2. Calculate the axis of symmetry.

$$x = (-2 + 3)/2 = 1/2$$

alternate solution:

$$x = -b/(2a) = --1/2 = 1/2$$

Step 3. $a > 0$, so the curve opens upward. Calculate the minimum point.

$$y = x^2 - x - 6$$
$$y = (1/2)^2 - 1/2 - 6$$
$$y = 1/4 - 1/2 - 6$$
$$y = 1/4 - 2/4 - 24/4$$
$$y = -25/4$$

$(1/2, -25/4)$ minimum point

Step 4. Graph the y-intercept: $y = c = -6$. Use symmetry to place a corresponding point on the opposite side of the axis of symmetry.

Step 5. Choose one more value of x and calculate the corresponding value of y. Use symmetry to place a corresponding point on the opposite side of the axis of symmetry.

Let $x = 5$.
$$y = x^2 - x - 6$$
$$y = 5^2 - 5 - 6$$
$$y = 14 \quad (5, 14)$$

See Figure 15-2.

FIGURE 15-2
Example 15.11

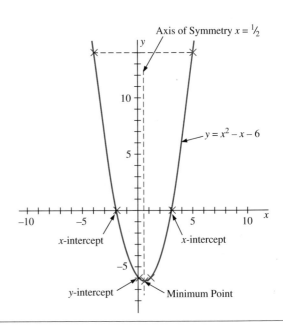

EXAMPLE 15.12 Graph $y = -x^2 - 8x + 20$.

Solution $a = -1$, $b = -8$, $c = +20$.

Step 1. Calculate the roots by factoring.
$$-x^2 - 8x + 20 = 0$$
$$x^2 + 8x - 20 = 0$$
$$(x + 10)(x - 2) = 0$$
$$x + 10 = 0 \quad x - 2 = 0$$
$$x = -10 \quad x = 2$$

Step 2. Calculate the axis of symmetry.
$$x = (-10 + 2)/2 = -8/2 = -4$$

Alternate solution:
$$x = -b/(2a) = --8/(2 \cdot -1) = -4$$

Step 3. $a<0$, so the curve opens downward. Calculate the maximum point.
$$y = -x^2 - 8x + 20$$
$$y = -(-4)^2 - 8(-4) + 20$$
$$y = -16 + 32 + 20$$
$$y = 36$$
$(-4, 36)$ maximum point

FIGURE 15-3
Example 15.12

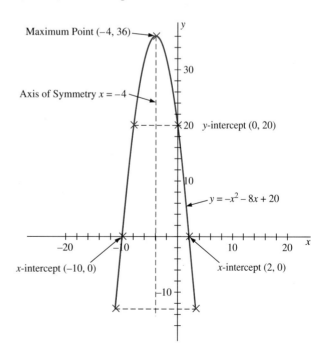

Step 4. Graph the y-intercept: $y = c = 20$. Use symmetry to place a corresponding point on the opposite side of the axis of symmetry.

Step 5. Choose one more value of x and calculate the corresponding value of y. Use symmetry to place a corresponding point on the opposite side of the axis of symmetry.

$$\text{Let } x = 3.$$
$$y = -x^2 - 8x + 20$$
$$y = -3^2 - 8(3) + 20$$
$$y = -13 \ (3, -13)$$

See Figure 15-3.

EXAMPLE 15.13 Graph $y = 0.01x^2 - 0.1x - 2$.

Solution $0.01x^2 - 0.1x - 2 = 0$

Step 1. Multiply both sides of the equation by 100.

$$x^2 - 10x - 200 = 0$$

Find the roots by factoring.

$$(x - 20)(x + 10) = 0$$
$$x - 20 = 0 \quad x + 10 = 0$$
$$x = 20 \quad\quad x = -10$$

Step 2. Find the axis of symmetry.

$$x = (20 - 10)/2$$
$$x = 5$$

Step 3. $a > 0$, so the curve opens upward. Calculate the minimum point.

$$y = 0.01x^2 - 0.1x - 2$$
$$y = 0.01(5^2) - 0.1(5) - 2$$
$$y = .25 - .5 - 2$$
$$y = -2.25$$
$$(5, -2.25) \text{ minimum point}$$

Step 4. Graph the y-intercept: $y = c = -2$. Use symmetry to place a corresponding point on the opposite side of the line of symmetry.

Step 5. Choose one more value of x and calculate the corresponding value of y. Use symmetry to place a corresponding point on the opposite side of the axis of symmetry.

Let $x = 30$
$$y = 0.01x^2 - 0.1x - 2$$
$$y = 0.01 \cdot 30^2 - 0.1 \cdot 30 - 2$$
$$y = 9 - 3 - 2 = 4$$
$$(30, 4)$$

See Figure 15-4.

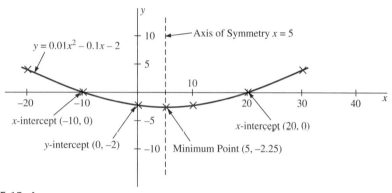

FIGURE 15-4
Example 15.13

EXAMPLE 15.14 Graph $y = -2x^2 + 6x - 7$.

Solution $a = -2$, $b = 6$, $c = -7$.

Step 1. Calculate the x-intercepts using the quadratic formula.
$$x = \frac{-b \pm \sqrt{b^2 - 4ac}}{2a}$$
$$x = \frac{-6 \pm \sqrt{(6)^2 - 4(-2)(-7)}}{2(-2)}$$
$$x = \frac{-6 \pm \sqrt{-20}}{-4}$$

The discriminant is negative; the parabola does not cross the x-axis.

Step 2. Calculate the axis of symmetry.
$$x = -b/(2a)$$
$$x = -6/(2 \cdot -2)$$
$$x = -6/-4 = 3/2$$

Step 3. a is negative, so the curve opens downward. Calculate the maximum point.

$$y = -2x^2 + 6x - 7$$
$$y = -2(3/2)^2 + 6(3/2) - 7$$
$$y = -2(9/4) + 18/2 - 7$$
$$y = -18/4 + 36/4 - 28/4$$
$$y = -10/4 = -5/2$$

$(3/2, -5/2)$ maximum point

Step 4. Graph the y-intercept: $y = c = -7$. Use symmetry to locate the corresponding point on the opposite side of the axis of symmetry.

Step 5. Let $x = -1$

$$y = -2x^2 + 6x - 7$$
$$y = -2(-1)^2 + 6(-1) - 7$$
$$y = -2 - 6 - 7 = -15$$

See Figure 15-5.

FIGURE 15-5
Example 15.14

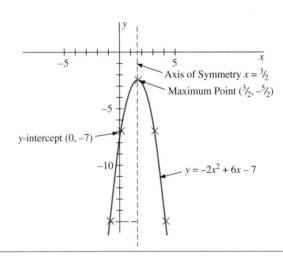

SECTION 15.5 EXERCISES

Graph these quadratic equations. If you have access to a graphing calculator, use it to confirm your results.

1. $y = 3x^2 + 7x - 4$
2. $y = -0.2x^2 + 3x - 52$
3. $y = 1/4x^2 - 3x + 10$
4. Graph these quadratic equations on the same graph. (They are quadratics with $b = c = 0$.)

A. $y = x^2$
B. $y = 2x^2$
C. $y = 4x^2$
D. $y = 1/2x^2$
E. $y = 1/4x^2$

5. From the graphs drawn in Problem 4, discuss the effect of *a* on the parabola.
6. Graph these quadratic equations on the same graph.
   A. $y = -x^2$
   B. $y = -2x^2$
   C. $y = -4x^2$
   D. $y = -1/2x^2$
   E. $y = -1/4x^2$
7. From the graphs drawn in Problems 4 and 6, discuss the effect of the sign of *a* on the parabola.
8. True or False. If *a* is negative, the parabola opens downward.
9. True or False. If *a* is positive, the parabola has a maximum point.
10. Graph these quadratic equations on the same graph.
    A. $y = (x - 3)^2$
    B. $y = (x + 3)^2$
    C. $y = x^2$
11. Graph these quadratic equations on the same graph.
    A. $y = -(x - 4)^2$
    B. $y = -(x + 4)^2$
    C. $y = -x^2$

15.6 ■ APPLICATIONS OF QUADRATIC EQUATIONS

We have graphed a wide variety of parabolas: open upward, open downward, open out wide, open narrow, two *x*-intercepts, one *x*-intercept, and no *x*-intercepts. So, what is the same about all these curves; what causes them all to be called parabolas?

Figure 15-6 shows the graph of the parabola $y = 1/8x^2$. The minimum point is labeled the **vertex**. Above the vertex, on the axis of symmetry, is a point called the **focus**. An equal distance below the vertex is a line perpendicular to the axis of symmetry called the **directrix**. In all parabolas, the following fact is true about any point on the parabola. **The distance from a point on the curve to the focus is the same as the perpendicular distance from the directrix to that point.**

To find the location of the focus, first rearrange the quadratic into **standard form**,

$$x^2 = 4py$$

where *p* is the distance between the focus and vertex, and between the vertex and directrix.

FIGURE 15-6
Focus, Vertex, and Directrix

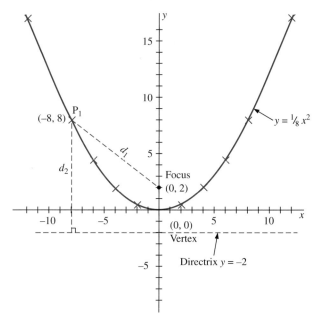

For the parabola in Figure 15-6,

$$y = 1/8 x^2$$
$$x^2 = 8y$$

In this case, $4p = 8$, $p = 2$. The focus is 2 units from the vertex. The focus and directrix are each 2 units from the vertex. See Figure 15-6.

Point $P_1(-8,8)$ lies on the parabola. Its distance to the focus $(0,2)$ is labeled d_1. Its distance to the directrix is labeled d_2. According to the definition of a parabola, d_1 should equal d_2. d_2 is the y value of P_1, 8, plus the distance from the x-axis down to the directrix, 2. $d_2 = 8 + 2 = 10$. Let's calculate d_1 to ensure that it also is 10.

$$d_1 = [(y_2 - y_1)^2 + (x_2 - x_1)^2]^{1/2}$$
$$d_1 = [(8 - 2)^2 + (-8 - 0)^2]^{1/2}$$
$$d_1 = [36 + 64]^{1/2}$$
$$d_1 = 10 \quad d_1 = d_2$$

Figure 15-7 again shows the parabola $y = 1/8x^2$. α and β represent electromagnetic waves (such as microwaves or light waves) entering the parabola parallel to the axis of symmetry. Wave α strikes the parabola at point B. Line AC also touches the parabola at point B. AC is **tangent** to the parabola at point B. Line DB is drawn perpendicular to AC at point B. It is the **normal** to the parabola at point B. The entering wave α forms an angle of incidence, θ_i, with the normal DB. The angle of reflection, θ_r equals the angle of incidence. The reflected wave passes through the focus.

Wave β strikes the parabola at point F. Line EG also touches the parabola at point F. EG is tangent to the parabola at point F. Line HF is drawn perpendicular to EG at point

FIGURE 15-7
Electromagnetic Waves Reflected through the Focus

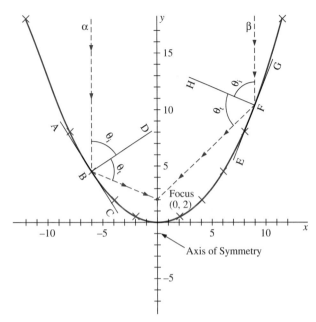

F. Line HF is the normal to the parabola at point F. The entering wave β forms an angle of incidence, θ_i, with the normal HF. The angle of reflection, θ_r, equals the angle of incidence. The reflected wave passes through the focus.

It is a property of the parabola that any wave entering parallel to the axis of symmetry will be reflected to the focus of the parabola. Conversely, waves being emitted at the focus will be reflected by the parabola parallel to the axis of symmetry.

If the parabola in Figure 15-7 is rotated about the y-axis, a three dimensional "dish" called a **paraboloid** is formed. A paraboloid is used for auto headlights and microwave antennas (dishes). In the case of the headlight, the filament for the bright beam is located at the focus. Light emitted is reflected off the paraboloid, parallel to the axis of symmetry, and leaves the headlight as parallel light rays.

In the case of the dish antenna, the receiver or transmitter element is located at the focus. Another property of the dish makes it a high-gain transmitter or receiver. See Figure 15-8.

Two arbitrary points on the parabola, P_1 and P_2, have been selected. The distance from the mouth of the parabola to the directrix is labeled d. We know that the distance from P_1 to the focus is equal to the distance from P_1 to the directrix. This distance is labeled d_1. The distance from P_1 to the mouth of the parabola is $d - d_1$. Suppose a microwave signal is being emitted from the focus toward P_1. How far does it travel getting to the mouth of the parabola? It travels d_1 units to P_1 and $d - d_1$ units on to the mouth.

$$\text{distance} = d_1 + (d - d_1) = d$$

Now consider P_2. A wave emitted from the focus travels d_2 units to P_2 and another $d - d_2$ units on to the mouth.

$$\text{distance} = d_2 + (d - d_2) = d$$

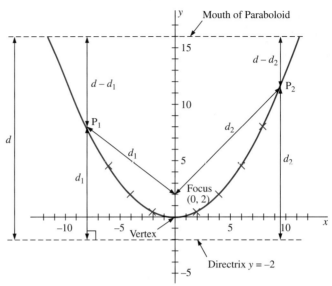

FIGURE 15-8
Distance from Focus to Mouth of Paraboloid

In the specific example in Figure 15-8, it is 16 units from the vertex to the mouth of the paraboloid and another 2 units from the vertex to the directrix. All waves emitted from the focus that reflect off the surface of the paraboloid travel 18 units to get to the mouth.

Since the distance from the focus to the mouth is the same for any wave reflected off of the paraboloid, the waves arrive at the mouth of the paraboloid traveling parallel to each other and in-phase with each other. There is no phase shift between the waves, and they do not tend to cancel each other along the axis of symmetry. This property makes the "dish" a high-gain receiver or transmitter.

SECTION 15.6 EXERCISES

1. For the parabola $y = 1/16x^2$, determine the focus, vertex, and directrix.
2. For the parabola $y = -2x^2$, determine the focus, vertex, and directrix.
3. Sketch the parabola $y = 1/16x^2$. Limit the height to 10.
4. Locate the focus and directrix.
5. Choose two points on the parabola. Show that each point is equidistant from the focus and directrix.
6. Draw a wave entering the mouth of the parabola parallel to the axis of symmetry. At the point where the wave contacts the parabola, draw the tangent and normal. Measure the angle of incidence and draw the angle of reflection. Does the reflected wave pass through the focus?
7. For the wave in Problem 6, calculate the distance traveled by the wave from the point where it enters the mouth to the focus.

SUMMARY

1. Equations of the form $y = ax^2 + bx + c$ are called quadratic equations.
2. The graph of a quadratic is called a parabola.
3. The x-intercepts are called the zeros or roots of the quadratic equation. The process of determining the x-intercepts is called finding the roots of the equation.
4. The roots of a quadratic can sometimes be found by factoring.

   Step 1: Set $y = 0$. $ax^2 + bx + c = 0$.
   Step 2: Factor the equation. $(x + d)(x + e) = 0$.
   Step 3: Set each factor equal to zero.
   $$x + d = 0 \quad x + e = 0$$
   Step 4: Solve each equation for x.
   $$x = -d \quad x = -e$$
   Step 5: Check each solution by substituting back into the original equation. y should equal zero for each of the roots.

5. The roots of a quadratic can be found by completing the square.

   Step 1: Set $y = 0$.
   Step 2: If the coefficient of the x^2 term is other than 1, divide each side of the equation by that coefficient.
   Step 3: Transpose the constant to the other side of the equation.
   Step 4: Take 1/2 the coefficient of the x term, square it, and add it to each side of the equation.
   Step 5: The left side of the equation is now a perfect square; factor it.
   Step 6: Take the square root of each side.
   Step 7: Solve for x.
   Step 8: Check each root.
   $$x = \frac{-b \pm \sqrt{b^2 - 4ac}}{2a}$$

6. The roots of a quadratic equation can always be found by using the quadratic formula.
7. In the quadratic formula, the quantity under the radical, $b^2 - 4ac$, is called the discriminant. If the discriminant is positive, the quadratic equation has two real roots. Its graph intersects the x-axis in two places. If the discriminant is 0, the quadratic has one real root. Its graph touches the x-axis in one place. If the discriminant is negative, there are two imaginary roots. The graph of the corresponding quadratic does not cross the x-axis.
8. To graph a parabola, follow this procedure.

Step 1. Calculate the roots by factoring, completing the square, or by applying the quadratic formula. Plot the roots.

Step 2. Calculate the axis of symmetry by averaging the roots or by plugging values for a and b into the equation $x = -b/(2a)$.

Step 3. Calculate the maximum or minimum point by plugging the value $x = -b/(2a)$ into the original equation. Plot the maximum or minimum point.

Step 4. Plot the y-intercept, c. Use symmetry to locate the corresponding point on the opposite side of the axis of symmetry.

Step 5. Choose more values of x as needed to provide a good spread of values and to help determine how wide or narrow the parabola is. Locate the corresponding points on the opposite side of the axis of symmetry.

9. The maximum point or minimum point of a parabola is called a vertex. The parabola wraps around a point on the axis of symmetry called the focus. The focus is p units from the vertex, where $x^2 = 4py$. p units on the other side of the focus is a line perpendicular to the axis of symmetry called the directrix.

10. In all parabolas, the distance from a point on the curve to the focus is the same as the perpendicular distance from the directrix to that point.

11. If the parabola is rotated about the y-axis, a three dimensional "dish" called a paraboloid is formed.

12. Any wave entering parallel to the axis of symmetry will be reflected to the focus of the parabola. Conversely, waves being emitted at the focus will be reflected by the parabola parallel to the axis of symmetry.

13. The distance from the focus to the mouth of a paraboloid is the same for any wave reflected off of the paraboloid. The waves arrive at the mouth of the paraboloid traveling parallel to each other and in-phase with each other. There is no phase shift between the waves, and they do not tend to cancel each other along the axis of symmetry. This property makes the "dish" a high-gain receiver or transmitter.

■ SOLUTIONS TO CHAPTER 15 SECTION EXERCISES

Solutions to Section 15.2, 15.3, and 15.4 Exercises

Problems 1 through 5 are the same in each exercise.

1. $x = 2$ $x = -4$
2. $x = -6$ $x = -10$
3. $x = 8$ $x = 5$
4. $x = 3/2$ $x = -6$
5. $x = 1/3$ $x = 5/4$

Solutions to Section 15.4 Exercises, Problems 6-11

6. $28 > 0$, two real roots, two x-intercepts
7. $68 > 0$, two real roots, two x-intercepts
8. 0, one real root, one x-intercept

9. −3 < 0, two imaginary roots, no x-intercepts
10. −8 < 0, two imaginery roots, no x-intercepts
11. 116 > 0, two real roots, two x-intercepts

Solutions to Section 15.5 Exercises

1. See Figure 15-9.

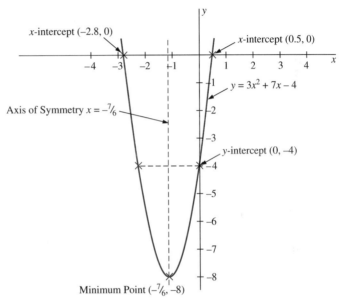

FIGURE 15-9
Solutions to Section 15.5 Exercises, Problem 1

2. See Figure 15-10.

FIGURE 15-10
Solutions to Section 15.5 Exercises, Problem 2

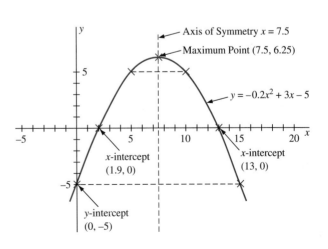

3. See Figure 15-11.

FIGURE 15-11
*Solutions to Section 15.5
Exercises, Problem 3*

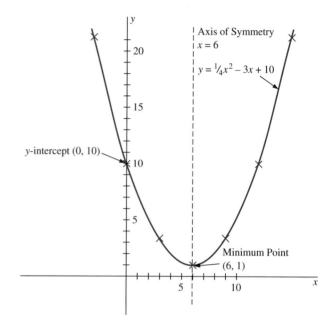

4. See Figure 15-12.

FIGURE 15-12
*Solutions to Section 15.5
Exercises, Problem 4*

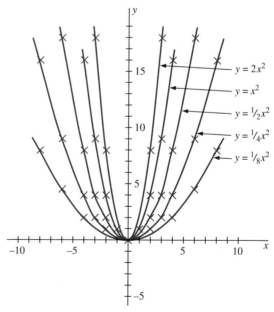

5. As *a* gets smaller, the parabola opens more widely.

FIGURE 15-13
Solutions to Section 15.5 Exercises, Problem 6

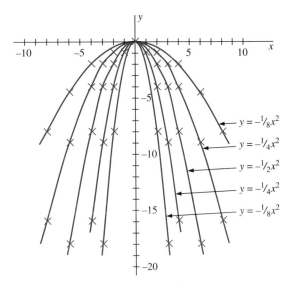

6. See Figure 15-13.
7. If *a* is negative, the parabola opens downward; if *a* is positive, the parabola opens upward.
8. True.
9. False. If *a* is positive, the parabola has a minimum point.
10. See Figure 15-14.

FIGURE 15-14
Solutions to Section 15.5 Exercises, Problem 10

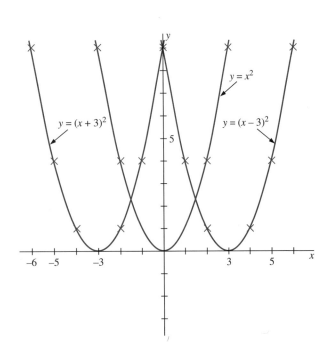

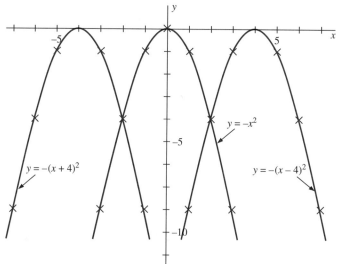

FIGURE 15-15
Solutions to Section 15.5 Exercises, Problem 11

11. See Figure 15-15.

Solutions to Section 15.6 Exercises

1. The curve opens upward. Vertex (0,0), focus (0,4), directrix $y = -4$.

2. The curve opens downward. Vertex (0,0), focus (0,−1/8), directrix $y = 1/8$.

3–6. See Figure 15-16.

7. 14

FIGURE 15-16
Solutions to Section 15.6 Exercises, Problems 3 through 6

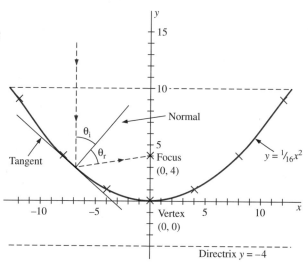

CHAPTER 15 PROBLEMS

Find the roots of each of these quadratics by factoring.

1. $y = x^2 - 2x - 99$
2. $y = x^2 + 10x + 24$
3. $y = x^2 - 10x + 21$
4. $y = 4x^2 - 4x - 15$
5. $y = -35x^2 + 64x - 21$
6. $y = -6x^2 - 7x + 28$

Find the roots of each of these quadratics by completing the square.

7. $y = x^2 - 2x - 99$
8. $y = x^2 + 10x + 24$
9. $y = x^2 - 10x + 21$
10. $y = 4x^2 - 4x - 15$
11. $y = -35x^2 + 64x - 21$
12. $y = -6x^2 - 7x + 28$

Use the quadratic formula to find the roots of each of these quadratic equations.

13. $y = x^2 - 2x - 99$
14. $y = x^2 + 10x + 24$
15. $y = x^2 - 10x + 21$
16. $y = 4x^2 - 4x - 15$
17. $y = -35x^2 + 64x - 21$
18. $y = -6x^2 - 7x + 28$

Graph these quadratic equations.

19. $y = x^2 - 2x - 99$
20. $y = x^2 + 10x + 24$
21. $y = x^2 - 10x + 21$
22. $y = 4x^2 - 4x - 15$
23. $y = -1/2x^2 - 3x + 8$
24. $y = -x^2 + 2x - 10$
25. Graph these quadratic equations on the same graph.
    A. $y = x^2 + 10$
    B. $y = x^2 - 10$
    C. $y = -x^2 + 10$
    D. $y = -x^2 - 10$
26. Graph these quadratic equations on the same graph.
    A. $y = x^2$
    B. $y = (x - 5)^2$

C. $y = (x + 5)^2$
D. $y = -(x - 5)^2$
E. $y = -(x + 5)^2$
F. $y = -x^2$

27. Graph these quadratic equations on the same graph.
    A. $y = -(x - 5)^2$
    B. $y = -2(x - 5)^2$
    C. $y = -3(x - 5)^2$
    D. $y = -1/3(x - 5)^2$
    E. $y = 1/9\, x^2$

28. Graph these quadratic equations on the same graph.
    A. $y = -(x - 5)^2$
    B. $y = -(x - 5)^2 + 2$
    C. $y = -(x - 5)^2 - 2$
    D. $y = -(x - 5)^2 + 4$
    E. $y = -(x - 5)^2 - 4$

29. Describe the effect of the magnitude of a on the parabola.

30. Describe the effect of the magnitude of c on the parabola.

31. For the parabola $y = -x^2/12$, determine the focus, vertex, and directrix. Graph and label.

32. For the parabola $y = -4x^2$, determine the focus, vertex, and directrix. Graph and label.

For $y = -1/12\, x^2$:

33. Sketch the parabola. Limit the height to 12.

34. Locate and label the focus, vertex, and directrix.

35. Choose two points on the parabola. Show that each point is equidistant from the focus and directrix.

36. Draw two waves entering the mouth of the parabola parallel to the axis of symmetry. At the point where each wave contacts the parabola, draw the tangent and normal. Measure the angle of incidence and draw the angle of reflection. Do the reflected waves pass through the focus?

37. For the waves in Problem 36, calculate the distance traveled by the waves from the point where they enter the mouth to the focus.

16
Number Systems

CHAPTER OUTLINE

- **16.1** Counting in Binary
- **16.2** Converting Binary to Decimal
- **16.3** Converting Decimal to Binary
- **16.4** Counting in Hexadecimal
- **16.5** Converting Binary to Hexadecimal
- **16.6** Converting Hexadecimal to Binary
- **16.7** Binary Addition
- **16.8** Binary 1's Complement Subtraction
- **16.9** Binary 2's Complement Subtraction
- **16.10** Signed 2's Complement Numbers
- **16.11** Binary Coded Decimal (BCD)

OBJECTIVES

After completing this chapter, you should be able to

1. count in binary.
2. convert binary numbers to decimal.
3. convert decimal numbers to binary.
4. count in hexadecimal.
5. convert binary numbers to hexadecimal.
6. convert hexadecimal numbers to binary.
7. add binary numbers.
8. subtract binary numbers using 1's complement.
9. subtract binary numbers using 2's complement.
10. use signed 2's complement numbers to jump ahead or back in a computer program.
11. convert decimal numbers to binary coded decimal.
12. convert binary coded decimal to decimal.
13. add binary coded decimal numbers.

NEW TERMS TO WATCH FOR

binary
bit
least significant bit (LSB)
most significant bit (MSB)
radix
hexadecimal
half adder
full adder
complement method

1's complement
overflow
end around carry (EAC)
2's complement
signed 2's complement
address
program counter
binary coded decimal (BCD)

16.1 ■ COUNTING IN BINARY

Binary numbers use only two digits, 0 and 1. Because of this, binary numbers are easily adapted to electronics. For example, a 1 can be represented by a high voltage and a 0 by a low voltage; a 1 by a closed switch and a 0 by an open switch; a 1 by a transistor that is turned off and a 0 by a transistor that is turned on; a 1 by a light-emitting diode (LED) that is turned on and a 0 by an LED that is turned off; and so on.

Let's review how we count in decimal and compare it to counting in binary. Decimal is a base 10 number system, which means that we have 10 digits to work with, 0–9. To count in decimal start in the first column with 0 and count up to 9. No more digits are available, so reset the first column to 0 and add 1 to the next column to the left. After 9 comes 10. Now start counting in the first column while leaving the new column at 1. After 9 comes 10, 11, 12, and so on until 19 is reached. Reset the first column and add 1 to the next column to the left. After 19 comes 20, 21, 22, and so on. When the first two columns are maximized at 99, reset both of the columns and add on the third column to the left. After 99 comes 100. Now start again on the first column while the second and third remain fixed at 10. After 99 comes 100, 101, 102, and so on.

FIGURE 16-1
Counting in Binary

```
0000
0001  ⎰ Maximum count in column 1.
0010  ⎱ Reset to zero. Add 1 to column 2.
0011  ⎰ Maximum count in columns 1 and 2.
0100  ⎱ Reset to zero. Add 1 to column 3.
0101  ⎰ Maximum count in column 1.
0110  ⎱ Reset to zero. Add 1 to column 2.
0111  ⎰ Maximum count in columns 1, 2, and 3.
1000  ⎱ Reset to zero. Add 1 to column 4.
1001  ⎰ Maximum count in column 1.
1010  ⎱ Reset to zero. Add 1 to column 2.
1011  ⎰ Maximum count in columns 1 and 2.
1100  ⎱ Reset to zero. Add 1 to column 3.
1101  ⎰ Maximum count in column 1.
1110  ⎱ Reset to zero. Add 1 to column 2.
1111    Maximum count in columns 1–4.
```

16.1 COUNTING IN BINARY

Binary is a base 2 number system, which means that we have 2 digits to work with, 0 and 1. Refer to Figure 16-1 as we count in binary.

To count in binary, start in the first column with 0 and count up to 1. That is all the digits available, so reset the first column to 0 and add 1 to the next column to the left. After 1 comes 10. It is better to call this number "one zero" or "one oh" rather than ten, because ten is taken as a decimal number. Now start counting in the first column while leaving the new column at 1. After 1 comes 10, 11. We have run out of digits in the first two columns very quickly. Reset the first two columns and add one to the next column to the left. After 11 comes 100. Now start again on the first column while the second and third columns remain fixed at 10. After 100 comes 101, 110, 111. After 111 comes 1000, 1001.

EXAMPLE 16.1 Count in binary from 10101 to 11111.

Solution

10101
10110
10111
11000
11001
11010
11011
11100
11101
11110
11111

In Figure 16-1, note that the first column (right column) changes (toggles) every time. The second column toggles every second time; the third column every fourth time; the fourth column every eighth time. This pattern continues and provides an excellent way to check your counting in binary.

Binary dig**it** is contracted into **bit**. 1001 is a four bit binary number. The right-most bit is called the **least significant bit** (**LSB**) and the left-most bit is called the **most significant bit** (**MSB**). See Figure 16-2.

FIGURE 16-2
LSB and MSB

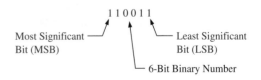

SECTION 16.1 EXERCISES

1. Count in binary from 0 to 10000.
2. Count in binary from 1011010 to 1100011.

16.2 ■ CONVERTING BINARY TO DECIMAL

In a base 10 number system, each decimal place to the left is worth 10 times the preceding place value: ..., 1000, 100, 10, 1. In a base 2 number system, each binary place to the left is worth 2 times the preceding place value: ..., 64, 32, 16, 8, 4, 2, 1. Figure 16-3 contrasts these two systems.

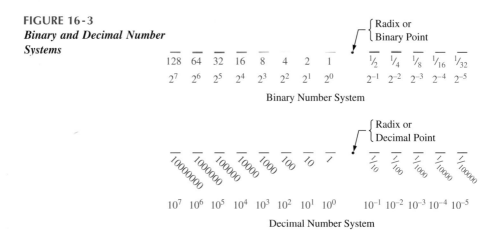

FIGURE 16-3
Binary and Decimal Number Systems

The place values can also be written in exponential form. In each number system, the first place to the left of the **radix** (decimal point or binary point) is the base number to the zero power or 1. Increment the exponent for each step to the left and decrement the exponent for each step to the right. Exponents for the places to the right of the radix are negative. Knowing the value of each place gives us a logical, straightforward method for converting binary numbers into their equivalent decimal number.

EXAMPLE 16.2   Convert binary 110001 into decimal.

Solution   Write the place value beneath each bit.

$$1\ 1\ 0\ 0\ 0\ 1$$
$$32\ 16\ 8\ 4\ 2\ 1$$

A 1 in a binary place means that value is contained in the decimal equivalent, and a zero means it is not. This binary number has a 32, a 16, no 8, no 4, no 2, and a 1.

$$32 + 16 + 1 = 49$$

Subscripts 2 and 10 are used to distinguish binary numbers from decimal numbers.

$$11001_2 = 49_{10}$$

EXAMPLE 16.3 Convert 1000110000_2 into decimal.

Solution Write the place value beneath each bit.

$$1 \; 0 \; 0 \; 0 \; 1 \; 1 \; 0 \; 0 \; 0 \; 0.$$
$$512 \; 256 \; 128 \; 64 \; 32 \; 16 \; 8 \; 4 \; 2 \; 1$$
$$512 + 32 + 16 = 560$$
$$1000110000_2 = 560_{10}$$

With a 2-bit number we can count to 11_2 or 3_{10}, and with a 3-bit number we can count to 111_2 or 7. What about an 8-bit number or a 16-bit number? There is a pattern here.

$$N = 2^n - 1$$

where n is the number of bits in the binary number and N is the maximum count.

EXAMPLE 16.4 How high can you count with an 8-bit binary number?

Solution Given: $n = 8$.
Find: N.
$$N = 2^n - 1$$
$$N = 2^8 - 1 = 255$$

EXAMPLE 16.5 How high can you count with a 16-bit binary number?

Solution Given: $n = 16$.
Find: N.
$$N = 2^n - 1$$
$$N = 2^{16} - 1 = 65\,535$$

With a 3-bit number we count from 000 to 111—eight different combinations counting all zeros. With a 4-bit number we count from 0000 to 1111—sixteen different combinations counting all zeros. The pattern here is

$$N = 2^n$$

where n is the number of bits and N is the number of possible combinations, including all zeros.

EXAMPLE 16.6 How many numbers are available using 8 binary bits?

Solution Given: $n = 8$.
Find: N.

$$N = 2^n$$
$$N = 2^8 = 256$$

EXAMPLE 16.7 How many numbers are available using 16 binary bits?

Solution Given: $n = 16$.
Find: N.

$$N = 2^n$$
$$N = 2^{16} = 65\,536$$

So, with 8 bits we can count to 255, and 256 numbers are represented. With 16 bits we can count to 65 535, and 65 536 numbers are represented.

SECTION 16.2 EXERCISES Convert into decimal.

1. 10101_2
2. 110011001_2
3. How high can you count using 7-bit binary numbers?
4. How many different combinations are there using 7-bit binary numbers?
5. How high can you count using 20-bit binary numbers?
6. How many different combinations are there using 20-bit binary numbers?

16.3 ■ CONVERTING DECIMAL TO BINARY

Two methods of converting decimal to binary are presented here.

Method 1 Begin at the binary point and label the values of each place to the left until the decimal number is surpassed. Place a 1 in the highest value that does not exceed the decimal number. Subtract that number from the decimal number to see how much is left to convert. Place a 1 in the largest binary place that does not exceed the difference. Continue until there is nothing left.

EXAMPLE 16.8 Convert 34_{10} to binary.

Solution Count up in binary until 34 is surpassed.

__ __ __ __ __ __ __
64 32 16 8 4 2 1

Place a 1 in the 32's place.

$$\frac{\quad 1 \quad \_\_ \_\_ \_\_ \_\_ \_\_}{64 \; 32 \; 16 \; \; 8 \; \; 4 \; \; 2 \; \; 1}$$

Subtract 32 to see how much remains.

$$34 - 32 = 2$$

Place a 1 in the 2's place. The result does not contain a 16 or 8 or 4. Place zeros in those places.

$$\frac{\quad 1 \; \; 0 \; \; 0 \; \; 0 \; \; 1 \quad \_\_}{64 \; 32 \; 16 \; \; 8 \; \; 4 \; \; 2 \; \; 1}$$

Subtract 2 from the remainder to see how much is left.

$$2 - 2 = 0$$

The process is complete. Place a zero in the 1's slot.

$$\frac{\quad 1 \; \; 0 \; \; 0 \; \; 0 \; \; 1 \; \; 0}{64 \; 32 \; 16 \; \; 8 \; \; 4 \; \; 2 \; \; 1}$$

$$34_{10} = 100010_2$$

Check: The result has a 1 in the 2's place and a 1 in the 32's place. $2 + 32 = 34$. It checks!

Method 2: Repetitive Division by 2 A more automated, less tedious method for converting decimal to binary is the repetitive division by 2 method. Divide the decimal number by 2 and list the remainder. This remainder will be the LSB of the result. Divide the quotient (don't worry about the remainder for a moment) by 2 and again list the remainder. Continue the division process until the quotient is 0. The remainders form the result with the last remainder as the MSB.

EXAMPLE 16.9 Convert 47_{10} to binary.

Solution

$$47 \div 2 = 23 \quad \text{R 1 LSB}$$
$$23 \div 2 = 11 \quad \text{R 1}$$
$$11 \div 2 = 5 \quad \text{R 1}$$
$$5 \div 2 = 2 \quad \text{R 1}$$
$$2 \div 2 = 1 \quad \text{R 0}$$
$$1 \div 2 = 0 \quad \text{R 1 MSB}$$

Read the result from MSB to LSB (up).

$$47_{10} = 101111_2$$

Check: Convert the result back to decimal.

$$1 + 2 + 4 + 8 + 32 = 47$$

EXAMPLE 16.10 Express 149_{10} as an 8-bit binary number.

Solution

$$
\begin{array}{rll}
149 \div 2 = 74 & \text{R 1 LSB} \\
74 \div 2 = 37 & \text{R 0} \\
37 \div 2 = 18 & \text{R 1} \\
18 \div 2 = 9 & \text{R 0} \\
9 \div 2 = 4 & \text{R 1} \\
4 \div 2 = 2 & \text{R 0} \\
2 \div 2 = 1 & \text{R 0} \\
1 \div 2 = 0 & \text{R 1 MSB} \\
149_{10} = 10010101_2
\end{array}
$$

All 8 bits were needed to express 149 in binary. Otherwise, zeros would have been used in the upper bits.

Check: $1 + 4 + 16 + 128 = 149$

SECTION 16.3 EXERCISES

Convert to binary using method 1.

1. 12_{10}
2. 14_{10}
3. 49_{10}
4. 148_{10}
5. 478_{10}

Convert to binary using method 2.

6. 12_{10}
7. 14_{10}
8. 49_{10}
9. 148_{10}
10. 478_{10}

16.4 ■ COUNTING IN HEXADECIMAL

Hexadecimal is a base 16 number system, which means we have 16 digits to work with: 0, 1, 2, 3, 4, 5, 6, 7, 8, 9, A, B, C, D, E, F. It is awkward at first, having letters A–F mixed in with the numerals. Refer to Figure 16-4 as we count in hex.

To count in hexadecimal, start in the first column with 0 and count up until all the digits have been used. Reset the first column and add 1 to the next column on the left. After 9 comes A, B, C, D, E, F, 10, 11, 12, 13 . . . When all the digits have been used in

16.4 COUNTING IN HEXADECIMAL

```
 0                               1A
 1                               1B
 2                               1C
 3                               1D
 4                               1E
 5                               1F  ⎰ Maximum Count Column 1.
 6                               20  ⎱ Reset to Zero. Add 1 to Column 2.
 7                               21
 8
 9
 A                               FF
 B                              100  ⎰ Maximum Count Columns 1 and 2.
 C                              101  ⎱ Reset to Zero. Add 1 to Column 3.
 D                              102
 E                              103
 F  ⎰ Maximum Count Column 1.
10  ⎱ Reset to Zero. Add 1 to Column 2.
11
12
13
14
15
16
17
18
19
```

FIGURE 16-4
Counting in Hexadecimal

the first column, reset to zero and add one to the next column to the left. After 1F comes 20. After FF comes 100.

EXAMPLE 16.11   Count in hex from 3FE8 to 4000.

Solution

$$3FE8$$
$$3FE9$$
$$3FEA$$
$$3FEB$$
$$3FEC$$
$$3FED$$
$$3FEF$$
$$3FF0$$
$$3FF1$$
$$3FF2$$

3FF3
3FF4
3FF5
3FF6
3FF7
3FF8
3FF9
3FFA
3FFB
3FFC
3FFD
3FFE
3FFF
4000

SECTION 16.4 EXERCISES

1. Count in hexadecimal from 0 to 20.
2. Count from 4ADC to 4B00.

16.5 ■ CONVERTING BINARY TO HEXADECIMAL

Hexadecimal has 16 different digits, 0–9 and A–F, where F represents 15. A 4-bit binary number has 16 different combinations and we can count to 1111 or 15. Hexadecimal and a 4-bit binary number are quite compatible. See Figure 16-5.

FIGURE 16-5
Hexadecimal versus 4-Bit Binary

Hexadecimal	4-Bit Binary
0	0000
1	0001
2	0010
3	0011
4	0100
5	0101
6	0110
7	0111
8	1000
9	1001
A	1010
B	1011
C	1100
D	1101
E	1110
F	1111
Maximum Count Column 1	Maximum Count Columns 1–4

To convert binary to hexadecimal, start at the binary point and work to the left, marking off groups of four. Convert each group of four bits individually to the corresponding hex digit.

EXAMPLE 16.12 Convert 101010000010_2 to hexadecimal.

Solution Start at the binary point and mark off groups of four.

$$1010\ 1000\ 0010.$$

Convert each group to its corresponding hex digit.

$$A\ \ 8\ \ 2$$
$$101010000010_2 = A82_{16}$$

EXAMPLE 16.13 Represent 1110110000011011_2 in hexadecimal.

$$1110\ 1100\ 0001\ 1011$$
$$E\ \ \ C\ \ \ 1\ \ \ B$$
$$1110110000011011_2 = EC1B_{16}$$

SECTION 16.5 EXERCISES

Convert these 8-bit binary numbers to hexadecimal.

1. 10010001
2. 11011100
3. 00101111
4. 10101011
5. 11100011

Convert these 16-bit numbers to hexadecimal.

6. 10000101 01111100
7. 11111111 11111111
8. 11001000 01100011
9. 01101011 11011110
10. 01001110 00110111

16.6 ■ CONVERTING HEXADECIMAL TO BINARY

Converting hexadecimal to binary is just as easy as converting binary to hexadecimal. Replace each hex digit with its 4-bit binary code. Refer to Figure 16-5 as needed.

EXAMPLE 16.14 Convert $F8B9_{16}$ into binary.

Solution

$$F\ \ \ \ 8\ \ \ \ B\ \ \ \ 9$$
$$1111\ 1000\ 1011\ 1001$$
$$F8B9_{16} = 1111100010111001_2$$

Binary is used in electronics; hexadecimal is used to represent or handle binary numbers.

We have learned to convert decimal to binary, binary to decimal, hexadecimal to binary, and binary to hexadecimal as indicated in Figure 16-6. An arrow pointing from one system to another indicates that we have covered that conversion. Binary-coded decimal is covered later in this chapter.

FIGURE 16-6
Number System Conversions

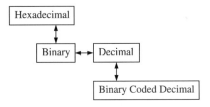

So how do you convert from hex to decimal? Procedures exist for doing that directly, but Figure 16-6 shows an indirect path to follow. Converting hex to binary is easy, and we already know how to convert binary to decimal.

EXAMPLE 16.15 Convert 153_{10} to hexadecimal.

Solution First convert 153_{10} to binary.

$$153 \div 2 = 76 \quad R\ 1 \quad LSB$$
$$76 \div 2 = 38 \quad R\ 0$$
$$38 \div 2 = 19 \quad R\ 0$$
$$19 \div 2 = 9 \quad R\ 1$$
$$9 \div 2 = 4 \quad R\ 1$$
$$4 \div 2 = 2 \quad R\ 0$$
$$2 \div 2 = 1 \quad R\ 0$$
$$1 \div 2 = 0 \quad R\ 1 \quad MSB$$
$$10011001_2$$

Now convert 10011001_2 to hexadecimal.

$$\begin{array}{cc} 1001 & 1001 \\ 9 & 9 \end{array}$$

$$153_{10} = 99_{16}$$

Check: If your calculator has the number systems built into it, it can be used to check these problems. Set your calculator to decimal mode if it is not already in it. Enter 153. Now set the calculator to hexadecimal mode. The display should read 99.

SECTION 16.6 EXERCISES

Convert these hexadecimal numbers to binary.

1. 38
2. 7A
3. 9C
4. F0DB
5. C0DA
6. 18FE

Convert.

7. $A7_{16}$ to decimal.
8. 97_{10} to hexadecimal.

16.7 ■ BINARY ADDITION

In Figure 16-7, the rectangle represents a circuit called a **half adder.** A half adder adds two bits and produces two outputs, a "sum" and a "carry" to the next column. The truth table for the circuit shows how the circuit will behave for every possible combination of inputs A and B, the two bits being added. The first three lines do not produce a carry: $0 + 0 = 0 \quad 0 + 1 = 1 \quad 1 + 0 = 1$. In the last line, $1 + 1$ is 10 or 2: a sum of 0 and a carry of 1.

FIGURE 16-7
Half Adder

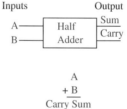

In Figure 16-8, the rectangle represents a circuit called a **full adder.** A full adder adds three bits, A and B and a carry from a previous addition called C_0 or C_{in}. Just like the half-adder, it produces two outputs, a sum and a carry to the next column. The truth table for the circuit shows how the circuit will behave for every possible combination of inputs A and B, the two bits being added. The last line is the trickiest: $1 + 1 + 1$ is 11 or 3, a sum of 1 and a carry of 1.

FIGURE 16-8
Full Adder

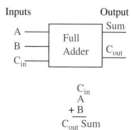

EXAMPLE 16.16 Add these binary numbers.

$$\begin{array}{r} \text{Carry} \quad 1011 \\ 1011 \\ +\ \ 1001 \\ \hline 10100 \end{array}$$

Check: 11 + 9 = 20.

EXAMPLE 16.17 Add these binary numbers.

$$\begin{array}{r} \text{Carry} \quad 1100 \\ 1110 \\ +\ \ 1101 \\ \hline 11011 \end{array}$$

Check: 14 + 13 = 27.

SECTION 16.7 EXERCISES

Add these binary numbers.
1. 1011 + 0010 =
2. 0100 + 1110 =
3. 1011 + 0111 =
4. 1010 + 1010 =

Convert to binary and add. Express the result in hexadecimal.

5. $9AEB_{16} + 3FCD_{16} =$
6. $73DC_{16} + 58AA_{16} =$
7. $6B3B_{16} + 41DC_{16} =$

16.8 ■ BINARY 1'S COMPLEMENT SUBTRACTION

Computers handle subtraction by a **complement method.** A complement method offers two advantages over regular subtraction.

1. In the complement approach to subtraction, the process is the same whether the minuend is smaller or larger than the subtrahend. Extra circuitry is not needed to determine which is larger because it does not matter.
2. The subtraction problem is converted into an addition problem. Adder circuits can be used for subtraction, minimizing circuitry.

The **1's complement** of 1 is 0; the 1's complement of 0 is 1. In other words, to take the 1's complement of a binary number, change each bit. To subtract using 1's complement, we need the concept of **overflow.** If two 4-bit numbers are being added, and a carry into the 5th column is produced, overflow has occured.

To subtract using 1's complement, add the 1's complement of the subtrahend to the minuend. If overflow occurs, the answer is positive and an **end around carry** (EAC) must be performed. To perform an EAC, add the overflow to the first column.

16.8 BINARY 1'S COMPLEMENT SUBTRACTION

EXAMPLE 16.18 Use the 1's complement method to subtract: $1010 - 1001 =$.

Solution The 1's complement of 1001 is 0110. Add 0110 to 1010.

$$\begin{array}{r} 1010 \\ -1001 \\ \hline \text{overflow} \rightarrow \end{array} \quad \begin{array}{r} 1010 \\ +0110 \\ \hline 1\,0000 \end{array} \quad \text{1's complement}$$

$$\begin{array}{r} \text{EAC} \quad +1 \\ \hline 0001 \end{array}$$

Two 4-bit numbers are being added. The carry into the fifth column is overflow. The answer is positive. Do an EAC.

$$1010 - 1001 = 0001$$

Check: $10 - 9 = 1$.

EXAMPLE 16.19 Use the 1's complement method to subtract: $11011101 - 1100010 =$.

Solution $11011101 - 1100010$

The minuend is an 8-bit number and the subtrahend is a 7-bit number. Add a leading zero to the subtrahend to make two 8-bit numbers. That leading zero becomes a 1 when the 1's complement is formed.

$$\begin{array}{r} 11011101 \\ -\ 1100010 \\ \hline \text{fill in leading zero} \end{array} \nearrow \quad \begin{array}{r} 11011101 \\ -01100010 \\ \hline \text{overflow} \rightarrow \end{array} \quad \begin{array}{r} 11011101 \\ +10011101 \\ \hline 1\,01111010 \end{array} \leftarrow \text{1's complement}$$

$$\begin{array}{r} \text{EAC} \quad +1 \\ \hline 01111011 \end{array}$$

$$11011101 - 1100010 = 01111011$$

Check: $221 - 98 = 123$.

If no overflow occurs, the answer is negative. Take the 1's complement to get the true magnitude of the answer.

EXAMPLE 16.20 Use the 1's complement method to subtract: $101 - 10100 =$.

Solution Even though it is obvious that the answer is going to be negative, the process is the same as before.

$$\begin{array}{r} 101 \\ -10100 \\ \hline \text{no overflow} \rightarrow \end{array} \quad \begin{array}{r} 00101 \\ +01011 \\ \hline 0\,10000 \end{array} \leftarrow \text{1's complement}$$

$$01111 \leftarrow \text{1's complement}$$

$$101 - 10100 = -01111$$

Check: $5 - 20 = -15$.

EXAMPLE 16.21 Use the 1's complement method to subtract: $00110110 - 11100110$.

Solution

$$\begin{array}{r} 00110110 \\ -11100110 \\ \hline \text{no overflow} \rightarrow \end{array} \quad \begin{array}{r} 00110110 \\ +00011001 \\ \hline 0\ 01001111 \end{array} \leftarrow \text{1's complement}$$

$$10110000 \leftarrow \text{1's complement}$$

$$00110110 - 11100110 = -10110000$$

Check: $54 - 230 = -176$.

SECTION 16.8 EXERCISES

Subtract using 1's complement.

1. $1100 - 0101 =$
2. $1101 - 0110 =$
3. $101 - 1110 =$
4. $1001 - 1111 =$
5. $C9_{16} - 8A_{16} =$
6. $6BD5_{16} - DDF_{16} =$
7. $9A_{16} - B7_{16} =$
8. $BD04_{16} - E008_{16} =$

16.9 ■ BINARY 2'S COMPLEMENT SUBTRACTION

There are two methods for taking the **2's complement** of a binary number.

Method 1: Take the 1's complement and add 1. In other words, to take the 2's complement, change each bit and add 1.

Method 2: Start at the binary point and work to the left. Leave each bit unchanged until the first 1 digit is passed. After the first 1 digit, change each bit.

EXAMPLE 16.22 Write the 2's complement of 1011000.

Solution Method 1: The 1's complement of 1011000 is 0100111. Add 1 to 0100111.

$$0100111 + 1 = 0101000$$

Method 2: 10111000 Working to the left, leave each bit unchanged until the first 1 is passed, in this case the first four bits. Change the remaining three. The 2's complement of 1011000 is 0101000.

To subtract using the 2's complement method, add the 2's complement of the subtrahend to the minuend. If overflow occurs, the answer is positive.

EXAMPLE 16.23 Use the 2's complement method to subtract: $1010 - 1001 = $.

Solution The 2's complement of 1001 is 0111. Add 0111 to 1010.

$$\begin{array}{r} 1010 \\ -1001 \\ \hline \text{overflow} \rightarrow 1\,0001 \end{array} \qquad \begin{array}{r} 1010 \\ +0111 \leftarrow \text{2's complement} \\ \hline \end{array}$$

Two 4-bit numbers are being added. The carry into the fifth column is overflow. The answer is positive.

$$1010 - 1001 = 0001$$

Check: $10 - 9 = 1$.

EXAMPLE 16.24 Use the 2's complement method to subtract: $11011101 - 1100010 = $.

Solution $11011101 - 1100010$
The minuend is an 8-bit number and the subtrahend is a 7-bit number. Add a leading zero to the subtrahend to make two 8-bit numbers. That leading zero becomes a 1 when the 2's complement is formed.

$$\begin{array}{r} 11011101 \\ -\ 1100010 \\ \hline \text{fill in leading zero} \nearrow \end{array} \qquad \begin{array}{r} 11011101 \\ -01100010 \\ \hline \text{overflow} \rightarrow \end{array} \qquad \begin{array}{r} 11011101 \\ +10011110 \\ \hline 1\,01111011 \end{array}$$

Overflow occurred; the answer is positive.

$$11011101 - 1100010 = 01111011$$

Check: $221 - 98 = 123$.

If no overflow occurs, the answer is negative. Take the 2's complement of the preliminary result to get the true magnitude of the answer.

EXAMPLE 16.25 Use the 2's complement method to subtract: $101 - 10100 = $.

Solution Even though it is obvious that the answer is going to be negative, the process is the same as before.

$$\begin{array}{r} 101 \\ -10100 \\ \hline \text{no overflow} \rightarrow \end{array} \qquad \begin{array}{r} 00101 \\ +01100 \leftarrow \text{2's complement} \\ \hline 0\,10001 \end{array}$$

$$01111 \leftarrow \text{2's complement}$$
$$101 - 10100 = -0.1111$$

Check: $5 - 20 = -15$.

EXAMPLE 16.26 Use the 2's complement method to subtract: $00110110 - 11100110$

Solution

$$\begin{array}{r} 00110110 \\ -11100110 \\ \hline \text{no overflow} \end{array} \rightarrow \begin{array}{r} 00110110 \\ +00011010 \\ \hline 001010000 \end{array} \leftarrow \text{2's complement}$$

$$10110000 \quad \text{2's complement}$$

$$00110110 - 11100110 = -10110000$$

Check: $54 - 230 = -176$.

SECTION 16.9 EXERCISES

Subtract using 2's complement.

1. $1100 - 0101 =$
2. $1101 - 0110 =$
3. $101 - 1110 =$
4. $1001 - 1111 =$
5. $C9_{16} - 8A_{16} =$
6. $6BD5_{16} - DDF_{16} =$
7. $9A_{16} - B7_{16} =$
8. $BD04_{16} - E008_{16} =$

16.10 ■ SIGNED 2'S COMPLEMENT NUMBERS

Microprocessors use **signed 2's complement** numbers. Signed numbers means that one bit has been reserved to represent the sign of the number, usually the most significant bit. If the most significant bit is 1, the number represented in the remaining bits is negative, and the number is written in 2's complement form. If the most significant bit is 0, the number represented in the remaining bits is positive, and the number is written in true magnitude form (what you see is what you get). See Figure 16-9.

FIGURE 16-9
Signed 2's Complement

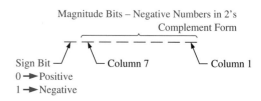

EXAMPLE 16.27 The following numbers are expressed in 5-bit signed 2's complement form, with the MSB as the sign bit. Write the equivalent decimal number: 01100 00011 01111 10011 10000 11111 11000.

Solution

$$01100 = +1100 = +(8 + 4) = +12$$
$$00011 = +0011 = +(2 + 1) = +3$$
$$01111 = +1111 = +(8 + 4 + 2 + 1) = +15$$

16.10 SIGNED 2'S COMPLEMENT NUMBERS

The MSB of the remaining numbers is 1. The numbers are negative in 2's complement form. Take the 2's complement of the complete number to get the true magnitude.

Number	2's Complement	Decimal Equivalent
10011	01100	$-(8+4) = -12$
10000	10000	-16
11111	00001	-1
11000	01000	-8

To express a negative number in signed 2's complement form, write the magnitude of the number in binary. Take the 2's complement. The MSB will turn out negative (1) as long as the magnitude is not too large.

EXAMPLE 16.28 Write these decimal numbers in 5-bit signed 2's complement form. $+9$ $+4$ $+0$ $+14$ $+15$ $+16$ -5 -2 -12 -14 -15 -16 -17.

Solution

Decimal Number	Magnitude in Binary	2's Complement Form
+9	01001	01001
+4	00100	00100
+0	00000	00000
+14	01110	01110
+15	01111	01111
+16	10000	

+16 is too large for our 5-bit number system because a 1 has appeared in the most significant bit, making the number appear negative. We learned in the last example that 10000 is -16, not $+16$. In our 5-bit system, positive numbers can range from 0 to 15 inclusive.

-5	00101	11011
-2	00010	11110
-12	01100	10100
-14	01110	10010
-15	01111	10001
-16	10000	10000
-17	10001	

17 is too large for our 5-bit 2's complement system. When the 2's complement is taken (01111), the MSB is 0, making the number appear positive. In our system, negative numbers can range from -1 to -16 inclusive.

SECTION 16.10 EXERCISES, Part 1

The following questions apply to a 6-bit signed 2's complement system (1 sign bit followed by 5 magnitude bits).
Write the equivalent decimal number.

1. 011011
2. 011111
3. 111111
4. 100000
5. 111000
6. What range of positive numbers can be represented by this system?
7. What range of negative numbers can be represented by this system?

Write these decimal numbers in 6-bit signed 2's complement form.

8. $+16$
9. -16
10. $+30$
11. -30

These questions refer to an 8-bit signed 2's complement system with the MSB as the sign bit. Negative numbers are written in 2's complement form. Write these numbers in 8-bit signed 2's complement form.

12. 98
13. 47
14. -25
15. -76
16. -127

Convert these 8-bit signed 2's complement numbers to decimal.

17. 01101100
18. 00000111
19. 10110110
20. 11111000
21. AE_{16}
22. FC_{16}

16.10 SIGNED 2'S COMPLEMENT NUMBERS ■ 533

In microcomputers, each location in memory has its own **address,** its own identifying number. Each memory location contains a binary number (1s and 0s) that can represent an instruction of a program the computer is running or that can represent data to be used in the program.

Here is a list of memory locations and the contents of those locations. The memory addresses are 16-bits wide (contain 16 bits), and each of the locations in memory contains 8 bits of information (8 1s and 0s). The addresses are listed in hex to keep from writing out 16 bits. Four hex digits are needed to represent each address. The contents of memory are listed in binary.

Address (Hex)	Contents (Binary)
0934	10010010
0935	01001000
0936	11001001
0937	01101111
0938	11100011
0939	00100000
093A	10100111
093B	00110011
093C	11000001
093D	00111001
093E	10011001
093F	10010010
0940	00001000
0941	11111000

Memory location $093B_{16}$ contains 00110011_2.

A **program counter** keeps track of where the computer is in its program by storing the memory address that contains the next instruction to be executed. To cause the computer to jump back to execute some sequence over again or to jump forward to avoid a section of program that does not need to be run at this time, a signed 2's complement number is added to the program counter. If the signed 2's complement number is negative, the computer will jump backward to a previous part of the program. If the number is positive, the computer will jump forward to a new part of the program.

Suppose the program counter contains 093C. When the computer is ready for its next instruction, it will take the data from memory location 093C, in this case 11000001, and decode it to see what to do. Once that instruction is read by the computer, the program counter will be incremented to contain the address of the next instruction.

EXAMPLE 16.29 The program counter contains 093D. The next instruction would normally be taken from that location in memory, but the program causes 0004 to be added to the program counter. From which address will the next instruction be taken and what will that instruction be?

Solution The new contents of the program counter will be 093D+0004. Let's add these in binary.

$$\begin{array}{r} 0000\ 1001\ 0011\ 1101 \\ +\ 0000\ 0000\ 0000\ 0100 \\ \hline 0000\ 1001\ 0100\ 0001 \end{array}$$

The program counter will contain 0941. The next instruction is the contents of 0941, which is 11111000.

EXAMPLE 16.30 The program counter contains 093D. What number should be added to the program counter to cause the program to jump backwards to location 0935? The number should be in signed 2's complement form.

Solution The program counter will have to step back 8 steps to get to 0935. Write 8 as a 16-bit binary number and take the 2's complement.

$$8_{10} = 00000000\ 00001000_2$$
$$\phantom{8_{10} = }11111111\ 11111000 \quad \text{2's complement}$$

FFF8 should be added to the program counter.

Check: Add FFF8 to 093D and see if it yields 0935.

$$\begin{array}{r} 0000\ 1001\ 0011\ 1101 \\ +\ 1111\ 1111\ 1111\ 1000 \\ \hline 0000\ 1001\ 0011\ 0101 = 0935_{16} \end{array}$$

EXAMPLE 16.31 The program instructs the computer to add the contents of memory location 0936 to the contents of location 093A and store the results in location 0940. What will be stored in location 0940 after the instruction has been run?

Solution 0936 contains 11001001 and 093A contains 10100111.

$$\begin{array}{r} 1111 \\ 11001001 \\ +\ 10100111 \\ \hline 1\ 01110000 \end{array}$$

An overflow occurred. The contents of memory location 0940 is changed to 01110000 even though it is not the correct sum.

16.11 BINARY CODED DECIMAL (BCD) — 535

SECTION 16.10 EXERCISES, Part 2

Here is a section of memory.

Address (Hex)	Contents (Binary)
3A87	01010010
3A88	01001000
3A89	10011011
3A8A	11101101
3A8B	01100011
3A8C	00101110
3A8D	11100001
3A8E	10110101
3A8F	10011101
3A90	11001001
3A91	11010011
3A92	10110000
3A93	11001001
3A94	10011000

23. If the contents of memory shown is all data, how many negative numbers are listed?
24. What is the contents of memory location 3A8F?
25. What is the contents of memory location 3A8C?
26. If the program tells the central processing unit (CPU) to add the contents of 3A8F to the contents of 3A8C and store the sum in 3A8B, what will be stored in 3A8B after the instruction has been run?
27. If the program counter contains 3A88 and 0006 is added to the program counter, what will be the next instruction read?
28. If the program counter contains 3A92 and FFFC is added to the program counter, what will be the next instruction read?
29. If the program counter contains 3A89 and FFFF is added to the program counter, what will be the next instruction read?
30. If the program counter contains 3A90, what number must be added to the program counter so that the next instruction comes from 3A87?
31. If the program counter contains 3A94, what number must be added to the program counter so that the next instruction comes from 3A87?

16.11 ■ BINARY CODED DECIMAL (BCD)

Decimal numbers can be represented with 1s and 0s in a manner that is different from the binary system we have already learned. In a system called **binary coded decimal (BCD)**, each decimal digit is represented by four 1s and 0s according to the 8 4 2 1 weighing that you have already learned.

Decimal	BCD
0	0000
1	0001
2	0010
3	0011
4	0100
5	0101
6	0110
7	0111
8	1000
9	1001
	1010 ⎫
	1011 ⎪
	1100 ⎬ These six combinations
	1101 ⎪ are not needed in BCD numbers.
	1110 ⎪
	1111 ⎭

EXAMPLE 16.32 Express 285_{10} in BCD.

Solution

$$2 \quad 8 \quad 5$$
$$0010 \quad 1000 \quad 0101$$
$$001010000101_{BCD} = 285_{10}$$

The first two leading digits are 0s. The result can be written without them (1010000101_{BCD}), but the other 0s are necessary as placeholders.

To convert BCD back to decimal, start at the right and mark off groups of four. Convert each group of four to decimal.

EXAMPLE 16.33 Convert $1111001000000010101_{BCD}$ to decimal.

Solution Start at the BCD point and mark off groups of four.

$$111 \quad 1001 \quad 0000 \quad 0001 \quad 0101$$

Convert each group into decimal.

$$7 \quad 9 \quad 0 \quad 1 \quad 5$$
$$1111001000000010101_{BCD} = 79015_{10}$$

16.11 BINARY CODED DECIMAL (BCD)

When counting in BCD, care must be taken to avoid the six unused states listed previously. When a column reaches 1001, that column must be reset to zero on the next count and the next column to the left must be incremented.

EXAMPLE 16.34 Count in BCD from 1000 1000 to 1 0000 0000.

Solution

$$
\begin{array}{c}
1000\ 1000 \\
1000\ 1001 \\
1001\ 0000 \\
1001\ 0001 \\
1001\ 0010 \\
1001\ 0011 \\
1001\ 0100 \\
1001\ 0101 \\
1001\ 0110 \\
1001\ 0111 \\
1001\ 1000 \\
1001\ 1001 \\
1\ 0000\ 0000
\end{array}
$$

BCD numbers can be added using the rules of binary addition, but the initial sum must be monitored to see that the result does not contain one of the six unused combinations.

EXAMPLE 16.35 Add $0101_{BCD} + 0111_{BCD} =$.

Solution

$$
\begin{array}{r}
0101 \\
+\ 0111 \\
\hline
1100
\end{array}
$$

The result is one of the unused combinations. To compensate for the six unused states, add six to the initial result.

$$
\begin{array}{r}
1100 \\
+\ 0110 \\
\hline
1\ 0010
\end{array}
$$

The final result is $1\ 0010_{BCD}$ or 12_{10}.

Check: $5 + 7 = 12$.

If the initial sum is a legitimate BCD number, but a carry to the next column is generated, the unused states were stepped through in the addition process. Six must be added to correct the initial sum.

EXAMPLE 16.36 Add $1001_{BCD} + 1000_{BCD} =$.

Solution

$$\begin{array}{r} 1001 \\ + \ 1000 \\ \hline 1\ 0001 \end{array}$$

The initial sum appears to be correct because 0001 is a legitimate BCD number. However, the carry to the next column indicates the six unused numbers have been involved in the addition process. Add six to compensate.

$$\begin{array}{r} 1\ 0001 \\ + \ 0110 \\ \hline 1\ 0111 \end{array}$$

Check: $9 + 8 = 17$.

To summarize, if the initial sum contains a non-BCD combination or if a carry to the next column is produced, add six to compensate.

EXAMPLE 16.37 Add $10\ 0101\ 1000_{BCD} + 100\ 0101\ 1000_{BCD} =$.

Solution

$$\begin{array}{r} 10\ 0101\ 1000 \\ + \ 100\ 0101\ 1000 \\ \hline 110\ 1011\ 0000 \end{array}$$

Six must be added to the first digit because a carry to the next column was produced. Six must be added to the middle digit because the initial sum is a non-BCD combination.

$$\begin{array}{r} 110\ 1011\ 0000 \\ + \ \ \ \ \ 0110\ 0110 \\ \hline 111\ 0001\ 0110 \end{array}$$

Check: $258 + 458 = 716$.

SECTION 16.11 EXERCISES

1. Count in BCD from 1000 1001 1001 to 1001 0010 0001.
2. List the six non-BCD combinations.
3. What two conditions must be watched for in BCD addition?

4. If one of the conditions listed in Problem 3 occur, what is done to compensate?
5. Add $0100_{BCD} + 0101_{BCD} =$
6. Add $0110_{BCD} + 1001_{BCD} =$
7. Add $1001_{BCD} + 1000_{BCD} =$
8. Add $01000111_{BCD} + 01010110_{BCD} =$
9. Add $10000001_{BCD} + 10010100_{BCD} =$

■ SUMMARY

1. Binary is a base 2 number system, which means there are two digits to work with, 0 and 1.
2. Binary digit is contracted into bit.
3. The right-most bit is called the least significant bit, and the left-most bit is called the most significant bit.
4. To convert binary to decimal, write the binary place values beneath each bit. Add the place values for the 1 bits.
5. The number N of binary numbers, including all zeros, that can be formed with n bits is

$$N = 2^n$$

6. Two methods of converting decimal into binary are presented in the chapter.
   Method 1: Begin at the binary point and label values of each place to the left until the decimal number is surpassed. Place a 1 in the highest value that does not exceed the decimal number. Subtract that number from the decimal number to see how much is left to convert. Place a 1 in the largest binary place that does not exceed the difference. Continue until there is nothing left.
   Method 2—Repetitive division by 2: Divide the decimal number by 2 and list the remainder. This remainder will be the LSB of the result. Divide the quotient (don't worry about the remainder for a moment) by 2 and again list the remainder. Continue the division process until the quotient is 0. The remainders form the result with the last remainder as the MSB.
7. Hexadecimal is a base 16 number system, which means we have 16 digits to work with: 0, 1, 2, 3, 4, 5, 6, 7, 8, 9, A, B, C, D, E, F.
8. To convert binary to hexadecimal, start at the binary point and work to the left, marking off groups of four. Convert each group of four bits individually to the corresponding hex digit.
9. To convert hexadecimal to binary, replace each hex digit with its 4-bit binary code.
10. A half adder adds two bits and produces two outputs, a sum and a carry to the next column.
11. A full adder adds three bits, A and B and a carry from a previous addition called C_0 or C_{in}. Just like the half adder, it produces two outputs, a sum and a carry to the next column.

12. A complement method offers two advantages over regular subtraction.
    a. In the complement approach to subtraction, the process is the same whether the minuend is smaller or larger than the subtrahend. Extra circuitry is not needed to determine which is larger because it does not matter.
    b. The subtraction problem is converted into an addition problem. Adder circuits can be used for subtraction, minimizing circuitry.
13. The binary 1's complement is formed by subtracting each bit from 1 (change each bit).
14. To subtract using 1's complement, add the 1's complement of the subtrahend to the minuend. If overflow occurs, the answer is positive and an end around carry (EAC) must be performed. To perform an EAC, add the overflow to the first column. If no overflow occurs, the answer is negative. Take the 1's complement to get the true magnitude of the answer.
15. There are two methods for taking the 2's complement.
    Method 1: Take the 1's complement and add 1. In other words, to take the 2's complement change each bit and add 1.
    Method 2: Start at the binary point and work to the left. Leave each bit unchanged until the first 1 digit is passed. After the first 1 digit, change each bit.
16. To subtract using the 2's complement method, add the 2's complement of the subtrahend to the minuend. If overflow occurs, the answer is positive. If no overflow occurs, the answer is negative. Take the 2's complement of the preliminary result to get the true magnitude of the answer.
17. In a signed 2's complement number, the most significant bit is reserved to represent the sign of the number. If the most significant bit is 1, the number represented in the remaining bits is negative, and the number is written in 2's complement form. If the most significant bit is 0, the number represented in the remaining bits is positive, and the number is written in true magnitude form (what you see is what you get).
18. In microcomputers, each location in memory has its own address. Each memory location contains a binary number (1s and 0s) that can represent an instruction of a program the computer is running or that can represent data to be used in the program.
19. A program counter keeps track of where the computer is in its program by storing the memory address that contains the next instruction to be executed. If a negative signed 2's complement number is added to the program counter, the computer will jump backward to a previous part of the program. If the number is positive, the computer will jump forward to a new part of the program.
20. In binary coded decimal (BCD), each decimal digit is represented by four 1s and 0s according to the 8 4 2 1 weighing.
21. To convert BCD back to decimal, start at the right and mark off groups of four. Convert each group of four to decimal.
22. These six combinations are not needed in BCD numbers: 1010 1011 1100 1101 1110 1111.
23. In BCD addition, the initial sum must be monitored to see whether a correction factor of 6 must be added.

SOLUTIONS TO CHAPTER 16 EXERCISES

Solutions to Section 16.1 Exercises

1. 00000
00001
00010
00011
00100
00101
00110
00111
01000
01001
01010
01011
01100
01101
01110
01111
10000

2. 1011010
1011011
1011100
1011101
1011110
1011111
1100000
1100001
1100010
1100011

Solutions to Section 16.2 Exercises

1. 21   **2.** 409   **3.** 127
4. 128   **5.** 1 048 575   **6.** 1 048 576

Solutions to Section 16.3 Exercises

1. 1100   **2.** 1110   **3.** 110001
4. 10010100   **5.** 111011110   **6.–10.** Same as 1–5.

Solutions to Section 16.4 Exercises

1. 00 10
01 11
02 12
03 13
04 14
05 15
06 16
07 17
08 18
09 19
0A 1A
0B 1B
0C 1C
0D 1D
0E 1E
0F 1F
    20

2. 4ADC 4AED 4AFE
4ADD 4AEE 4AFF
4ADE 4AEF 4B00
4ADF 4AF0
4AE0 4AF1
4AE1 4AF2
4AE2 4AF3
4AE3 4AF4
4AE4 4AF5
4AE5 4AF6
4AE6 4AF7
4AE7 4AF8
4AE8 4AF9
4AE9 4AFA
4AEA 4AFB
4AEB 4AFC
4AEC 4AFD

Solutions to Section 16.5 Exercises

1. 91
2. DC
3. 2F
4. AB
5. E3
6. 857C
7. FFFF
8. C863
9. 6BDE
10. 4E37

Solutions to Section 16.6 Exercises

1. 00111000
2. 01111010
3. 10011100
4. 1111000011011011
5. 1100000011011010
6. 0001100011111110
7. 167
8. 61

Solutions to Section 16.7 Exercises

1. 1101
2. 10010
3. 10010
4. 10100
5. $DA68_{16}$
6. $CC86_{16}$
7. $AD17_{16}$

Solutions to Section 16.8 Exercises

1. 0111
2. 0111
3. −1001
4. −0110
5. 00111111
6. 5DF6
7. −1D
8. −2304

Solutions to Section 16.9 Exercises

1. +111
2. +111
3. −1001
4. −110
5. $3F_{16}$
6. $5DF6_{16}$
7. $-1D_{16}$
8. -2304_{16}

Solutions to Section 16.10 Exercises, Parts 1, 2, and 3

1. +27
2. +31
3. 111111 = −1
4. −32
5. −8
6. 0 − 31
7. −1 through −32
8. 010000
9. 110000
10. 011110
11. 100010
12. 01100010
13. 00101111
14. 11100111
15. −76=10110100
16. 10000001
17. +108
18. +7

19. −74
20. −8
21. −82
22. −4
23. 10
24. 10011101
25. 00101110
26. 11001011
27. 3A8E 10110101
28. 3A8E 10110101
29. 3A88 01001000
30. FFF7$_{16}$
31. FFF3$_{16}$

Solutions to Section 16.11 Exercises

1. 1000 1001 1001
   1001 0000 0000
   1001 0000 0001
   1001 0000 0010
   1001 0000 0011
   1001 0000 0100
   1001 0000 0101
   1001 0000 0110
   1001 0000 0111
   1001 0000 1000
   1001 0000 1001
   1001 0001 0000
   1001 0001 0001
   1001 0001 0010
   1001 0001 0011
   1001 0001 0100
   1001 0001 0101
   1001 0001 0110
   1001 0001 0111
   1001 0001 1000
   1001 0001 1001
   1001 0010 0000
   1001 0010 0001
2. 1010, 1011, 1100, 1101, 1110, 1111
3. Non-BCD combination and overflow.
4. Add six to the initial sum.
5. 1001$_{BCD}$
6. 1 0101$_{BCD}$
7. 1 0111$_{BCD}$
8. 1 0000 011$_{BCD}$
9. 1 0111 0101$_{BCD}$

CHAPTER 16 PROBLEMS

1. Count in binary from 1010 to 10000.
2. Count in binary from 11011010 to 11100100.

Convert into decimal:

3. 11001_2
4. 101000111_2
5. How high can you count using 3-bit binary numbers?
6. How many different combinations are there using 3-bit binary numbers?
7. How high can you count using 24-bit binary numbers?
8. How many different combinations are there using 24-bit binary numbers?

Convert to binary using method 1.

9. 21_{10}
A. 36_{10}
B. 52_{10}
C. 231_{10}
D. 255_{10}

Convert to binary using method 2.

E. 26_{10}
F. 31_{10}
10. 48_{10}
11. 129_{10}
12. 234_{10}
13. 100_{10}
14. Count in hexadecimal from 8 to 2E.
15. Count from AFEC to B001.

Convert these 8-bit binary numbers to hexadecimal.

16. 11100010_2
17. 10111100_2
18. 01110101_2
19. 10101101_2
1A. 11111010_2
1B. 10110100_2
1C. 10111101_2
1D. 10101001_2

Convert these 16-bit binary numbers to hexadecimal.

1E. $01110001\ 10000110_2$
1F. $11110110\ 00110101_2$
20. $10101010\ 01010101_2$
21. $00001100\ 10101011_2$
22. $11110101\ 11101001_2$

Convert these hexadecimal numbers to binary.

23. $4E_{16}$
24. 82_{16}
25. $D0_{16}$
26. AE_{16}
27. $C5_{16}$
28. BA_{16}
29. ED_{16}
2A. 43_{16}
2B. $67F_{16}$
2C. $2A4D_{16}$
2D. $3FA7_{16}$
2E. $11DE_{16}$

Convert.

2F. $C2_{16}$ to decimal.
30. 82_{10} to hexadecimal.

Add these binary numbers.

31. $0101_2 + 1101_2 =$
32. $1110_2 + 0010_2 =$
33. $0101_2 + 1011_2 =$
34. $1000_2 + 1100_2 =$
35. $10000011_2 + 01011101_2 =$
36. $01011000_2 + 00110011_2 =$

Convert to binary and add. Express the result in hexadecimal.

37. $6ABD_{16} + 4BA4_{16} =$
38. $378D_{16} + 2E4F_{16} =$
39. $89A3_{16} + 21DF_{16} =$

Subtract using 1's complement.

3A. $0111_2 - 0100_2 =$
3B. $1011_2 - 0111_2 =$
3C. $100111_2 - 11011_2 =$
3D. $110_2 - 1001_2 =$
3E. $100_2 - 1111_2 =$
3F. $11110000_2 - 11001011_2 =$
40. $BB_{16} - AA_{16} =$
41. $53DA_{16} - EAC_{16} =$
42. $CA_{16} - FE_{16} =$
43. $AA00_{16} - FEEF_{16} =$

Subtract using 2's complement.

44. $1011_2 - 0110_2 =$
45. $1001_2 - 0101_2 =$
46. $110011_2 - 10101_2 =$
47. $110_2 - 1001_2 =$
48. $1010_2 - 1011_2 =$
49. $11001010_2 - 01011011_2 =$
4A. $DC_{16} - 9A_{16} =$
4B. $6D3C_{16} - AAA_{16} =$
4C. $8E_{16} - AF_{16} =$
4D. $D4CB_{16} - EEF2_{16} =$

The following questions apply to a 6-bit signed 2's complement system (1 sign bit followed by 5 magnitude bits). Write the equivalent decimal number.

4E. 001110
4F. 010101
50. 010001
51. 011101
52. 101010

53. 111101
54. 101011
55. 110101
56. What range of positive numbers can be represented by a 16-bit signed 2's complement system?
57. What range of negative numbers can be represented by a 16-bit signed 2's complement system?

Write these decimal numbers in 8-bit signed 2's complement form.

58. +93   **59.** −64   **60.** +127
61. −128 **5A.** +68   **5B.** −51
5C. +100 **5D.** −120 **5E.** +1
5F. −1

Convert these 8-bit signed 2's complement numbers into decimal.

62. 00100101   **63.** 01101001   **64.** 00100100
65. 10011010   **66.** 10101101   **67.** 11001101
68. $5D_{16}$ **69.** 60_{16} **6A.** $8F_{16}$
6B. EC_{16}

Here is a section of memory. Data is stored in signed 2's complement form.

Address (Hex)	Contents (Binary)
AE0B	10110000
AE0C	10010101
AE0D	00010010
AE0E	11000101
AE0F	11010100
AE10	01101100
AE11	11101011
AE12	11000101
AE13	11010100
AE14	01101001
AE15	00010011
AE16	10110110
AE17	11011001
AE18	11011000
AE19	00100101
AE20	01101000

6C. If the contents of memory shown is all data, how many negative numbers are listed?
6D. What is the contents of memory location AE14?

6E. What is the contents of memory location AE15?

6F. If the program tells the central processing unit (CPU) to add the contents of AE14 to the contents of AE15 and store the sum in AE16, what will be stored in AE16 after the instruction has been run?

70. If the program counter contains AE19 and FFF4 is added to the program counter, what will be the next instruction read?

71. If the program counter contains AE0D and 000A is added to the program counter, what will be the next instruction read?

72. If the program counter contains AE10 and FFFF is added to the program counter, what will be the next instruction read?

73. If the program counter contains AE18, what number must be added to the program counter so that the next instruction comes from AE0A?

74. If the program counter contains AE18, what number must be added to the program counter so that the next instruction comes from AE0D?

75. Using 16 bits for memory addresses, how many locations in memory can be addressed?

76. Count in BCD from 001101100111 to 010000000101.

77. Why is 6 sometimes added to correct BCD sums?

78. Add $0101_{BCD} + 0011_{BCD} =$

79. Add $0111_{BCD} + 1001_{BCD} =$

7A. Add $1000_{BCD} + 1000_{BCD} =$

7B. Add $01010110_{BCD} + 00110111_{BCD} =$

7C. Add $10010101_{BCD} + 00100010_{BCD} =$

7D. Add $01111000_{BCD} + 10000011_{BCD} =$

7E. Add $01000011_{BCD} + 10001001_{BCD} =$

Appendix

■ SOLUTIONS TO CHAPTER 1 PROBLEMS

1. True
3. False   Subtraction is not commutative.
5. False   The answer to a subtraction is called the difference.
7. False   Numbers being subtracted are called subtrahend.
9. True
11. False   41 is not divisible by 5.
13. False   Including 1, there are 11 prime numbers less than 30.
15. True
17. False   The least common multiple for 54, 21, and 14 is 756.
19. False $8/5 - 3/2 = 1/10$
21. True
23. True
25. False   To subtract unlike fractions, first find the LCD.
27. True
29. True
31. False $8^{2/3} = 1/8^{-2/3}$
33. True
35. $11 \cdot 5 \cdot 3$
37. $3 \cdot 3 \cdot 3 \cdot 3 \cdot 10 \cdot 10 \cdot 10$
39. $13 \cdot 11 \cdot 5$
41. $13 \cdot 17 \cdot 19$
43. 2/3
45. 11/13
47. 23/29
49. 169/143
51. 600/2040

SOLUTIONS TO CHAPTER 1 PROBLEMS ■ 549

53. 225/2325

55. 51 450/38 850

57. LCM = $11 \cdot 3 \cdot 3 \cdot 2 \cdot 2 = 396$

59. LCM = $5 \cdot 5 \cdot 5 \cdot 3 \cdot 7 \cdot 3 = 7875$

61. LCM = $2 \cdot 3 \cdot 5 \cdot 7 \cdot 9 = 1890$

63. $9/3 = 3$

65. $(2 + 1 + 4)/5 = 7/5$

67. 11/10

69. $\dfrac{158}{15}$

71. $\dfrac{271}{120}$

73. $\dfrac{43}{24}$

75. 3/5

77. $\dfrac{13}{40}$

79. $\dfrac{19}{24}$, LCD = $2 \cdot 2 \cdot 2 \cdot 3 = 24$

81. $\dfrac{3}{7}$

83. $\dfrac{20 \times \cancel{5}}{\cancel{3} \times \cancel{9}} \times \dfrac{\cancel{3}}{\cancel{5}} \times \dfrac{\cancel{9}}{\cancel{20}} = \dfrac{1}{1} = 1$

85. $\dfrac{3 \times \cancel{7}}{4 \times \cancel{5}} \times \dfrac{\cancel{5} \times 5}{\cancel{7} \times 7} = \dfrac{3 \times 5}{4 \times 7} = \dfrac{15}{28}$

87. $\dfrac{5 \times \cancel{7}}{3 \times \cancel{17}} \times \dfrac{5 \times \cancel{13}}{3 \times \cancel{7}} \times \dfrac{5 \times \cancel{17}}{3 \times \cancel{13}} = \dfrac{5 \times 5 \times 5}{3 \times 3 \times 3} = \dfrac{125}{27}$

89. $\dfrac{3}{5} \times \dfrac{9}{5} = \dfrac{3 \times 9}{5 \times 5} = \dfrac{27}{25}$

91. $\dfrac{49}{72} \times \dfrac{36}{35} = \dfrac{\cancel{7} \times 7}{2 \times \cancel{2} \times \cancel{2} \times \cancel{3} \times \cancel{3}} \times \dfrac{\cancel{2} \times \cancel{3} \times \cancel{2} \times \cancel{3}}{\cancel{7} \times 5} = \dfrac{7}{2 \times 5} = \dfrac{7}{10}$

93. $\dfrac{11}{17} \times \dfrac{34}{121} = \dfrac{\cancel{11}}{\cancel{17}} \times \dfrac{\cancel{17} \times 2}{\cancel{11} \times 11} = \dfrac{2}{11}$

95. $\dfrac{\tfrac{42 - 15}{35}}{\tfrac{14 + 10}{35}} = \dfrac{\tfrac{27}{35}}{\tfrac{24}{35}} = \dfrac{27}{24} = \dfrac{9 \times \cancel{3}}{8 \times \cancel{3}} = \dfrac{9}{8}$

97. $\dfrac{1}{2 - \dfrac{1}{\tfrac{6}{6} - \tfrac{1}{6}}} = \dfrac{1}{2 - \dfrac{1}{\tfrac{5}{6}}} = \dfrac{1}{2 - \tfrac{6}{5}} = \dfrac{1}{\tfrac{10}{5} - \tfrac{6}{5}} = \dfrac{1}{\tfrac{4}{5}} = \dfrac{5}{4}$

99. $3 \cdot 3 \cdot 3 \cdot 3 \cdot 3 = 243$

101. 1

103. $1/(5 \cdot 5 \cdot 5) = 1/125$

105. $8^3 = 512$

107. $1/12^2 = 1/144$

109. $5^{2+3+1} = 5^6 = 15\,625$

111. $7^{5-2} = 7^3 = 343$

113. $10^{7+3-5-4} = 10^1 = 10$

115. $3^2 \times 10^{4 \times 2} = 9 \times 10^8$

117. $2^{3 \times 2} / 3^{2 \times 2} = 2^6 / 3^4 = 64/81$

119. $(9^2 \cdot 8^2 / 25^2) = 81 \cdot 64 / 625 = 5184/625$

121. 8.9466

123. 0.000 002 17

125. 0.1479

127. 0.010 27

129. $(10 \cdot 10 \cdot 10 \cdot 10)^{1/3} = 10(10)^{1/3}$

131. $(2 \cdot 2 \cdot 2 \cdot 2 \cdot 2 \cdot 2 \cdot 2 \cdot 2 \cdot 2 \cdot 2 \cdot 2)^{1/4} = 2 \cdot 2 \cdot (2 \cdot 2 \cdot 2)^{1/4} = 4(8)^{1/4}$

133. 1/10

135. $8^{3/4}/32$

137. $2 + 24 - 6 = 20$

139. 18 1/7

141. 51 7/32

143. 154%

145. 60%

147. 39.5%

149. 81.90%

151. 0.0078

153. $38/100 = 19/50$

155. 3419.56

157. $\$10.33 \cdot 105.5\% = \$10.33 \cdot 1.055 = \$10.90$

159. 820 000 Ω

161. 380 000 Ω

163. 39 000 Ω · 95% = 37 050 Ω   39 000 Ω · 105% = 40 950 Ω

165. 26.5 Ω · 99% = 26.235 Ω   26.5 Ω · 101% = 26.765 Ω

Unscramble these letters:

M S U   S(U)M
A B E S   B A (S) E
D D D N E A   (A) D D E (N) D
D D D I V E I N   (D) I V I D (E) (N) D
U T E Q N O I T   Q U O T (I) E N (T)
L I M T L I P R U E   M (U) L T (I) P L I E (R)
U L I P L I A M T C D N   (M) U L T I P L I C A (N) (D)

Now unscramble the circled letters to answer the question.

What's the difference?

M I N U E N D   M I N U S   S U B T R A H E N D

SOLUTIONS TO CHAPTER 2 PROBLEMS

1. F   To the left.
3. F   Is always between 1 and 9.99...
5. F   Is not used to hold a decimal place.
7. F   Accuracy.
9. F   Accuracy is used.
11. F   Means 10^3.
13. T
15. T
17. 0.001 08
19. 38 900
21. 1×10^{-2}
23. 2.4689×10^4
25. 3.625×10^{-6} upper; 3.615×10^{-6} lower
27. 1650 upper; 1550 lower
29. 1340
31. 16 950
33. 15.9
35. 5.6×10^4

37. 119
39. 10
41. 1.9
43. 2×10^{-4}
45. 12.1
47. 0.000 6
49. 0.000 002 004
51. 0.000 007 9
53. 5 660
55. 46 800
57. 2007   4
59. 0.0006   1
61. 0.000 600   3
63. 3.50 mA   3
65. 2007   units
67. 0.0006   ten thousandths
69. 0.000 600   millionths
71. 3.50 mA   hundred thousandths

Unscramble these letters:

AEMG M(E)GA
IOCMR M(I)C(R)O
IOPC (P)(I)(C)O
EART T(E)(R)A
EOTMF (F)EM(T)O

Now unscramble the circled letters to solve the puzzle.

~~U.S. CUSTOMARY~~   ~~POST~~   ~~BREAK~~

M E T R I C   P R E F I X

SOLUTIONS TO CHAPTER 3 PROBLEMS

1. <
3. <
5. >
7. <

9. $<$
11. 11
13. -11
15. 87
17. -56
19. -108
21. 176
23. 0
25. -11
27. -11
29. 13
31. 38
33. -117
35. -4
37. 14
39. -15
41. -68
43. -3
45. 46
47. 0
49. 0
51. 28
53. -30
55. 32
57. -1260
59. 1
61. 3
63. -4
65. 17
67. 5
69. -3
71. 39
73. 79
75. 100
77. 25
79. 1/49
81. 116
83. 13
85. 100
87. 0
89. 0
91. $11u^2 + 13v - 7t^2$
93. $20m^2n^3 - 3m^2 - 7n^2$
95. $22t^2 + 2t$
97. $-32x + 8y$
99. $-6x^3y^2z^2 - 12x^3y^3z^3$
101. $-20t^5$
103. $24t^9u^{14}v^4$
105. $120x^{10}y^8z^{10}$
107. $48v^6$
109. $54a^{10}b^7c^9d^{13}$
111. 10^4
113. 10^{-1} or $1/10$
115. $3a^5b^{-2}c^{-6}$ or $3a^5/(b^2c^6)$
117. $v^{-2}w^6/3$ or $w^6/(3v^2)$
119. $10c^9d^{-3}e^{-2}$ or $10c^9/(d^3e^2)$
121. $8a^9b^{12}c^{15}$
123. $256p^8q^{12}r^{16}$
125. $20x^4y^2z^5$
127. $5m^4n^5o^2(2)^{1/3}$
129. $r^4s^{11}t^6(4)^{1/4}$
131. $t^2 - 1$
133. $t^2 - 4t + 4$
135. $6s^2 + 9s - 42$
137. $8x^2y^2 - 8xy - 30$
139. $8r^2 + 2rs - 6s^2$
141. $16u^2 - 40uv + 25v^2$
143. $16x^2 - 25y^2$
145. $16x^4y^6 - 48x^2y^3 + 36$
147. $20x^6y^8 - 7x^3y^4 - 6$
149. $16g^8 - 40g^4h^3 + 25h^6$
151. $20a^2 + 12ab - 15ac - 8b^2 + 6bc$

153. $30a^2 + 14ab - 42ac - 8b^2 + 14bc$
155. $6t^6 + 15t^4u - 12t^4u^2 + 8t^3u^2 - 30t^2u^3 - 10tu^3 - 8u^4 - 24tu^4$
157. $5XY^2(X^2 - 3)$
159. $13(r + s)(r - s)$
161. $5(y^2 + 3z^2)$
163. $(7r + 8t)(7r - 8t)$
165. $(x + 3)(x - 2)$
167. $(u + 7)(u - 3)$
169. $(w + 3)(w + 7)$
171. $(y - 3)(y - 4)$
173. $(3t - 5)(2t + 3)$
175. $(6v + 3)(3v - 4)$
177. $3(5x - 2)(x + 1)$
179. $(5s - 3t)(2s + 3t)$
181. $(7w + 3x)(4w - 5x)$
183. $a - b + c$
185. $(x - y)/(2x - 3y)$
187. $63x^7y^2/(21x^5y^6)$
189. $-75w^5y^4z^4$
191. $47s/(5t^2)$
193. $(5bc - 5ac + 5ab)/(abc)$
195. $(12m^3 + 21n^3)/(28m^2n^2)$
197. $5/(12p^3q^2)$
199. $4st^2u^3(t - 6)/(t + 1)$
201. $22p^3s^3/(21r^2q)$
203. $5(2u + 3v)/[2(2u - 3v)]$
205. $x/(x + 1)$
207. $(a + 1)/(a - 1)$
209. $(1 - x)/x$

■ SOLUTIONS TO CHAPTER 4 PROBLEMS

15. -22
17. -25
19. $8/5$
21. $31/7$
23. 0
25. 3
27. $32\,765$
29. $19/4$
31. $67/2$
33. $22/5$
35. $-5/2$
37. $-105/64$
39. $5/7$
41. $C_T - C_1 - C_2$
43. $V_B - V_E$
45. $f_r/(f_2 - f_1)$
47. c/λ
49. $f_r/(f_2 - f_1)$
51. $L_M(L_1L_2)^{1/2}$
53. N_2V_1/V_2
55. $V_{CC} - I_BR_B$
57. $(V_{CC}R_3 - V_3R_2 - V_3R_3)/V_3$
59. $\beta(V_{EE} - V_{BE})/I_E - \beta R_E$
61. $[Z^2 - (X_L - X_C)^2]^{1/2}$
63. $1/(4\pi^2f_c^2R_AC_AC_B)$
65. $Q^2/[C(Q^2 + 1)(2\pi f_r)^2]$
67. $(R_2R_A + R_2R_C)/(R_C - R_2)$
69. $(R_AR_2 - R_1R_2)/(R_1 + R_2)$
71. $1/(1/R_T - 1/R_1 - 1/R_3)$
73. $(V_{CC} \cdot R_3)/V_3 - R_2 - R_3$ or $(V_{CC}R_3 - V_3R_2 - V_3R_3)/V_3$
75. $\beta[V_{EE} - V_{BE})/I_E - R_E]$ or $\beta(V_{EE} - V_{BE})/I_E - \beta R_E$

SOLUTIONS TO CHAPTER 5 PROBLEMS

1. False m^2
3. True
5. True
7. False   7.481 gallons
9. False   distance
11. False   kilogram
13. False   distance per time^2
15. True
17. True
19. True
21. True
23. True
25. False   farads
27. False   Resistance is the reciprocal of conductance.
29. False   electromotive force
31. True
33. False
35. True

	Quantity	Symbol	Units US	SI
37.	Force	F	pound	newton
39.	Length	s	foot	meter
41.	Power	P	horsepower	watts
43.	Area	A	ft^2	m^2
45.	Energy	W	Btu	joule
47.	Temperature	T	Fahrenheit	Celsius
49.	Acceleration	a	ft/sec^2	m/sec^2

51. g	**53.** a	**55.** d	**57.** k
59. i	**61.** p	**63.** x	**65.** w
67. c	**69.** u	**71.** s	**73.** j
75. v			

	Quantity	Symbol	Unit
77.	Charge	Q	coulombs
79.	Capacitance	C	farads
81.	Current	I	amps
83.	Energy	W	joules

85. e
89. f
93. c
97. 132 mi
101. 18.5 ton
105. 1.000×10^{-3} km²
109. 28 000 km/m³
113. 3500 mbar
117. 289 kwatts

87. j
91. b
95. 1.4×10^{10} A
99. 114 m
103. 24 m/sec²
107. 11 000 liters
111. 77.7 mph
115. 20.7 MPa
119. 0.3619 kwatt-hr

Unscramble these letters:

E Y H N R H E N R Y
M A G R G R A M
E N O W N T N E W T O N
A A D R F F A R A D
O S D P U N P O U N D S
O J L U E J O U L E

Now unscramble the circled letters to solve the puzzle.

The United States leads all other countries in the consumption of this.

E N E R G Y

SOLUTIONS TO CHAPTER 6 PROBLEMS

1. 158 Ω	**3.** 24 kΩ	**5.** orange blue red
7. 101 kPa	**9.** 400 kV	**11.** 2.0 Ω
13. 48 ft per sec	**15.** 174 pF	**17.** 0.96 µH
19. 75 ft	**21.** 26.5 kΩ	**23.** 3.6 MΩ
25. 9.55 MΩ	**27.** 770 Ω	**29.** 60 ft per sec
31. 8190 N	**33.** 100°C	**35.** 8430 lb
37. 420 mb	**39.** 390 THz	**41.** 190 Mm/sec
43. 630 THz	**45.** 230 Mm/sec	

SOLUTIONS TO CHAPTER 7 PROBLEMS

1. α, φ	**3.** DEF, β	**5.** No
13. 48″	**15.** No	**17.** 9.6 in²
19. 13.2 in²	**21.** 8.49 acres	**23.** 13 300 ft²
25. 29.30 m	**27.** 6.28 acres	**29.** 12 500 ft²
31. 74°49′48″	**33.** 64.6608°	**35.** 62°2′14″
37. 1.1 rad	**39.** 5π/6	**41.** 6.8 ft²
43. 399 ft	**45.** 188.5 rad per sec	**47.** 2.2 sec
49. 30 ft/sec, 7.5 ft/sec	**51.** 38 in²	**53.** 1.41 rad
55. 30.00 rev/sec	**57.** 0.38 sec	**59.** 4480 ft²
61. 164 yd³	**63.** 20 yd³	

SOLUTIONS TO CHAPTER 8 PROBLEMS, PART 1

1. (1.050, 0.650)	**3.** 0.671 in	**5.** 1.700 in
7. 1.972 in	**9.** 1.393 in	

SOLUTIONS TO CHAPTER 8 PROBLEMS, PART 2

1. See Figure 8-56.

3. No

5. AB (1/2, 3), AC (−6, 11/2), AD (−3/2, 1/2), BC (3/2, 9/2), BD (6,−1/2), CD (−1/2, 2)

7. $y - 4 = -3(x - 6)$

FIGURE 8-56
Solution to Chapter 8 Problems, Part 2, Problem 1

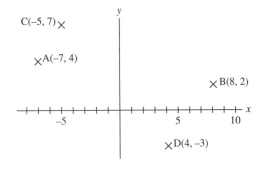

9. $y = 3x - 5$
11. $y - 4 = -2(x + 3)$
13. $y + 2 = 1/3(x - 3)$
15. A. $x = 4$, B. $y = -4$, C. $y = x + 6$, D. $y = -1/3x + 4$, E. $y = 8/3(x - 4)$, F. $y = -3/5x - 6$
17. See Figure 8-57.

FIGURE 8-57
Solution to Chapter 8 Problems, Part 2, Problem 17

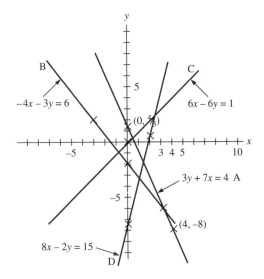

19. See Figure 8-58.
21. See Figure 8-59.

SOLUTIONS TO CHAPTER 9 PROBLEMS

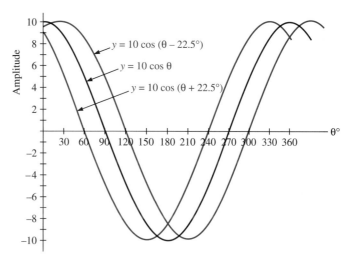

FIGURE 8-58
Solution to Chapter 8 Problems, Part 2, Problem 19

FIGURE 8-59
Solution to Chapter 8 Problems, Part 2, Problem 21

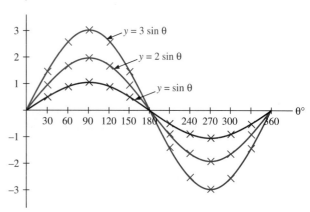

■ SOLUTIONS TO CHAPTER 9 PROBLEMS

1. Figure 9-24.

FIGURE 9-24
Solution to Chapter 9, Problem 1

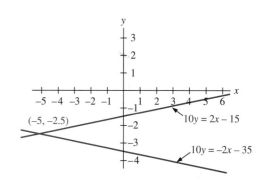

3. Figure 9-25.

FIGURE 9.25
Solution to Chapter 9, Problem 3

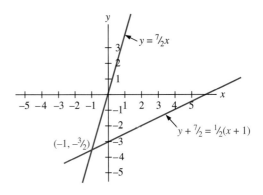

5.

Resistor (Ω)	100	470	560	330	110
Voltage drop (V)	2.6	12.2	14.6	20.7	6.9
Current (mA)	10.0	10.0	34.4	24.4	24.4

7.

Resistor (kΩ)	33	47	56
Voltage (V)	2.55	7.45	4.54
Current (μA)	77.3	158.5	81.2

■ SOLUTIONS TO CHAPTER 10 PROBLEMS

1. $b = (c^2 - a^2)^{1/2}$
3. 4.5 ft

Complete the chart.

	sin A	cos A	tan A	sin B	cos B	tan B
5. Figure 10-39.	0.478	0.879	0.544	0.879	0.478	1.839

Complete the chart.

	Angle A	Angle B
7. Figure 10-39.	28.5°	61.5°

9. 35.7° and 54.3°
11. 81 km and 31 km
13. 45.0 m
15. Not possible—hypotenuse must be longest side
17. 0.53

SOLUTIONS TO CHAPTER 10 PROBLEMS ▪ 561

19. 1.64
21. 0.68
23. 19.0°
25. 52.9°
27. 37.2°
29. 56°
31. 40.3°
33. See Figure 10-41. $n_t = 1.52$

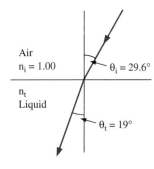
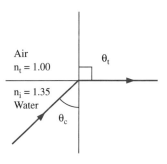

FIGURE 10-41
Solutions to Chapter 10, Problems 33 and 35

35. See Figure 10-41. $\theta_c = 47.8°$
37. See Figure 10-42. $x = 0.774"$ $\theta_{t\ air} = 25.0°$

FIGURE 10-42
Solution to Chapter 10, Problem 37

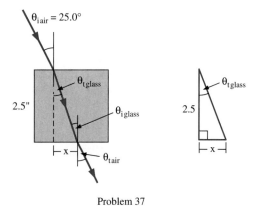

39. See Figure 10-43. $h = 22.7$ ft
41. See Figure 10-44. $BC = 113$ ft

FIGURE 10-43
Solutions to Chapter 10, Problems 39 and 41

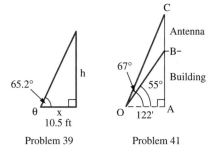

Problem 39   Problem 41

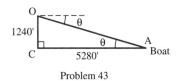   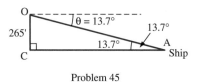

Problem 43   Problem 45

FIGURE 10-44
Solutions to Chapter 10, Problems 43 and 45

43. See Figure 10-44. θ = 13.2°
45. See Figure 10-44. AC = 1090 ft
47. See Figure 10-45. S54.6°E, 567 ft

FIGURE 10-45
Solutions to Chapter 10 Problems 47, 49, and 51

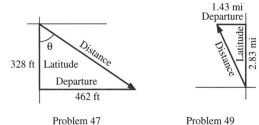

Problem 47   Problem 49

Problem 51

49. See Figure 10-45. N26.8°W, 3.17 mi

51. See Figure 10-45. S46.6°W, 101 km

53. See Figure 10-46.
lat = 14 km N
dep = 9.0 km W

FIGURE 10-46
Solutions to Chapter 10, Problems 53, 55, and 57

Problem 53

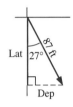
Problem 55

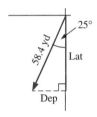
Problem 57

55. See Figure 10-46.
lat = 78 ft S
dep = 39 ft E

57. See Figure 10-46.
lat = 52.9 yd S
dep = 24.7 yd W

■ SOLUTIONS TO CHAPTER 11 PROBLEMS

1. False
3. True
5. False
7. True
9. True
11. False
13. True

15. False
17. False
19. False
21. True
23. False
25. True
27. False
29. False
31. 55 ∠2° lb
33. 109 ∠20° lb
35. 39 ∠11° lb
37. 60 ∠88° lb
39. −31, −58
41. −33, 10
43. 212 ∠30° lb
45. 144 ∠133° lb
47. 145 ∠−132° lb
49. 84 ∠47° lb
51. 94 ∠6° lb
53. 39 ∠−169° lb
55. 64 ∠−56° lb
57. 128 ∠−110° lb
59. 128 ∠−152° lb
61. a. True
    b. False
    c. False
    d. True
    e. True
63. a. True
    b. False
    c. False
    d. True
    e. True
65. See Figure 11-54.
67. See Figure 11-55.
69. $y_A = 100 \sin \theta$ V
    $y_B = 50 \sin (\theta - \pi/4)$ V

FIGURE 11-54
Solution Chapter 11, Problem 65

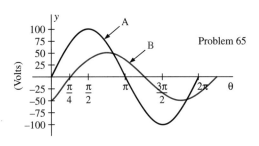

FIGURE 11-55
Solution to Chapter 11, Problem 67

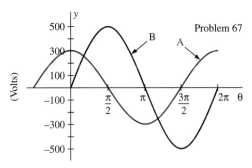

71. $y_A = 300 \sin(\theta + \pi/4)$ V
$y_B = 500 \sin \theta$ V

73. **a.** True
    b. False
    c. False
    d. True
    e. False

75. **a.** False
    b. True
    c. True
    d. False
    e. True
    f. True
    g. True
    h. False
    i. False
    j. True
    k. False
    l. False
    m. False
    n. True
    o. True
    p. False
    q. False
    r. False

77.–79. See Figure 11-56.

FIGURE 11-56
Solutions to Chapter 11, Problems 77 and 79

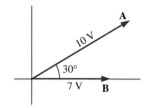

Problem 77

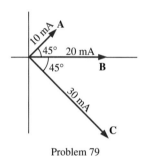

Problem 79

81. $y_A = 10 \sin(\theta + 30°)$ V
$y_B = 7 \sin \theta$ V

83. $y_A = 10 \sin(\theta + 45°)$ mA
$y_B = 20 \sin \theta$ mA
$y_C = 30 \sin(\theta - 45°)$ mA

85. a. True
  b. True
  c. False
  d. True
  e. True

87.–89. See Figure 11-57.

91. $y_D = 30 \sin(\theta - 80°)$ V
$y_E = 10 \sin(\theta - 55°)$ V
$y_F = 20 \sin \theta$ V

93. a. False
  b. True
  c. True
  d. False
  e. False
  f. True
  g. False

FIGURE 11-57
Solutions to Chapter 11, Problems 87 and 89

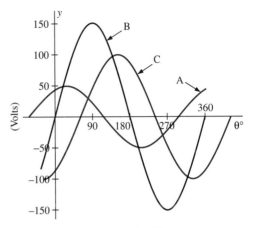

Problem 87

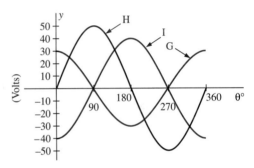

Problem 89

    h. False
    i. True
    j. True

95. a. True
    b. True
    c. True
    d. True
    e. True

97. See Figure 11-58.

99. $y_A = 50 \sin \theta$ V
    $y_B = 30 \sin (\theta + 25°)$ V
    $y_C = 40 \sin (\theta - 35°)$ V

101. $y_A = 50 \sin \theta$ V
    $y_B = 50 \sin (\theta - 120°)$ V
    $y_C = 50 \sin (\theta + 120°)$ V

103. 933.33 Hz

FIGURE 11-58
Solution to Chapter 11, Problem 97

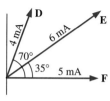

Problem 97

■ SOLUTIONS TO CHAPTER 12 PROBLEMS

1 and 3. See Figure 12-23.

FIGURE 12-23
Solutions to Chapter 12, Problems 1 and 3

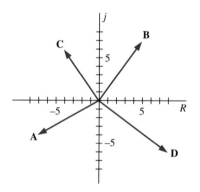

5. $110 \angle 49°$
7. $25 \angle 135°$
9. $23.5 + j7.40$ kips
11. $-7 + j9$
13. $60 + j52$ or $79 \angle 41°$
15. $90.4 + j145$ or $171 \angle 58°$
17. $-31 - j11$ or $33 \angle -161°$
19. 30
21. $807 \angle -46°$
23. 80
25. 53
27. $(-51 - j33)/90$

29. $0.434 \angle 38.9°$

31. $49 \angle 27°$

33. $Z = 78 \angle 44°$ Ω. See Figure 12-24.
$I = 320 \angle 0°$ mA
$V_R = 18 \angle 0°$ V
$V_C = 7.0 \angle -90°$ V
$V_L = 24 \angle 90°$ V

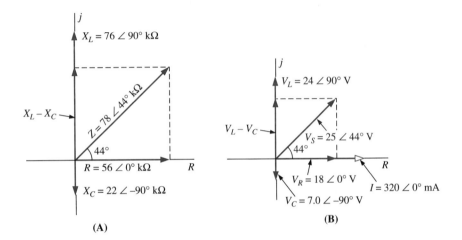

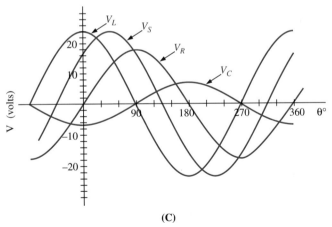

FIGURE 12-24
Solutions to Chapter 12, Problem 33

35. See Figure 12-25. $X = 79 \angle 90°$ Ω (inductive).

FIGURE 12-25
Solutions to Chapter 12, Problem 35

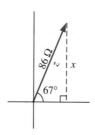

■ SOLUTIONS TO CHAPTER 13 PROBLEMS

Surveying

1. N15°E, S15°W
3. S62°E
5. 124 m at N76°E
7. **a.** 6.5 ft, **b.** 25.6 ft

Mechanics

1. Force at pinned end, 218 lb down
   Force at support, 557 lb up
3. 3.01 kips at 29.7° and 3.13 kips at 33.5°

Navigation

1. 272°/66 kn
3. 065°/243 kn
5. 149 kn

■ SOLUTIONS TO CHAPTER 14 PROBLEMS

1. 3
3. 1
5. −1
7. 3.7706
9. 1.7706
11. −0.2294
13. 598.8
15. 3.874
17. 1.4022×10^{-3}
19. log 10 000 = 4
21. $\log_5 125 = 3$
23. $\log_2 1024 = 10$
25. $10^{-2} = 0.01$
27. $10^{-2.750} = 1.78 \times 10^{-3}$
29. $2^{20} = 1\ 048\ 576$

31. 98.37 dB
33. 20 mW/m^2
35. 25.4 dBm
37. 158 μW
39. 21.25 dB
41. −3.7 dB
43. 1400 W
45. 110 μW
47. 28 μW
49. 1.322
51. −0.146
53. 1.799
55. 4.322
57. 2.651
59. 22.69 dB
61. 29 V
63. 51mV
65.
    A. 14 kHz
    B. 26 kHz
    C. 74 kHz
    D. 150 kHz
    E. 240 kHz
    F. 360 kHz
    G. 560 kHz
    H. 760 kHz
    I. 960 kHz
    J. 1.25 MHz
    K. 2.12 MHz
    L. 2.8 MHz
    M. 4.2 MHz
    N. 5.8 MHz
67. See Figure 14-12.
69. See Figure 14-13.
71. 100
73. 5.8539
75. ln 9000 = 9.1049

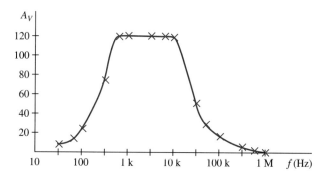

FIGURE 14-12
Solutions to Chapter 14, Problem 67

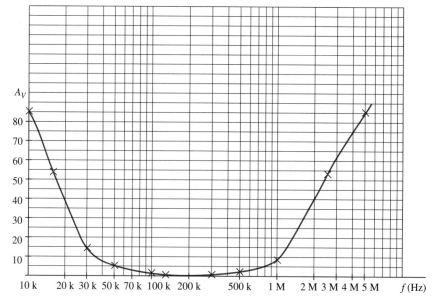

FIGURE 14-13
Solutions to Chapter 14, Problem 69

77. $\epsilon^{12.000\,000} = 162\,755$
79. 2.0 msec
81. 10.0 msec
83. 63% of 35 V = 0.63 · 35 V = 22 V
85. 97% of 35 V = 0.97 · 35 V = 33 V
87. 26 V
89. 70 msec
91. 30 V
93. 1.3 sec

SOLUTIONS TO CHAPTER 15 PROBLEMS

1., 7., 13. $x = 11 \quad x = -9$
3., 9., 15. $x = 7 \quad x = 3$
5., 11., 17. $x = 3/7 \quad x = 7/5$
19. See Figure 15-17.

FIGURE 15-17
Solutions to Chapter 15, Problem 19

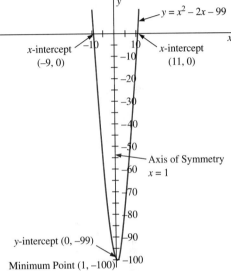

FIGURE 15-18
Solutions to Chapter 15, Problem 21

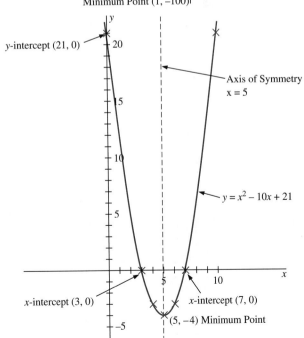

FIGURE 15-19
Solutions to Chapter 15, Problem 23

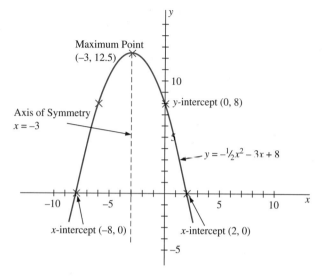

FIGURE 15-20
Solutions to Chapter 15, Problem 25

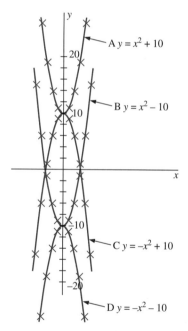

21. See Figure 15-18.
23. See Figure 15-19.
25. See Figure 15-20.
27. See Figure 15-21.
29. If $a<0$, the parabola opens downward; if $a>0$, the parabola opens upward.

FIGURE 15-21
Solutions to Chapter 15, Problem 27

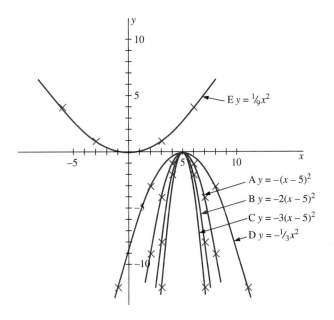

31. See Figure 15-22.

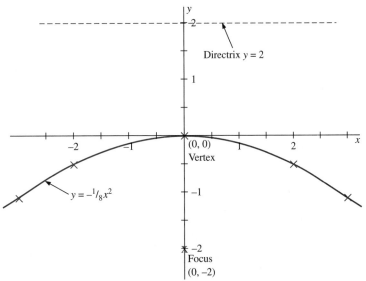

FIGURE 15-22
Solutions to Chapter 15, Problem 31

33 and 35. See Figure 15-23.

35. $d_1 = d_2, d_3 = d_4$

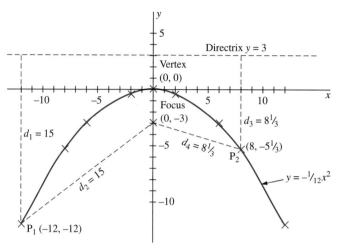

FIGURE 15-23
Solutions to Chapter 15, Problems 33 and 35

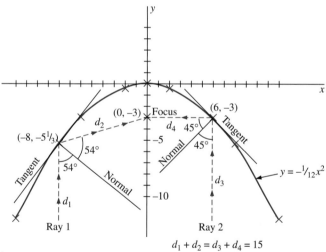

FIGURE 15-24
Solutions to Chapter 15, Problem 37

37. See Figure 15-24.

$$d_1 = 12 - 5\ 1/3 = 6\ 2/3$$
$$d_2 = [(8 - 0)^2 + (-5\ 1/3 + 3)^2]^{1/2}$$
$$d_2 = [64 + 49/9]^{1/2} = [576/9 + 49/9]^{1/2} = [625/9]^{1/2} = 25/3$$
$$d_2 = 8\ 1/3$$

Total distance traveled by ray 1 = 8 1/3 + 6 2/3 = 15.
$$d_3 = 12 - 3 = 9$$
$$d_4 = [(6-0)^2 + (-3+3)^2]^{1/2} = 6$$
Total distance traveled by ray 2 = 9 + 6 = 15.

■ SOLUTIONS TO CHAPTER 16 PROBLEMS

1. 1010, 1011, 1100, 1101, 1110, 1111, 10000
3. 25
5. 7
7. $16\,777\,215_{10}$
9. 10101_2
B. 110100_2
D. 11111111_2
F. 11111_2
11. 10000001_2
13. 1100100_2
15. AFEC, AFED, AFEE, AFEF, AFF0, AFF1, AFF2, AFF3, AFF4, AFF5, AFF6, AFF7, AFF8, AFF9, AFFA, AFFB, AFFC, AFFD, AFFE, AFFF, B000, B001
17. BC_{16}
19. AD_{16}
1B. $B4_{16}$
1D. $A9_{16}$
1F. $F635_{16}$
21. $0CAB_{16}$
23. 01001110_2
25. 11010000_2
27. 11000101_2
29. 11101101_2
2B. 011001111111_2
2D. 0011111110100111_2
2F. 194_{10}
31. 10010_2
33. 10000_2
35. 11100000_2
37. $B661_{16}$

39. $AB82_{16}$
3B. 100_2
3D. -11_2
3F. 100101_2
41. $452E_{16}$
43. $-54EF_{16}$
45. 100_2
47. -11_2
49. 1101111_2
4B. 6292_{16}
4D. $-1A27_{16}$
4F. $+21_{10}$
51. $+29_{10}$
53. -3_{10}
55. -11_{10}
57. -1 through $-32\,768$
59. 11000000
5B. 11001101
5D. 10001000
5F. 11111111
61. 10000000
63. $+105_{10}$
65. -102_{10}
67. -51_{10}
69. 96_{10}
6B. -20_{10}
6D. 01101001
6F. 01111100
71. 11011001
73. $FFF2_{16}$
75. $65\,536_{10}$
77. To compensate for the six binary numbers that are not used in BCD math.
79. $0001\ 0110_{BCD}$
7B. $1001\ 0011_{BCD}$
7D. $0001\ 0110\ 0001_{BCD}$

Index

A
$a^{b/c}$ key, 10
Absolute pressure, 172–173
Absolute value, 74
Acceleration (a), 150–151
Accuracy, 59–61
Acute angle, 194, 195
Acute triangle, 196
Addend, 2
Addition, 2–3
  binary, 525–526
  complex numbers, 402–403
  fractions, 7, 94–96
  monomials, 79
  precision, 62
  signed numbers, 74–75
  vectors, 345–350, 359–363
Addition method, for systems of linear equations, 287–289
Addition property of equality, 113
Additive inverse, 113
Air speed, true, 438
Airplane, stall speed, 184
Alternate interior angles, 330
Alternating current (ac) circuit, series, analysis of, 407–415
Amp (A), 154, 173
Angle(s), 194–195
  central, 207
  critical, 327
  of depression, 329–332
  of elevation, 329–332
  interior, alternate, 330
  precision, 320
  sides, 194
Angle of incidence, 323, 324, 327
Angle of transmission, 323, 324
Angstrom (A), 144
Angular displacement (θ), 216
Angular velocity (Ω), 215
Arc length, 214–215
Area (A), 146–147
  circle, 207–208
  quadrilateral, 204–207
  sector, 213–214
  triangle, 200–202
Arithmetic, 2–4
  scientific notation, 49–50
Arrow, vector, 345
Augend, 2
Axis of symmetry, 494

B
Base, 14
Base number, 451
Bearing, 332–335
Binary coded decimal, 535–539
Binary numbers, 514–515
  addition, 525–526
  decimal conversion, 516–518
  hexadecimal conversion, 522–525
  signed, 530–535
  subtraction, 526–530
Binomials, 79
  multiplication, 84–86
Bit, 515
B&K Precision Electronic Multimeter, 53–54

C
Calculator, 47–50
  +/− key, 15, 47
  degree mode, 312
  EXP key, 47–48
  inverse trigonometric functions, 319
  parentheses, 291
  P>R key, 397
  R>P key, 400
  x^2 key, 14–15
  x^3 key, 15
  y^x key, 14
Capacitance (C), 157, 158
Capacitive reactance, of series ac circuit, 407–409
Capacitor, 157, 158
  charge, 470
  series ac circuit, 407–409
Caret (ˆ) key, 14
Cartesian coordinate system, 234–236
Celsius (°C), 154
Central angle, 207
Charge (Q), 154
  capacitor, 470
Chord, 207
Circles, 207–211
  arc, 214–215
  radius, 212–213
  sector, 213–214
Circuit
  series, 171–172
  tank, 175–177
Circumference, 210
Coefficient matrix, 291–292
Coefficient of friction, 177–178
Common factor, 128
Common logarithms, 452–453
Common multiple, 7
Common terms, collecting, 128–132
Complex numbers, 393–394
  addition, 402–403
  division, 406–407
  graph, 394–395
  multiplication, 405–406
  polar notation, 396–401
  rectangular notation, 395–401
  subtraction, 403–405
Complex plane, 394
Composite number, 4
Conductance (G), 157
Constant term, 79
Conversion, 144
Coordinates
  polar, 396–401
  rectangular, 395–401
Cosecant (csc), 264, 310
Cosine (cos), 260, 310
Cotangent (cot), 264, 310
Coulomb (C), 154
Course, true, 438

579

Cramer's rule, 292–295
Critical angle, 327
Cube root, 16
Current (I), 154
   peak, of sine wave, 372
Cycle, 159
Cycles per second, 159

D
Daisy chain method, 345–347, 349
Decibel (dB), 455
Decimal notation, 25, 43, 44–45, 51–52
   binary coded, 535–539
   binary conversion, 518–520
Degree mode, calculator, 312
Degrees (°), 211
Degrees Celsius (°C), 154, 183–184
Degrees Fahrenheit (°F), 183–184
Degrees kelvin (K), 154
Denominator, 4
   rationalizing, 21
Density (ρ), 152
Departure, 332–335
Dependent variable, 254
Depression, angle of, 329–332
Determinant, 292
Diameter, 207
Dielectric, 157, 158
Difference, 3
   precision, 63
Difference of two squares, 85
Digits, significant, 52–63
Dimensional analysis, 144–146
Directrix, 501, 502
Displacement, angular, 216
Distance, between two points, 236–237
Distributive property, 128
Dividend, 4
Division, 4
   complex numbers, 406–407
   fractions, 11–12, 97–98
   monomials, 81–82
   signed numbers, 78
Divisor, 4
Drift correction angle, 438
Dynamics, 429
Dyne, 151

E
EE (enter exponent) key, 47–48
Electromagnetic waves, 186–189
Electromotive force (EMF), 154–156
   alternating current, 156
   direct current, 155
Electron, 154
Elevation, angle of, 329–332
Elimination method, for systems of linear equations, 287–289
Energy (w), 159, 161
Energy, vs. power, 162
Engineering notation, 50–52
Equality
   addition property, 113
   multiplication property, 115
Equations
   inverted, 132–135
   linear, 245–248
      distance between two points, 236–237
      general form, 252–254
      graphing, 254–259, 282–284
      point-slope form, 249–250
      slope of a line, 238–240
      slope-intercept form, 250–252
      systems, 282–284
         addition method, 287–289
         substitution method, 284–287
      two, with two unknowns, 289–291
   literal, 112
      collecting common terms, 128–132
      inverting both sides, 132–135
      raising both sides to a power, 123–128
      removing a factor, 115–119
      removing a term, 113–115
      variable within parentheses, 120–123
   loop, 295–301
   quadratic, 486–487
      applications, 501–504
      graph, 494–501
      roots
         by completing the square, 488–491
         by factoring, 487–488
         by quadratic formula, 491–494
      standard form, 501–502
Equilateral triangle, 196
EXP (exponent) key, 47–48
Exponent(s)
   negative, 15–16
   positive, 13–15
   rational, 16–18
Exponential expressions
   accuracy, 60–61
   division, 18–19
   multiplication, 18
   radical expression, 123
   raising a power to a power, 19–20
   raising a product to a power, 20–21
   raising a quotient to a power, 21–22
Exponential notation, 451
Exponential quantity, 14

F
Factor, common, 128
Factoring, 87–89
   trinomials, 89–91
Fahrenheit (°F), 154
Farad (F), 157, 158
Fathom, 144
Feet, square, 146
Five-cycle semilog graph paper, 465
Focus, of parabola, 501, 502
FOIL, 84–85
Foot, 144
Foot-pound, 159
Force (F), 151–152
   moment of, 432–436
Formula, 112
Fractions
   addition, 7, 94–96
   complex, 12–13, 98–100
   division, 11–12, 97–98
   least common denominator, 8–10
   like, 7
   multiplication, 10–11, 96–97

percent to, 25
raising, 6–7, 93
reducing, 5–6, 91–92
subtraction, 7, 94–96
unlike, 8
Free body diagram, 430
Frequency (f), 159, 161
   electromagnetic wave, 186
   measurement, 186
   sine wave, 377
Friction, coefficient of, 177–178
Full adder, 525
Furlong, 144

G
Gage pressure, 172–173
Grade, 243
Gradient, 243
Graphs, 254–259
   complex numbers, 394–395
   linear equations, 254–259, 282–284
   log, 463–466
   quadratic equations, 494–501
   sine waves, 373–375
   trig functions, 260–263
Greater than (>) sign, 74
Ground speed, 438

H
Half adder, 525
Henry (H), 159
Hertz (Hz), 159, 186, 377
Hexadecimal numbers, 520–525
Hypotenuse, 198

I
Imaginary number, 393
Impedance (Z), 408
Inch, 144
Independent variable, 254
Index, 123
Index of refraction, 187, 323
Inductance (L), 159, 160
Inductive reactance, of series ac circuit, 407–409
Inductor, 159, 160
   of series ac circuit, 407–409
Instantaneous value, 368

Integers
   negative, 15–16
   positive, 13–15
Intensity, of sound wave, 455
Interior angles, alternate, 330
International System of units, 144
Inverse trigonometric functions, 317–323
Isosceles triangle, 196

J
Joule (J), 159

K
Kelvin (K), 154, 183–184
Kilopound (kip), 152, 345

L
Latitude, 332–335
Least common denominator, 8–10
Least common multiple, 7–8, 93–94
Least significant bit, 515
Least significant digit, 57
Length (s), 144–146
   arc, 214–215
Less than (<) sign, 74
Light, refraction, 187, 323–329
   from normal, 326
   toward normal, 325, 326
Like terms, 79
Line(s)
   parallel, 241–242
   perpendicular, 242–243
   slope, 238–240
   transversal, 330
Line segment, midpoint, 244–245
Linear equations, 245–248
   distance between two points, 236–237
   general form, 252–254
   graphing, 254–259, 282–284
   point-slope form, 249–250
   slope of a line, 238–240
   slope-intercept form, 250–252
   systems, 282–284
      addition method, 287–289
      substitution method, 284–287
   two, with two unknowns, 289–291

Linear velocity (v), 149–150, 216–217
Literal equations, 112
   collecting common terms, 128–132
   inverting both sides, 132–135
   raising both sides to a power, 123–128
   removing a factor, 115–119
   removing a term, 113–115
   variable within parentheses, 120–123
Log graph paper, 463–466
Logarithm(s), 451
   common, 452–453
   graph paper, 463–466
   natural, 467–468
   power gain application, 457–460
   properties, 460–461
   significant digits, 453–455
   sound application, 455–457
   universal time constant, 468–475
   voltage gain application, 461–463
Logarithmic notation, 451–452
Log-log graph paper, 465
Loop equations, 295–301
Lower bound, 57

M
Magnitude, 74
Mass (m), 151, 182–183
Matrix, 291–292
Measures
   acceleration, 150–151
   area, 146–147
   capacitance, 157, 158
   charge, 154
   conductance, 157
   current, 154
   density, 152
   energy, 159, 161
   force, 151–152
   frequency, 159, 161
   inductance, 159, 160
   length, 144–146
   mass, 151
   power, 161–162
   pressure, 152–153

resistance, 156–157
temperature, 154
time, 149
velocity, 149–150
voltage, 154–156
volume, 147–149
Mechanics, 429–437
Meter (m), 144, 146
Metric prefixes, 50, 52
Micron (μn), 144
Midpoint, of line segment, 244–245
Mile, 144
square, 146
Minuend, 3
Minutes ('), 211
Moment arm, 432
Moment of force, 432–436
Monomials, 79
division, 81–82
multiplication, 79–81
powers, 83–84
roots, 83–84
Most significant bit, 515
Multiplicand, 3
Multiplication, 3–4
accuracy, 59–60
binomials, 84–86
complex numbers, 405–406
fractions, 10–11, 96–97
monomials, 79–81
polynomials, 86–87
signed numbers, 77–78
Multiplication property of equality, 115
Multiplier, 3

N
Natural logarithms, 467–468
Nautical mile, 144
Navigation, 437–445
Negative integers, 15–16
Newton (N), 151
Newton-meter, 159
Numbers
accuracy, 57–61
binary, 514–518, 522–535
complex, 393–394
decimal, 25, 43, 44–45, 51–52, 516–520, 535–539

hexadecimal, 520–525
imaginary, 393
ordered pair, 235
precision, 61–62
signed, 74–78, 530–535
Numerator, 4

O
Obtuse angle, 194, 195
Obtuse triangle, 196
Ohm (Ω), 156, 173, 408
Ohm's law, 173–174
Operator, 394
Order of operations, 23–24, 63–65
Origin, 234

P
Pain threshold, 455
Parabola, 494–501
Paraboloid, 503–504
Parallel lines, 241–242
Parallelogram, 203
area, 205
Parallelogram method, of vector addition, 347–350
Parentheses
calculator, 291
nested, 23
variable within, 120–123
Pascal (Pa), 153
Peak current, 372
Peak voltage, 372
Percent, 24–31
Period, of electromagnetic wave, 186
Perpendicular lines, 242–243
Phasors, 368–378
diagram, 372–373
Planned true heading, 438
Plus/minus (+/−) key, 15, 47
Polar notation, 396
Polynomial, 79
multiplication, 85
Positive integers, 13–15
Potential difference, 154
Pound (lb), 152
Pounds per square inch (psi), 153
Power (P), 161–162
vs. energy, 162

Power gain, 457–460
Powers
monomials, 83–84
raising, 123–128
P>R key, 397
Precision, 61–62, 320
Prefixes, metric, 50, 52
Pressure (P), 144, 152–153
column of liquid, 185–186
gage, 172–173
Prime factors, 4–5
Prime number, 4
Problem solving, 170–171
Product, 3
Program counter, 533
Pythagorean theorem, 198–200

Q
Quadrants, Cartesian, 234–235
Quadratic equations, 486–487
applications, 501–504
graph, 494–501
roots
by completing the square, 488–491
by factoring, 487–488
by quadratic formula, 491–494
standard form, 501–502
Quadrilaterals, 203
area, 204–207
Quotient, 4

R
Radian measure, 212–213
Radical, 16, 123
Radical expression, 123
Radicand, 16, 123
Radius, 207
Radix, 30
Reciprocal, 11, 115
Rectangle, 203
Rectangular notation, 395–396
Reduction, 144
Refractive index, 187
Resistance (R), 156–157
series ac circuit, 407–409
Resistors
color coding, 27–30
parallel, 178–181

precision, 29
reliability, 30
series ac circuit, 407–409
Rhombus, 203
Right angle, 194, 195
Rise, line, 238
R>P key, 400
Run, line, 238

S
Scalar quantities, 345
Scalene triangle, 196
Scientific notation, 44–50
   arithmetic operations, 49–50
   decimal conversion, 44–46
Secant (sec), 207, 264, 310
Second (sec), 149
Second function, 15
Seconds (''), 211
Sector, area of, 213–214
Series circuit, 171–172
Side adjacent, 308
Side opposite, 308
Siemen (S), 157
Sign, 74
Signed numbers, 74
   addition, 74–75
   binary, 530–535
   division, 78
   multiplication, 77–78
   subtraction, 75–77
Significant digits, 52–63
   accuracy of, 59–61
   logarithmic notation, 453–455
   precision of, 61–62
   trigonometric functions, 314, 320
Sine (sin), 260, 310
Sine wave
   frequency, 377
   graph, 373–375
   peak voltage, 372
Slope, line, 238–240
Slug, 151
Snell's law, 323–329
Sound level, 455–457
Speed, stall, 184
Square, 203
Square root, 16
Stall speed, 184

Statics, 429
Subscripts, 73
Substitution method, systems of
    linear equations, 284–287
Subtraction, 3
   binary numbers, 526–530
   complement method, 526–530
   complex numbers, 403–405
   fractions, 7, 94–96
   monomials, 79
   precision, 63
   signed numbers, 75–77
   vectors, 350–352, 363–367
Subtrahend, 3
Sum, 2
   precision, 62
Superelevation, 181–182
Surveying, 332–335, 423–429
Système International d'unites (SI),
    144

T
Tangent (tan), 260, 310
Tank circuit, 175–177
Temperature (T), 154
Term(s), 79, 113
   collection, 128–132
   constant, 79
   factors, 115
   like, 79
   transposed, 113–114, 117
   variable, 79
Time (t), 149
Time constant (τ), 470
Torricelli's theorem, 174–175
Transversal, 330
Trapezoid, 203
   area, 205–206
Triangles, 196–197
   area, 200–202
Trigonometric functions, 308–317
   graphing, 260–263
   inverse, 317–323
   significant digits, 320, 324
Trigonometric identities, 263–264
Trinomials, 79
   factoring, 89–91
   multiplication, 85
True air speed, 438

True course, 438
True heading, planned, 438
Two points, distance between,
    236–237

U
Unit phasor, 368
United States Customary Units,
    144
Universal time constant chart,
    468–475
Upper bound, 57

V
Variable, 73, 79
   dependent, 254
   independent, 254
   within parentheses, 120–123
Variable term, 79
Vector(s), 345
   addition, 345–350, 359–363
     daisy chain method, 345–347,
      349
     parallelogram method, 347–350
   closure, 425–429
   horizontal components, 352–355
   mechanics application, 429–437
   navigation application, 437–445
   perpendicular, 423–425
   subtraction, 350–352, 363–367
   surveying application, 423–429
   vertical components, 352–355
Vector sum, 359
Velocity
   angular (Ω), 215
   linear (v), 149–150, 216–217
Vertex
   angle, 194
   parabola, 501, 502
   triangle, 196
Volt (V), 154–156, 173
Voltage, peak, 372
Voltage drop, 154
Voltage gain, 461–463
Volume (V), 147–149, 217–223

W
Watt (W), 161
Wavelength (λ), 186

Weight, 182–183
Wind correction angle, 438
Wind triangle, 438

X
x intercept, 247
x^2 key, 14–15
x^3 key, 15
x-axis, 234
x^y key, 14

Y
y intercept, 247
Yard, 144
y-axis, 234
y^x key, 14